纪念吴文俊先生诞辰100周年

国家出版基金项目
NATIONAL PUBLICATION FOUNDATION

The Complete Works of Wu Wen-Tsun

Mathematics Mechanization II

吴文俊全集·数学机械化卷 II

——Mechanical Theorem Proving in Geometries: Basic Principles

吴文俊 著

高小山 编订

科学出版社
龍門書局

北京

内 容 简 介

本卷收录了吴文俊的 *Mechanical Theorem Proving in Geometries: Basic Principles* 一书. 书中论述初等几何机器证明的基本原理, 证明了奠基于各种公理系统的各种初等几何, 只需相当于乘法交换律的某一公理成立, 大都可以机械化. 因此在理论上, 这些几何的定理证明可以借助于计算机来实施. 可以机械化的几何包括了多种有序或无序的常用几何、投影几何、非欧几何与圆几何等.

全书共分六章. 前两章是关于几何机械化的预备知识, 集中介绍了常用几何; 后四章致力于几何的机械化问题. 第 3 章为几何定理证明的机械化与 Hilbert 机械化定理, 第 4, 5 章分别为(常用)无序几何的机械化定理和(常用)有序几何的机械化定理, 第 6 章阐述各种几何的机械化定理.

本书可供数学工作者和计算机科学工作者以及高等院校有关专业的师生参考.

图书在版编目(CIP)数据

吴文俊全集. 数学机械化卷. II/吴文俊著; 高小山编订. —北京: 龙门书局, 2019.5

国家出版基金项目

ISBN 978-7-5088-5551-6

I. ①吴… II. ①吴… ②高… III. ①数学理论-文集 IV. ①O1-53

中国版本图书馆 CIP 数据核字(2019) 第 074593 号

责任编辑: 李 欣 赵彦超 / 责任校对: 邹慧卿

责任印制: 肖 兴 / 封面设计: 无极书装

科学出版社 出版

北京东黄城根北街 16 号

邮政编码: 100717

http://www.sciencep.com

北京通州皇家印刷厂印刷

科学出版社发行 各地新华书店经销

*

2019 年 5 月第 一 版 开本: 720 × 1000 1/16

2019 年 5 月第一次印刷 印张: 19

字数: 383 000

定价: 158.00 元

(如有印装质量问题, 我社负责调换)

编 者 序

中国现代数学的崛起，开始于 20 世纪初，经历了几代人坚苦卓绝的努力. 在这百年奋战中涌现出来的数学家中，吴文俊是最杰出的代表之一. 他早年留学法国，留学期间就已在拓扑学方面做出了杰出贡献，提出了后来以他的名字命名的“吴公式”和“吴示性类”. 回国后提出了“吴示嵌类”等拓扑不变量，发展了统一的嵌入理论. 他关于示性类与示嵌类的研究，已成为 20 世纪拓扑学的经典，至今还在前沿研究中使用. 20 世纪 70 年代以来，吴文俊院士在汲取中国古代数学精髓的基础上，开创了崭新的现代数学领域——数学机械化. 他发明的被国际上誉为“吴方法”的数学机械化方法，改变了国际自动推理的面貌，形成了自动推理的中国学派，已使中国在数学机械化领域处于国际领先地位. 上述工作无疑属于 20 世纪中国数学赶超国际先进水平的标志性成果，而吴文俊院士博大精深的科学研究，除了拓扑学与数学机械化以外，还跨越了代数几何、博弈论、中国数学史、计算图论、人工智能等众多领域，并在每个领域都留下了这位多能数学家的重要贡献.

吴文俊先生是一位具有强烈爱国精神的数学家. 自 1950 年谢绝法国师友的挽留回到祖国后，半个世纪如一日，为在他深爱的中华故土发展数学事业而鞠躬尽瘁. 除了第一流的科研成果，吴文俊先生长期身处中国数学界领导地位，在团结带领整个中国数学界赶超世界先进水平方面，也做出了不可磨灭的贡献. 特别是，吴文俊先生在担任中国数学会理事长期间，领导中国数学最终成功地加入了国际数学联盟，此举大大提高了我国数学界的国际地位，同时也为我国成功举办 2002 年国际数学家大会铺平了道路.

吴文俊治学严谨，学术思想活跃，无论获得多么高的声誉，他总是勤奋地在科研第一线工作，一生积极进取，锲而不舍，不断取得新的成就. 在开始从事机器证明时，他已近花甲之年，从零开始学习编写计算机程序，每天十多个小时在机房连续工作，终于在几何定理机器证明这一难题上取得成功.

吴文俊先生为中国现代数学的发展建立了丰功伟绩，而他本人却始终淡泊、谦逊. 他处事公正豁达，待人充满善意，受过他帮助的人可以说不计其数. 正因如此，这位有着崇高国际声望而平易近人的学者，受到了每一个认识他的人格外的爱戴与尊敬.

2019 年 5 月 12 日是吴文俊先生百年诞辰. 为了纪念这个特殊的日子，我们编辑出版了《吴文俊全集》，通过系统地收录、整理吴文俊先生的学术著作和论文，纪念吴先生的学术思想及学术成就. 全集共计 13 卷, 包括拓扑学 4 卷、数学机械化 5 卷以及数学史、博弈论与代数几何、数学思想各 1 卷; 同时, 全集还设有附卷, 收录吴文俊先生的同事、学生和其他社会各界人士发表过的与吴先生有关的各类文献资料.

最后, 我们对在全集编辑中给予帮助的各位同事表示衷心感谢; 感谢国家出版基金对于全集出版的资助; 感谢科学出版社编辑人员在出版全集时认真细致的专业精神; 感谢相关出版与新闻机构在版权方面提供的帮助.

李邦河　高小山　李文林

2019 年 3 月

Preface

There seems to be no doubt that geometry originates from such practical activities as weather observation and terrain survey. But there are different manners, methods, and ways to raise the various experiences to the level of theory so that they finally constitute a science. F. Engels said, "The objective of mathematics is the study of space forms and quantitative relations of the real world." During the time of the ancient Greeks, there were two different methods dealing with geometry: one, represented by the Euclid's "Elements," purely pursued the logical relations among geometric entities, excluding completely the quantitative relations, as to establish the axiom system of geometry. This method has become a model of deduction methods in mathematics. The other, represented by the relevant work of Archimedes, focused on the study of quantitative relations of geometric objects as well as their measures such as the ratio of the circumference of a circle to its diameter and the area of a spherical surface and of a parabolic sector. Though these approaches vary in style, have their own features, and reflect different viewpoints in the development of geometry, both have made great contributions to the development of mathematics.

The development of geometry in China was all along concerned with quantitative relations. For example, the measure of area and volume and the application of the Kou-Ku theorem[1] occupied a central position in ancient Chinese geometry. At that time, Chinese mathematicians did not work out a tedious stack of axioms but extracted a few general principles and took them as the basis of wide applications, deductions and proofs. Such principles include the out-in complementary principle, Liu-Hui's principle in volume theory, and Liu-Zu's principle. The last corresponds to Cavalieri's principle which appeared in Europe in the sixteenth century. The two different systems of Archimedes and Euclid both appeared in ancient Chinese geometry. The Kou-Ku theorem led naturally to the evaluation problem of square roots while the method of extracting square roots and cubic roots in ancient Chinese mathematics proceeded step by step with the out-in complementary principle as its geometric background. From the evaluation of square and cubic roots, the method and theory of solving quadratic equations and equations of higher degree were developed progressively, which led to the discovery of Thieh Yuan Shu[2] as well as the algebraization of geometry, polynomial operations, elimination methods and so forth. The mechanization problem of theorem proving in geometry elaborated in this book, from thoughts to methods, can be traced back at least to the Song-Yuan dynasties (960–1368).

---

1 Called the *Pythagorean theorem* in the West. [Transl.]

2 A method for solving algebraic equations of higher degree. [Transl.]

Though those ideas were very primitive, they are the main inspiration to the author's findings.

The various tendencies that arose due to different understandings in the development of ancient geometry were not united until the appearance of analytic geometry in the seventeenth century. The final union has made it possible to treat geometric problems using algebraic methods, so that one can avoid the high proof techniques involved in the synthetical method of Euclid. Even though the synthetical proof method is fascinating and absorbing by its intuition and had a period of renaissance in the nineteenth century, it is eventually unable to match the algebraic method of analytic geometry. In many important areas of modern geometry, such as differential geometry and algebraic geometry, a number system (generally a field, or, more particularly, the real field or the complex field) is usually assumed at the beginning. From this system one forms an affine or a projective space and then introduces geometric configurations such as curves and surfaces and their relations by using coordinates and algebraic relations among functions and their derivatives. The purely synthetical method starting from axioms occupies only a small, and mainly an elementary, corner. Moreover, the use of this method to give a detailed analysis of the logical relations among axioms is usual for considering foundations, but not for enriching the concrete contents of geometry. It seems to be the trend of the times to build geometry on the basis of numbers.

Nevertheless, we cannot always only consider the intuitive background and the origination of geometry, but also have to consider its foundations even though it can be established in the form of pure quantitative relations. How to refine axiom systems (or some principles) of geometry from the original figures in the real world and how to develop coordinate systems from axiom systems so that algebraic methods can play their role in geometry are all problems which should be explained. For the often seen and often used Euclidean geometry, a detailed exposition is given in the classical work "Grundlagen der Geometrie" by Hilbert (1899). One might think it easy to establish analytic geometry from geometric axioms by introducing coordinates. This is due to the illusion given by the application of all our knowledge about the real number system and elementary geometry. In fact, along the way there are quite a few difficulties and setbacks. This can be seen clearly from the manner in which the number system and then the coordinate system were introduced according to the axiom system in Hilbert's book. In this development, Desargues' two axioms and Pascal's axiom named by Hilbert play an extremely important role. As will be pointed out in Chaps. 2 and 6 of this book, there is a big difference in the introduction of number systems for projective geometry and for affine geometry. Though both depend only upon Desargues' axioms (and some simplest incidence axioms, etc.) so that one may introduce isomorphic number systems on lines, in affine geometry there is a *canonical* isomorphism relation among these number systems. In contrast to that, in projective geometry the isomorphism is *canonical* only under the assumption that the so-called Pappus' axiom holds. This seems to be neglected in popular geometry books. Perhaps one of the reasons for this is that in these books either Euclid's tradition is stressed, i.e., the logical relations among axioms are emphasized while leaving the number system and the introduction of coordinates to a secondary position, or the geometries are established directly from the number system without considering the real originality and axiomatic foundation of geometric objects. Of those which take the connection of axiom system and quantitative relations as the subject of exploration, Hilbert's book is still a good representative up to the present day. Also, one can see that the introduction of coordinates, even for the commonly known projective geometry and affine geometry, is not as simple as often imagined. For other geometries based on various axiom systems, the difficulty of introducing number systems and coordinate systems is not hard to imagine.

Moreover, it is usually not easy to completely reduce the proof-problems of geometric theorems to purely algebraic problems even if we have arrived at the coordinate system and established the corresponding analytical geometry, starting from the axiom system. First of all, the number of calculations for solving these algebraic problems is often so large that we are deterred from making further efforts. Secondly, the algebraic relations representing the geometric relations are usually disorderly and unsystematic and we are at a loss as to what to do. It often requires high techniques to find a way from these unorderly algebraic relations to achieve a proof. This can become clear as soon as we have a look at many books written in the past as, for example, Salmon (1879a, b). Now, in virtue of the emergence of the computer, we have efficient means to deal with complicated calculations. Therefore, putting the unorderly algebraic relations in perfect order so that the computer can display its full power becomes the key to the whole problem.

Up to this point, the problem of proving geometric theorems can be divided into three steps:

First, starting from the axiom system of a geometry, introduce a number system and a coordinate system to reduce the problem of proving geometric theorems to a purely algebraic problem.

Second, well-order the algebraic relations corresponding to the hypothesis of a geometric theorem and then verify whether the algebraic relations corresponding to the conclusion of the theorem can be deduced from the well-ordered hypothesis relations according to some specific decision procedure.

Third, implement the decision procedure mentioned in the second step and run it on a computer in order to confirm the truth of the theorem.

We shall call the first step the *algebraization* and *coordinatization*, and the second step the *mechanization* of geometry. For the third step, whether a computer can be used to finally verify the theorem depends completely on the possibility of mechanization in the second step. Since computers can recognize only a finite number of objects, a precondition is that the algebraic relations in the second step must appear in a finite form. Hence, if the algebraic relations in the second step involve notions such as continuity and limit or even appear in the form of transcendental functions, it will exclude the possibility of using a computer. On the contrary, if these algebraic relations appear in the form of polynomials with integers as coefficients, the use of a computer is only the programming of the procedure of the second step and will not lead to any essential difficulties. If one can perform these three steps (in fact, only the first two steps are required) to complete theorem proving for a geometry, we shall say that the geometry is *mechanizable* and call this conclusion a *mechanization theorem*. Whether the mechanization theorem holds depends on the realization of the first two steps, both of which are purely theoretical problems.

Whether or not a geometry is mechanizable is not obvious. For the various, usual geometries, in particular differential geometry, which cannot be separated from function and differential and extensively uses concepts like continuity and limit, it appears that we are unable to perform theorem proving using a computer. It is actually not so and the reason is quite deep. According to Hilbert's "Grundlagen der Geometrie," an important point is that, usually, the foundation of geometry can completely exclude such axioms as the axioms of continuity. In Rashevsky's preface to the Russian edition of Hilbert's book, the central idea was precisely pointed out. In fact, the development of geometry may be independent of the axioms of continuity. At the end of another article concerning Bolyai–Lobachevsky's non-Euclidean geometry, Hilbert (1903) also pointed out this fact. The geometric axioms, theorems, and proofs can, in fact, be stated and completed in a finite number of constructive steps. This also serves as a basis for the mechanizability of theorem proving in geometry. As to the knowledge of the author, one of the earliest real mechanization theorems appeared in Hilbert's book "Grundlagen der Geometrie." The third chapter of this book will be devoted to the proof of this mechanization theorem and we shall call it Hilbert's *mechanization theorem* (cf. Wu 1982b).

The same arguments apply to differential geometry as well. If we leave aside the essence of geometry and take only the logical relations for consideration (this is enough for theorem proving), those functions and derivatives occurring in the statement of various concepts and theorems in differential geometry may be completely formally treated, yet we need not consider whether they are related to the real continuity and limit process. Therefore, even in differential geometry there are possibilities for processing theorem proving using a computer through a finite number of formal constructive steps. It is actually so. We have already proved some theorems in differential geometry with the aid of computers (some experiments were made on microcomputers, cf. Wu 1979, 1982a).

We roughly identify geometries into two classes according to whether or not they involve the concept of differentiation. One class consists of *differential geometries*, and the other of the so-called *elementary geometries*. For convenience, we rename the elementary geometry that is usually called Euclidean geometry *ordinary geometry*. The author will elaborate (in succession) on the theory and methods of the mechanization of various geometries. In this book we shall only be concerned with the basic principles of the mechanization of elementary geometries. In its twist book "Theory, Methods and Practice of Mechanical Theorem Proving in Geometries (Part on Elementary Geometries)," we shall explain the concrete implementation of these methods, including programming, the estimation of computational complexity, proofs of concrete theorems, the discovery of new theorems and the improvements of efficiency by using geometric theory and methods as well as other diverse applications. In other relevant books by this author, we shall devote ourselves to expounding on the mechanization problem and related theoretical problems in differential geometry.

This book is divided into six chapters. The first two chapters are the preliminaries to the mechanization of geometry. As a concrete example, we concentrate on ordinary geometry, i.e., Euclidean geometry as it is usually called, and explain in detail the process of establishing a coordinate system starting from the axiom system. For the sake of simplicity, we consider only the planar case; and in order to have a foundation, all our discussions will be made around the five groups of axioms (axioms of incidence, axioms of order, axioms of congruence, axiom of parallels and axioms of continuity) proposed by Hilbert. In establishing ordinary geometry, Hilbert abandoned axioms of continuity, but the axioms of order and axioms of incidence run all through his system, and even the concept and axioms of incidence themselves cannot be independently stated without the concept of order. How to introduce perpendicularity, congruence, and other metric concepts earlier, not depending on the concept and axioms of order, seemingly was first proposed by Hjelmslev and had lasted until the 50–60s of this century, cf. Bachmann (1959a) and the works of Klingenberg, Lenz, Reidemeister, Schülte, Sperner, Winterniz, and others. Their studies focused on the logical dependency relations among different groups of axioms which reflect the major geometric concepts. As opposed to this character, this book aims at studying the mechanization rather than axiomatization of geometry. Problems concerning the dependence and independence of axioms are not what we are interested in. But as for the theme of mechanization, whether or not the axioms of order are involved has extreme relevance for both theory and methods. Moreover, in some modern geometries such as algebraic geometry, the considered fields are mainly the complex field and arbitrary fields of characteristic 0, so no order relation exists. Even in an ordered geometry, theorems which really involve the order relation do not occupy a major position. In view of this, an order relation is necessarily relegated to a secondary position in establishing geometry from the axiom system. For this reason, during the establishment of ordinary geometry in the first two chapters, the introduction of an order relation is postponed as long as possible in order to introduce various unordered geometries and ordered geometries in succession. The introduction of the number system associated with a geometry to complete the algebraization and coordinatization (as indicated before), though still based on the two axioms named after Desargues and Pascal (as in Hilbert's book), is not like Hilbert's mixing it with the axioms of order. Moreover, we add an axiom of infinity to exclude finite geometry, for the mechanizability of finite geometry is obvious.

The latter four chapters of this book are devoted to the mechanization problem of various geometries. The traditional manner of Euclid requires finding an individual proof method for an individual theorem, while every proof requires some new and usually ingenious ideas (see Kline 1972: pp. 307 f.; Chinese edition, II. 8). The method sought in the present book on the mechanization of theorem proving in geometry is going to be a general method which can be used not only for an individual theorem, but also for a whole class of theorems, or even all theorems in a geometry. Based on the methods described in this book, one can obtain unified proofs or disproofs for whole classes of theorems after a finite number of steps regardless of difficulty or ease. To achieve this goal, it is necessary to use algebraic techniques that take quantitative relations as their main objects, with the algebraization of geometry as a key step. This just follows the idea of Descartes in reducing the bulk of problem solving to the mechanization of deduction procedures by using algebraic methods. It is also like the words "by performing with the heaven element as to make clear the problem and elastic the method and to save labour many times" in Suan Hsüeh Chhi Mĕng[3] (1299) by Chu Shih-Chieh, the earlier Chinese creator of Thieh-Yuan Shu for the algebraization of geometry in the Yuan dynasty. In summary, if we can really reach the effective mechanization, "for doing a little bit to help build geometry there will be no need to have geniuses" – a phrase borrowed from Chasles (1889: p. 269).

Not every geometry nor every class of geometric theorems has mechanical proving methods. For those theorems which can be exactly proved mechanically, we consider them to be of three types, for which three corresponding mechanical methods will be discussed in Chaps. 3, 4, and 5, respectively. It is assumed that for each type of theorem the algebraization and coordinatization at the first step are already completed and the proof-problem is turned into a problem of manipulating algebraic expressions. Theorems of the first type are characterized by the fact that the algebraic expressions corresponding to the hypothesis are linear with respect to certain variables. This type of theorem includes the so-called pure point of intersection theorems for which a mechanical proof method appeared in Hilbert (1899); so we shall call this result *Hilbert's mechanization theorem*. Theorems of the second type are characterized by the fact that the algebraic expressions corresponding to the hypothesis and conclusion may be given as polynomial equations. A mechanical method for proving this type of theorem was proposed by Wu (1978) and has been experimented on using several computers. Theorems of the third type are characterized by the fact that the hypothesis and conclusion may be expressed as arbitrary polynomial equations or inequalities with coefficients in a real closed field, so that the original geometry should involve an order relation. A mechanical method for proving this type of theorem was given by Tarski (1948). We shall call the corresponding result Tarski's mechanization theorem. Each of the above three mechanization theorems has its own application domain, in particular, the second and the third do not contain one another.[4] Our mechanization theorem is not suited for those theorems whose conclusion involves an order relation, whereas Tarski's mechanization theorem can only be applied to the case in which the geometry-associated field is a real closed field, but not to the case in which the associated field is the complex field or some other general field. From the viewpoint of applicability, the relationships among the three (denoted by I, II, and III, respectively) are depicted below. As to the methods, the three differ from each other. Therefore, for the overlapping parts of the application domains, theorems can be proved by different mechanical methods. With respect

3 "Introduction to Mathematical Studies." [Transl.]

4 Dr. Hoon Hong has pointed out that the mechanization theorem presented in Chap. 4 of this book may be considered as a special case of the mechanization theorem of Tarski, because any complex number can be represented by means of two real numbers and thus the decision problem for the complexes can be reduced to that for the reals (cf. Tarski 1948: chap. 3). [Transl.]

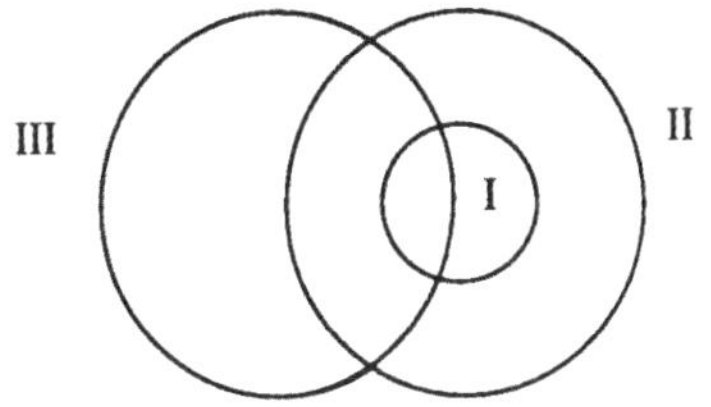

to efficiency, I is the most efficient; with it one can prove quite non-trivial theorems even with hand-calculation. Then comes II, the power of which cannot be seen with hand-calculation, though, one can prove quite non-trivial theorems even on a microcomputer. For III, the efficiency is rather low, and there is no report that any significant theorem has been proved up to this day. This is true even when the method runs on big, high speed computers.

Since the applicability of Hilbert's mechanical method is too narrow and the efficiency of Tarski's mechanical method is too low, our focal point will be Method II, which is presented in Chap. 4. The mechanical method presented in that chapter, in theory, requires the help of algebraic geometry. Though modern algebraic geometry is one of the most active branches of mathematics, it does not fit our requirement because of its existential character. For the demands of mechanization, the provided means must be constructive, and only in this case will we then have the possibility to implement the corresponding procedure on computer. Fortunately, such a constructive theory of algebraic geometry that satisfies our need had been developed by Ritt (1932, 1950) early on. Ritt used analytic methods in his argumentation, whereas concepts like continuity and limit are inconsistent with mechanical proving methods. It is therefore necessary to make a proper revision of Ritt's work and part of Chap. 4 will be devoted to this revision. Some concepts and methods like the reduction and well-ordering of polynomial sets are taken directly from Ritt and differ only slightly in our elaboration. Ritt's method may also be used for the mechanical proving of theorems in differential geometries for which a brief description may be found in Wu (1979, 1982a). The author will give a more detailed elaboration on this in relevant books.

Applying the methods presented in those chapters to investigate geometry, one can prove that theorem proving in numerous geometries – those occurring in the course of establishing ordinary geometry, starting from the so-called unordered Pascalian geometry – is all mechanizable, i.e., the mechanization theorems for those geometries hold. In Chap. 6, we shall also investigate other geometries such as projective geometry, hyperbolic and elliptic non-Euclidean geometries, and Möbiusian and Laguerrean circle geometries that are different from ordinary geometry, and prove similar mechanization theorems. Since in these geometries the process of coordinatization from axiomatization is rather tedious, we will not be able to argue in as much detail as in Chaps. 1 and 2.

Although this book exhibits many mechanizable geometries, not every geometry is mechanizable. What counts is whether Pappus' or Pascal's axiom holds, or whether the commutative law of multiplication in the geometry-associated number system holds. Concretely speaking, we have the following

Conjecture. Desarguesian geometry is non-mechanizable.

We are, however, unable to prove this conjecture. Mathematical logicians might give a correct answer.

Since the discovery (at the end of 1976 and beginning of 1977) of the mechanical method of theorem proving shown in Chap. 4 of this book, more than five years have quickly elapsed. In this period, the author has received much encouragement and sometimes unforeseen, warm help. Those to be acknowledged are too many to be listed here. Nevertheless, there are three who deserve special thanks.

For theoretical developments the author is deeply grateful to Prof. Shihua Hu and Prof. Hao Wang for their encouragement and support. The author's ideas and methods on the mechanization of geometry theorem proving conflict with the long-evolved traditional ones. Without first the consent of Prof. Hu from the viewpoint of mathematical logic at that time, it would have been possible for our ideas and methods to come to a premature end. Professor Wang's penetrative achievements in mechanical theorem proving have won him universal praise for a long time while his incisive arguments call for more deep thought. His arguments such as replacing qualitative difficulties by quantitative complexities and making differences between foundational proof and example proof in mechanical theorem proving all give the author deep inspirations for his research.

Mr. Fanshu Mong provided much convenience for the author to experiment on a table-computer, Great Wall 203. A few months of experiments on that computer convinced the author to purchase a more modern table computer to do more experiments. Those experiments demonstrated that our method can be put to practice, so that we do not become armchair strategists. We therefore express our deep thanks to Mr. Mong.

Wen-tsün Wu

Contents

Author's note to the English-language edition

Mechanical theorem proving (MTP) is a traditional topic of mathematical logic. In contrast to the usual approaches to MTP by mathematical logic, approaches using algebraic methods seem to be originated in the paper by the present writer (Wu 1978). In 1984 appeared the present book "Basic Principles of Mechanical Theorem Proving in Geometries" devoted to a systematic exposition of such algebraic methods for MTP. The book, written in Chinese and published in China, was little known beyond China. The method became widely known owing to papers by Wu (1978, 1984a) and Chou (1984) published in the anthology edited by Bledsoe and Loveland (1984). Since that time the algebraic methods of treating MTP have been developed rapidly and vigorously in various directions. For example, in 1986 there appeared several papers on MTP based on Gröbner basis method and others (e.g., Kutzler and Stifter 1986, Kapur 1986). In this note we shall give a brief review of the achievements of MTP in recent years restricted, however, to the methods as exhibited in the present book alone. Thus it may serve merely as complement and addendum to the original version of the book.

The book had a subtitle "Part I on Elementary Geometries." It was intended to be followed by a Part II on differential geometries. The basic principles underlying such MTP of differential geometries have been described in detail in the paper by Wu (1989b) and a survey article has also been published recently (Wu 1991). However, it seems that the time for writing such a book has not yet come. The computational difficulties are awful and up to now only a few results have been shown (Chou and Gao 1989b, 1991; Li 1991). We shall thus restrict ourselves below to the case of elementary geometries alone.

The present writer has emphasized on various occasions that our method of MTP may be considered as a particular application of our general method of polynomial equations-solving, which has an immense domain of applications besides MTP. Again, we shall restrict ourselves to MTP proper alone.

This book spends more than one-third of its length, viz. Chaps. 1 and 2, to re-visit Hilbert's axiomatization of Euclidean geometry, modified to meet the purpose of MTP. They are followed by Chap. 3, which is devoted to the Hilbert mechanization theorem, the earliest and also the simplest of such mechanization theorems. The theorem had already appeared in the first edition of Hilbert's classic "Grundlagen der Geometrie" in 1899, however, in a disguised form. The significance of the theorem seems to have been first recognized by the present writer when he taught a course on MTP at the Graduate School of the University of Science and Technology, associated with the Chinese Academy of Sciences, during the academic year 1979/1980. We remark in passing that Professor S. C. Chou, now renowned for his work on MTP, was at that time a graduate student at the school and attended the previously mentioned course, unnoticed, however, by the present writer.

The contents of Chap. 3 were first published in a separate paper (Wu 1982). Owing to the great importance of Hilbert's contributions, little recognized before or even now, it seems appropriate to reproduce below the opening passage of this paper:

"The great merit of Hilbert's classic "Grundlagen der Geometrie" of 1899 is universally recognized as being representative for axiomatization of mathematics, laying in particular a rigorous foundation of the Euclidean geometry. However, another great merit (perhaps greater in the opinion of the present author) of this classic seems hardly to be noticed up to the present. In fact, this classic is also representative for the mechanization of geometry, showing clearly at the same time the way to achieve it."

In another paper (Wu 1983), the present writer pointed out some contrasts between the MTP based on our algebraic methods and the ordinary theorem proving in Euclidean fashion based on logical deductions. We reproduce below also the opening passage of that paper:

"... almost all theorems in ordinary geometry (i.e. ordinary Euclidean geometry) are only *generically true*, or true only under some non-degeneracy conditions usually not explicitly described in the statement of the theorem. This fact is quite fatal to make the usual Euclidean fashion of proving theorems rigorous as one believes to be so, since it is difficult to state clearly the non-degeneracy conditions to be observed and to verify that the previous theorems to be applied in the proof of the theorem in question do not fall into the degeneracy cases under which the previous theorems to be applied might not be true. On the contrary our mechanical method of theorem-proving in geometry permits to discover automatically the non-degeneracy conditions to be observed and offers the means of dealing with such degeneracy cases in systematic mechanical way. It is in fact this crucial point which is responsible for the high efficiency of our method in proving quite non-trivial theorems without the defect pointed out above."

The domain of applicability of Hilbert's mechanization theorem is very limited: It concerns only pure intersection theorems. It was D. M. Wang who first noticed that the method may be extended to theorems of metric character, midpoint, orthogonality, etc. (Wang 1989b). However, the domain of applicability is still quite limited: For example, it does not work for most theorems involving distances or circles.

The heart of this book is Chap. 4. It deals with MTP for which both hypothesis and conclusion of a theorem to be proved may be expressed in the form of polynomial equations of arbitrary degree. Part of the contents of this chapter were also published in a separate paper (Wu 1984b). The notations used in both Chap. 4 and the previously mentioned paper are rather cumbersome. Of late (from 1985–86 onwards) these notations have been replaced by the naive notions of *zero-set*. Thus for any polset *PS* and any pol G in the polynomial ring $\boldsymbol{K}[X] = \boldsymbol{K}[x_1, \ldots, x_n]$ over a field $\boldsymbol{K}$ of characteristic 0, we denote by $\mathrm{Zero}(PS/G)$ the totality of zeros of *PS* which are not zeros of G. Here, by a *zero* we mean one in an arbitrary extension field of $\boldsymbol{K}$. In particular, we put

$$\mathrm{Zero}(PS/1) = \mathrm{Zero}(PS).$$

Besides, various notions and notations have been made precise in later developments. For example, the *remainder* of a polynomial G w.r.t. an ascending set (asc-set) *AS* as defined in Sect. 4.5 has been denoted as $\mathrm{Remdr}(G/AS)$. Observe that it is different from the notion of *pseudoremainder* usually adopted in the literature (see Wu 1986a, 1987a).

In this way the Lemma 1 of Sect. 4.5 as well as the accompanying formula

$$|\Lambda| = |\Omega_\Phi| \cup |\Lambda + I_1| \cup \cdots \cup |\Lambda + I_n| \tag{1}$$

has been reformulated and generalized as the following.

Well-ordering principle. There is an algorithm which permits one to determine for any polset *PS* an asc-set *CS* in a finite number of steps such that

$$\mathrm{Zero}(PS) \subset \mathrm{Zero}(CS) \quad \text{and} \tag{2}$$

$$\mathrm{Remdr}(P/CS) = 0, \quad \text{for all } P \text{ in } PS. \tag{3}$$

Moreover, let I_i be the initials of pols in *CS*, and J the product of all such initials; then we have

$$\mathrm{Zero}(CS/J) \subset \mathrm{Zero}(PS) \subset \mathrm{Zero}(CS), \tag{4}$$

$$\mathrm{Zero}(PS) = \mathrm{Zero}(CS/J) + \mathrm{UNION}_i\ \mathrm{Zero}(PS + \{I_i\}). \tag{I}$$

In (I) the asc-set *CS* corresponds to Φ of Lemma 1. It is one of particular interest associated to the given polset *PS* and is characterized by the two properties (2) and (3). It is in later developments called a *characteristic set* (char-set) of *PS*.

The algorithm for arriving at a char-set *CS* of a polset *PS* as given in Sect. 4.3 was turned in later developments into the following concise form:

$$\begin{array}{ccccccccc} PS & = & PS_1 & \cdots & PS_i & \cdots & PS_m & & \\ & & BS_1 & \cdots & BS_i & \cdots & BS_m & = & CS \\ & & RS_1 & \cdots & RS_i & \cdots & RS_m & = & \text{Empty}. \end{array} \tag{S}$$

In the scheme (S) each BS_i is an asc-set of lowest ordering contained in the polset PS_i, RS_i is the set of non-zero remainders of pols in $PS_i - BS_i$ w.r.t. BS_i, and

$$PS_i = PS_{i-1} + RS_{i-1}. \tag{5}$$

Consider a geometric theorem $T = \{\text{HYP}, \text{CONC}\}$ for which the hypothesis and conclusion are given respectively in the form of polynomial equations HYP $= 0$, CONC $= 0$ with HYP $\subset \boldsymbol{K}[X]$ and CONC $\in \boldsymbol{K}[X]$. Then a zero of HYP is nothing else but a geometrical configuration verifying the hypothesis of the theorem T, eventually in some fictitious extended space. With this understanding the mechanization theorem of unordered geometries as given in Sect. 4.1 has been reformulated and made precise as the following.

Special MTP-principle I. Given a theorem $T = \{\text{HYP}, \text{CONC}\}$ with HYP $\subset \boldsymbol{K}[X]$ and CONC $\in \boldsymbol{K}[X]$ let Zero(HYP) be decomposed as in (I) with corresponding char-set CS and initials I_i. Then we have (J is the product of all I_i)

$$\text{Zero}(\text{HYP}/J) \subset \text{Zero}(\text{CONC}) \tag{6}$$

if

$$\text{Remdr}(\text{CONC}/CS) = 0. \tag{7}$$

In other words, the theorem is generically true under the non-degeneracy conditions

$$J \neq 0 \quad \text{or} \quad I_i \neq 0 \text{ for all } i \tag{8}$$

if (7) is true. The condition (7) is also necessary for the above to hold in case CS is an irreducible asc-set.

From the well-ordering principle the following can easily be deduced.

Zero-decomposition theorem (special form). There is an algorithm which permits one to determine for any polset PS a finite number of asc-sets AS_j in a finite number of steps such that

$$\text{Zero}(PS) = \text{UNION}_j \ \text{Zero}(AS_j/J_j), \tag{II}$$

in which each J_j is the product of all initials of pols in the corresponding asc-set AS_j.

Corresponding to this zero-decomposition theorem we have then the following.

Special MTP-principle II. Given a theorem $T = \{\text{HYP}, \text{CONC}\}$ with HYP $\subset \boldsymbol{K}[X]$ and CONC $\in \boldsymbol{K}[X]$ let Zero(HYP) be decomposed as in (II). Then we have for each j (J_j is the product of all initials of pols in AS_j)

$$\text{Zero}(AS_j/J_j) \subset \text{Zero(CONC)} \tag{9}$$

if

$$\text{Remdr}(\text{CONC}/AS_j) = 0. \tag{10}$$

In other words, the theorem is generically true under the non-degeneracy condition

$$J_j \neq 0 \tag{11}$$

for those geometrical configurations verifying hypothesis as well as $AS_j = 0$ if (10) is true. The condition (10) is also necessary for the above to hold in case AS_j is an irreducible asc-set.

Suppose now *IRR* is an *irreducible* asc-set with a generic point *GZ*. The totality of all specializations of the generic point *GZ* is then an irreducible algebraic variety with *GZ* as a generic point in the ordinary sense which will be denoted by Spec(*GZ*), or Var[*IRR*] since *GZ* is determined by *IRR*. Note that here square brackets instead of parentheses have been used. In fact, Var[*IRR*] is a subvariety of the algebraic variety Zero(*IRR*) but usually only a proper one. The latter one is the variety associated with the ideal Ideal(*IRR*) with *IRR* as a basis which may well be denoted by Var(*IRR*) with *IRR* in parentheses. Furthermore, the finite basis of the ideal associated with the irreducible variety Var[*IRR*] may be determined by computations if required. This was announced as a theorem in Sect. 4.4 without indicating how such computations should be carried out. Such explicit computations based on the theory of Chow forms were given in Wu (1989c). An alternative, ingenious method of such computations based on Gröbner basis determination was given by Chou et al. (1985) and also by Wang (1989c).

With these notions, the main theorem in Sect. 4.5, viz., the irreducible decomposition theorem of algebraic varieties, has been refined to the following.

Zero-decomposition theorem (general form). There is an algorithm which permits one to determine for any polset *PS* a finite number of irreducible asc-sets IRR_k in a finite number of steps such that

$$\text{Zero}(PS) = \text{UNION}_k\ \text{Var}[IRR_k], \tag{III}$$

the union being uncontractible and unique up to order.

Corresponding to the above zero-decomposition theorem we have now the following.

General MTP-principle III. Given a theorem $T = \{\text{HYP}, \text{CONC}\}$ with HYP $\subset$ $\boldsymbol{K}[X]$ and CONC $\in \boldsymbol{K}[X]$ let Zero(HYP) be decomposed as in (III). Then we have for each k

$$\text{Var}[IRR_k] \subset \text{Zero(CONC)} \tag{12}$$

if and only if

$$\mathrm{Remdr}(\mathrm{CONC}/IRR_k) = 0. \tag{13}$$

In other words, the theorem T is true on the whole component $\mathrm{Var}[IRR_k]$ if and only if (13) is verified.

For these decomposition theorems and the corresponding MTP principles we refer again to Wu (1986a, 1987a) as well as a survey paper (Wu 1992a). Among the MTP principles, Principle III is clearly the most general and also the complete one. However, MTP principle I, though less general and not so precise, is already powerful enough and has been applied with great success in recent years. In fact, up to now most of the geometric theorems which have been proved are based on the special principle I alone. The most far-reaching scale of experiments on the method, with some variations and improvements, was carried out by Chou and has been published in book form (Chou 1988). In addition, Wang and Gao also proved a lot of theorems which have been collected in a summary article (Wang and Gao 1987). The method has also been applied to the discovery of new theorems. As examples we may cite the Pascal conic theorem already given in this book's original version, the theorems for Leisenring lines by D. M. Wang (1989a), a generalization of Gauss mid-line theorem of a complete quadrilateral to a 6-pole configuration in planar kinematics (see Wu 1989a), and an interesting theorem for regular pentagonal planar sections of square pyramids (D. K. Wang 1990, 1992).

The general MTP principle III shows that, for MTP, the notions of *non-degeneracy conditions* and *generic truth* of a geometric theorem are not at all indispensable and may be avoided if we like. It is merely owing to the extreme computational complexity involved in a decomposition like (III) that we would prefer to satisfy ourselves in proving the mere *generic* truth of a theorem using Principle I.

The decomposition of Zero(PS) in the form of (III) for a general polset PS is usually too complicated to be carried out. However, in the case of PS being the hypothesis set of a geometric theorem, we may sometimes take advantage of the inherent geometric properties to render the decomposition easy to attain. Thus, the present writer introduced the notions of oriented lines, oriented circles, etc. to simplify the hypothesis equations of a geometric theorem in order to bring about such decompositions in an easy way (Wu 1987b). As an example we may cite the theorem of Feuerbach of which a proof has already been given in Sect. 4.8. By the new method, it is shown that Zero(PS) with PS the corresponding hypothesis set will be decomposed into 8 irreducible components, for which the theorem will be true on the whole of 4 of them but not so for the other 4. The theorem becomes even more precise in that sense of contact of circles will be clarified. As a further example, let us consider a conjecture by Thebault in 1938. The conjecture was first proved by Taylor in 1982 in ordinary Euclidean fashion and then by Chou in 1986 by means of the special MTP principle I. On the other hand, the present writer gave a proof in showing that the corresponding algebraic variety will be decomposed into 4 irreducible components, for which the Thebault conjecture will be true on one of them but not so on the other three. No non-degeneracy conditions will be involved and the statement of the theorem is quite precise, for which the usual proofs may be lacking. The same method has also been applied to give a more precise and exact answer to an enumerative problem of algebraic geometry (Wu 1992d). It seems that the method may be successively applied to other circumstances, e.g., the case of Appolonius and Hart configurations of 3 or 4 circles for which the Thebault–Taylor–Chou theorem is probably just a very special result.

Though the computer can distinguish real variables $+x$ and $-x$ of opposite sign, it has no means to decide whether the variable x itself is positive or negative. For example, it is impossible to realize the non-negative distance between two points in a simple way on a computer. The MTP method in Chap. 4 and the MTP principles described above are all restricted to unordered geometries, for which order relations or algebraically relations like > 0, < 0, ≥ 0, ≤ 0 are neglected. In this respect, we may cite an ingenious device by Gao (1987, 1990) by which the difficulties can be partially resolved. A lot of interesting theorems have been proved in this way. In general, according to a device of Seidenberg, inequalities can be turned into equalities by introducing new variables. Thus, $x > 0$ is equivalent to the *existence* of a variable y such that $y^2 \cdot x = 1$, similarly for the others. With this device the hypotheses and conclusions of geometry theorems involving order relations will all be turned into polynomial equations and, apparently, the method of Chap. 4 or the above MTP principles may again be applied. Chapter 5 is devoted to its exposition. Just in this way Chou and the present writer in collaboration were able to prove the delicate bisector theorem that triangles with equal internal bisectors are isosceles (Wu and Lu 1985). The applicability of the method is, however, quite limited. In fact, usually it will arrive at the problem of deciding whether an algebraic expression is always positive or not, a problem out of reach of the MTP methods described in this book and those described above.

Early in 1950, Tarski already gave a general MTP method for proving all theorems involving both equations and inequations as well as inequalities. The effectiveness of Tarski's method has been highly raised in later years mostly due to the invention of the CAD method by Collins. In various ways the CAD method has been improved by students of Collins and in recent years has been applied to prove some interesting geometry theorems involving inequalities. On the other hand, Chou and Gao (1989a) have combined the CAD method with our MTP method to prove a lot of non-trivial geometry theorems involving inequalities too. Besides, the present writer has discovered a general optimization method and reduced the problem of proving such geometry theorems to problems of optimization. A number of such non-trivial theorems have been proved in this way (Wu 1988, 1992b, c). However, all the proofs of Chou, Gao, and the present writer will rely heavily on the ultimate decision of positiveness of algebraic expressions. More powerful methods remain to be sought.

The final Chap. 6 deals with some miscellaneous topics. For MTP of various kinds of geometries and for the role of transcendental functions we may mention the papers by Gao (1987, 1989, 1990). However, much more can be done and they are waiting to be further developed. For other ramifications which have been developed in later years but not touched on in this book, we may cite particularly automatic discovering of unknown relations, initiated in Wu (1986b, c). In this respect we may also cite an interesting paper by Gao and D. K. Wang (1992). The methods of proving by examples owing to Hong, Zhang, and Yang are exceedingly interesting and deserve special notice (e.g., Hong 1986, Yang et al. 1992; also D. M. Wang 1988, Shi 1989). Furthermore, this whole book is restricted to the case for which the basic field $\boldsymbol{K}$ is of characteristic 0. Lin and Liu (1992), however, have extended our MTP method to the case of fields of finite characteristic; interesting theorems have been proved which show clearly the difference between the cases of even and odd characteristic. Applications of our MTP methods have also been made to domains beyond mathematics, for which we may mention in particular the paper by Kapur and Mundy (1988).

The ordinary Euclidean geometry occupies a particularly important place in high school teaching. It is therefore natural that our MTP methods will have a large impact on education in mathematics. Unfortunately, the reformation of geometry teaching in this direction has never been realized in China, though it is one of the earliest desires of the present writer to usher in at least the Hilbert MTP in high schools. Most recently, at Zhang's initiative, there appeared papers by Chou et al. (1992a, b) which give MTP a traditional fashion. Here "traditional" means that the use of coordinates will be avoided so that each step of the proofs will have evident geometrical meaning. Both methods of MTP, traditional and not, are now under consideration in China so that perhaps they will actually be taught in our high schools. Clearly, it is a matter of highest importance.

The algorithms of MTP have been turned into programs of which the earliest and simplest one was first done by the present writer under the name of China Prover, in memory of the fact that our MTP method had its origin in the study of our ancient Chinese mathematics, as explained in the original preface of this book. The program is obsolete and has been replaced by those on polynomial equations solving which include MTP as a particular application. The original Prover may still have some interest in that the branching process had been taken into account which is peculiar to MTP but not so to general polynomial equations solving. Besides, almost all the members in the Mathematics Mechanization Center (MMC) here in China and those in close connection with MMC have their own programs, both for MTP and polynomial equations solving. We may cite in particular the program systems described in Chou and Gao (1992) and Gao et al. (1992). Packages about zero-set decompositions based on our methods have appeared also beyond China in various forms in software systems like REDUCE, SCRATCHPAD, etc. Perhaps the most complete and also the most powerful package is due to D. M. Wang and has been implemented in the software system MAPLE under the name of CharSets. This package includes also a rare method of factorization of multivariate polynomials in general algebraic field due to S. Hu and D. M. Wang (1986). This method of factorization based on the well-ordering principle was discovered by Hu and Wang when they were graduate students in the previously mentioned graduate school, attending a course given by the present writer on MTP in 1984.

The present writer owes much to many colleagues in reading and checking the original Chinese text. Particularly Dr. X. S. Gao has pointed out some gap in the proof of dimension theorem in Sect. 4.6 which has been accordingly filled in this English translation. Finally, the writer would like to express his gratitude to Ms. X. F. Jin and Dr. D. M. Wang who have undertaken the painstaking work of translation.

References

Bledsoe, W. W., Loveland, D. W. (eds.) (1984): Automated theorem proving: after 25 years. American Mathematical Society, Providence.

Chou, S. C. (1984): Proving elementary geometry theorems using Wu's algorithm. In: Bledsoe, W. W., Loveland, D. W. (eds.): Automated theorem proving: after 25 years. American Mathematical Society, Providence, pp. 243–286.

Chou, S. C. (1985): Proving and discovering geometry theorems using Wu's method. Ph.D. Thesis, University of Texas at Austin, Austin, Texas.

Chou, S. C. (1988): Mechanical geometry theorem proving. Reidel, Lancaster.

Chou, S. C., Gao, X. S. (1989a): On the mechanical proof of geometry theorems involving inequalities. University of Texas at Austin, Preprint TR-89-31.

Chou, S. C., Gao, X. S. (1989b): Mechanical theorem proving in differential geometry – I. Space curves. Math. Mech. Res. Preprints 4: 109–123.

Chou, S. C., Gao, X. S. (1991): Theorems proved automatically using Wu's method – part on differential geometry (space curves) and mechanics. Math. Mech. Res. Preprints 6: 37–55.

Chou, S. C., Gao, X. S. (1992): An algebraic system based on the characteristic set method. In: Proceedings IWMM '92, Beijing, pp. 1–18.

Chou, S. C., Schelter, W. F., Yang, J. G. (1985): Characteristic sets and Gröbner bases. University of Texas at Austin, Preprint.

Chou, S. C., Gao, X. S., Zhang, J. Z. (1992a): Automatic production of traditional proofs for theorems in Euclidean geometry – parts I–IV. Wichita State University, Preprints WSUCS-92-3,5-7.

Chou, S. C., Gao, X. S., Zhang, J. Z. (1992b): Automated geometry theorem proving by vector calculation. Wichita State University, Preprint.

Gao, X. S. (1987): Trigonometric identities and mechanical theorem proving in elementary geometries. J. Syst. Sci. Math. Sci. 7: 264–272 (in Chinese).

Gao, X. S. (1989): Mechanical theorem proving in Cayley–Klein geometries. Math. Mech. Res. Preprints 3: 39–50.

Gao, X. S. (1990): Transcendental functions and mechanical theorem proving in elementary geometries. J. Automat. Reason. 6: 403–417 [also in Math. Mech. Res. Preprints 2: 37–47 (1987)].

Gao, X. S., Wang, D. K. (1992): On the automatic derivation of a set of geometric formulae. Math. Mech. Res. Preprints 8: 26–37.

Gao, X. S., Li, Y. L., Lin, D. D., Lu, X. S. (1992): A geometric theorem prover based on Wu's method. In: Proceedings IWMM '92, Beijing, pp. 201–205.

Hong, J. W. (1986): Can we prove geometry theorem by computing an example? Sci. Sin. A29: 824–834.

Hu, S., Wang, D. M. (1986): Fast factorization of polynomials over rational number field or its extension fields. Kexue Tongbao 31: 150–156.

Kapur, D. (1986): Using Gröbner bases to reason about geometry problems. J. Symb. Comput. 2: 399–408.

Kapur, D., Mundy, J. L. (1988): Wu's method and its applications to perspective viewing. Artif. Intell. 37: 15–36.

Kutzler, B., Stifter, S. (1986): On the application of Buchberger's algorithm to automated geometry theorem proving. J. Symb. Comput. 2: 389–397.

Li, Z. M. (1991): Mechanical theorem proving of the local theory of surfaces. Math. Mech. Res. Preprints 6: 102–120.

Lin, D. D., Liu, Z. J. (1992): Some results on theorem proving in finite geometry. In: Proceedings IWMM '92, Beijing, pp. 222–235.

Shi, H. (1989): On the resultant formula for mechanical theorem proving. Math. Mech. Res. Preprints 4: 77–86.

Wang, D. K. (1990): A mechanization proving of a group of space geometry problems. Math. Mech. Res. Preprints 5: 66–81.

Wang, D. K. (1992): Mechanical solution of a group of space geometry problems. In: Proceedings IWMM '92, Beijing, pp. 236–243.

Wang, D. M. (1988): Proving-by-examples method and inclusion of varieties. Kexue Tongbao 33: 2015–2018.

Wang, D. M. (1989a): A new theorem discovered by computer prover. J. Geom. 36: 173–182.

Wang, D. M. (1989b): On Wu's method for proving constructive geometric theorems. In: Proceedings IJCAI-89, Detroit, pp. 419–424 [also in Math. Mech. Res. Preprints 3: 89–101 (1989)].

Wang, D. M. (1989c): A method for determining the finite basis of an ideal from its characteristic set with application to irreducible decomposition of algebraic varieties. Math. Mech. Res. Preprints 4: 124–140.

Wang, D. M., Gao, X. S. (1987): Geometry theorems proved mechanically using Wu's method – part on Euclidean geometry. Math. Mech. Res. Preprints 2: 75–106.

Wu, W.-t. (1978): On the decision problem and the mechanization of theorem-proving in elementary geometry. Sci. Sin. 21: 159–172 [also in Bledsoe, W. W., Loveland, D. W. (eds.): Automated theorem proving: after 25 years. American Mathematical Society, Providence, pp. 213–234 (1984)].

Wu, W.-t. (1982): Toward mechanization of geometry – some comments on Hilbert's "Grundlagen der Geometrie". Acta Math. Sci. 2: 125–138.

Wu, W.-t. (1983): Some remarks on mechanical theorem-proving in elementary geometry. Acta Math. Sci. 3: 357–360.

Wu, W.-t. (1984a): Some recent advance in mechanical theorem-proving of geometries. In: Bledsoe, W. W., Loveland, D. W. (eds.): Automated theorem proving: after 25 years. American Mathematical Society, Providence, pp. 235–242.

Wu, W.-t. (1984b): Basic principles of mechanical theorem proving in elementary geometries. J. Syst. Sci. Math. Sci. 4: 207–235 [also in J. Automat. Reason. 2: 221–252 (1986)].

Wu, W.-t. (1986a): On zeros of algebraic equations – an application of Ritt principle. Kexue Tongbao 31: 1–5.

Wu, W.-t. (1986b): A mechanization method of geometry and its applications – I. distances, areas and volumes. J. Syst. Sci. Math. Sci. 6: 204–216.

Wu, W.-t. (1986c): A mechanization method of geometry and its applications – I. distances, areas and volumes in Euclidean and non-Euclidean geometries. Kexue Tongbao 32: 436–440.

Wu, W.-t. (1987a): A zero structure theorem for polynomial-equations-solving. Math. Mech. Res. Preprints 1: 2–12.

Wu, W.-t. (1987b): On reducibility problem in mechanical theorem proving of elementary geometries. Chin. Q. J. Math. 2: 1–20 [also in Math. Mech. Res. Preprints 2: 18–36 (1987)].

Wu, W.-t. (1988): A mechanization method of geometry and its applications – III. mechanical proving of polynomial inequalities and equations-solving. Syst. Sci. Math. Sci. 1: 1–17 [also in Math. Mech. Res. Preprints 2: 1–17 (1987)].

Wu, W.-t. (1989a): A mechanization method of geometry and its applications – IV. some theorems in planar kinematics. Syst. Sci. Math. Sci. 2: 97–109.

Wu, W.-t. (1989b): On the foundation of algebraic differential geometry. Syst. Sci. Math. Sci. 2: 289–312 [also in Math. Mech. Res. Preprints 3: 1–26 (1989)].

Wu, W.-t. (1989c): On the generic zero and Chow basis of an irreducible ascending set. Math. Mech. Res. Preprints 4: 1–21.

Wu, W.-t. (1991): Mechanical theorem proving of differential geometries and some of its applications in mechanics. J. Automat. Reason. 7: 171–191 [also in Math. Mech. Res. Preprints 6: 1–22 (1991)].

Wu, W.-t. (1992a): A report on mechanical geometry theorem proving. Prog. Natural Sci. 2: 1–17.

Wu, W.-t. (1992b): On problems involving inequalities. Math. Mech. Res. Preprints 7: 1–13.

Wu, W.-t. (1992c): On a finiteness theorem about optimization problems. Math. Mech. Res. Preprints 8: 1–18.

Wu, W.-t. (1992d): A mechanization method of equations solving and theorem proving. Adv. Comput. Res. 6: 103–138.

Wu, W.-t. (1993): Mechanical theorem proving in differential geometries. In: Proceedings XXI DGM, Singapore (to appear).

Wu, W.-t., Lu, X. L. (1985): Triangles with equal bisectors. People's Education Press, Beijing (in Chinese).

Yang, L., Zhang, J. Z., Li, C. Z. (1992): A prover for parallel numerical verification of a class of constructive geometry theorems. In: Proceedings IWMM '92, Beijing, pp. 244–250.

1 Desarguesian geometry and the Desarguesian number system

1.1 Hilbert's axiom system of ordinary geometry

What we call *ordinary geometry* in this book is the usual Euclidean geometry.

The famous book "Grundlagen der Geometrie" by Hilbert (1899) first put forward a complete axiom system for ordinary geometry, so that since then it has had a truly rigorous foundation. The axiom system of Hilbert takes some *fundamental concepts*, which need not be defined, as the target of discussion. These fundamental concepts are divided into two classes – fundamental objects like *points*, *lines*, and *planes*, and fundamental relations among these objects like *belonging to*, *between*, and *congruent to*. They obey a certain number of axioms and are considered the starting point of logical deduction. Hilbert classified these axioms into five groups which together constitute a rigorous axiom system that suffices to completely capture ordinary geometry. The names of these five groups of axioms are as follows:

H I	Axioms of incidence (axioms of subordination);
H II	Axioms of order;
H III	Axioms of congruence;
H IV	Axiom of parallels;
H V	Axioms of continuity.

The main goal of Hilbert was to give a systematic logical analysis for the perception of space. To achieve this goal, he explored in detail the logical relationships among some axioms and introduced the concept of independence. However, when Hilbert actually set up his axiom system, he did not strictly abide by the requirement of independence of axioms. For example, the statement of the axioms of order H II must depend on the axioms of incidence H I while the statement of axioms of congruence H III must depend on the first and the second group of axioms. The axioms listed in the first edition of "Grundlagen der Geometrie" in 1899 are not completely independent of each other; some can be deduced from the others. Only in later editions did Hilbert make some revision of his axiom system so that the "superfluous" axioms do not appear any more, which in turn makes some axioms unnatural. Moreover, some intuitively quite self-evident theorems also have to be inferred from axioms while the inferences are often rather cumbersome and lengthy. In our opinion, the tendency to eliminate axioms in order to pursue independence at the cost of the simplicity of whole theory may make us lose more than what we wish to gain and is best avoided. As to the main subject of this book, the mechanization of geometry, we are even less concerned with the independence of axioms.

Actually, in contrast with the idea of axiomatization, the original book by Hilbert already contained the ideas and methods of mechanization consistent with the main subject of this book. It seems that this fact has not been observed previously. It is even hard to say whether Hilbert himself had definitely recognized this point. In this chapter and Chaps. 2 and 3, we shall explain the important role Hilbert's book has played on the mechanization of geometry.

Although for our purposes we do not think the axiom system of Hilbert is completely satisfactory, below we shall still list the revised axiom system given in the 8th edition of Hilbert's book (1956) as the basis of our discussions. The listed axioms will at the same time be restricted to the case of ordinary plane geometry in order to simplify our presentation. Thus, the fundamental objects only consist of two kinds, i.e., points and lines. In addition, to simplify the terminology, two, three, ... points or lines are always understood to be two, three, ... distinct points or lines unless stated otherwise.

H I Axioms of incidence (axioms of subordination)

The fundamental relation is that a point *belongs to* a line. We shall also follow the customary phraseologies such as a point *lies on* a line, a line *passes through* a point, a line *connects* two points, two lines *intersect* at a point. The axioms are:

I1. For every two points A, B there exists a line that passes through each of the points A, B.
I2. For every two points A, B there exists no more than one line that passes through each of the points A, B.
I3. There exist at least two points on a line. There exist at least three points that do not lie on a line.

By Axioms I1 and I2, the line uniquely defined by the two points A, B will be denoted by AB. In case the two lines l_1, l_2 intersect, we denote the corresponding point of intersection by $l_1 \wedge l_2$.

H II Axioms of order

The fundamental relation is that a point is *between* or lies *between* two points. In the planar case, there are four axioms including Pasch's axiom. By these axioms, one may define *segment*, *ray* or *half-line*, *polygonal line*, *angle* and derive some theorems such as a point on a line separates the line into two *sides*, a line in a plane separates the plane into two *sides* and an angle or a *polygon* separates the plane into *interior* and *exterior* parts. A more detailed recounting can be found in Sect. 2.5 and is omitted here.

H III Axioms of congruence

The fundamental relation is the *congruence* or *equality* of segments and angles. The axioms are:

III1. If A, B are two points on a line a and A' is a point on another line a', then it is always possible to find a point B' on a given side of line a' through A' such that segment AB is congruent or equal to segment $A'B'$. In symbols,

$$AB \equiv A'B'.$$

III2. If a segment $A'B'$ and a segment $A''B''$ are congruent to the same segment AB, then segment $A'B'$ is also congruent to segment $A''B''$.
III3. On a line a let AB and BC be two segments which except for B have no point in common. Furthermore, on the same or on another line a' let $A'B'$ and $B'C'$ be two segments which except for B' also have no point in common. In that case, if

$$AB \equiv A'B', \quad BC \equiv B'C',$$

then

$$AC \equiv A'C'.$$

III4. Let $\measuredangle(h, k)$ be an angle and a' a line and let a definite side of a' be given. Let h' be a ray on line a' that emanates from the point O'. Then there exists one ray k' emanating from O' such that angle $\measuredangle(h, k)$ is congruent or equal to the angle $\measuredangle(h', k')$ and at the same time all interior points of angle $\measuredangle(h', k')$ lie on the given side of a'. Symbolically,

$$\measuredangle(h, k) \equiv \measuredangle(h', k').$$

Every angle is congruent to itself, i.e.,

$$\measuredangle(h, k) \equiv \measuredangle(h, k)$$

is always true.
III5. If for two triangles ABC and $A'B'C'$ the congruences

$$AB \equiv A'B', \quad AC \equiv A'C', \quad \measuredangle BAC \equiv \measuredangle B'A'C'$$

hold, then the congruence

$$\measuredangle ABC \equiv \measuredangle A'B'C'$$

is also satisfied.

Starting from these axioms and Axioms H I–H II, one may define such concepts as right angle, perpendicularity, displacement of segments and of angles, addition of segments and of angles as well as quantitative comparison of segments and of angles.

By applying Axioms H I–H III, we may prove the theorems about congruent triangles, isosceles triangles, perpendicular lines, bisection of segments and of angles and so on in ordinary geometry. In addition, we can prove some theorems of inequalities such as *the exterior angle of a triangle is greater than any interior angle that is not adjacent to it* and *the sum of two sides of a triangle is greater than its third side* as was done in Euclid's "Elements." However, by axioms H I–H III one can only prove that the sum of the three angles of a triangle is less than or equal to two right angles, but cannot prove that it must be equal to two right angles.

H IV Axiom of parallels

This group of axioms consists only of one, the so-called Euclid's axiom: Let a be any line and A a point not on it. Then there is at most one line, determined by a and A, that passes through A and does not intersect a.

By this axiom and the axioms of congruence listed before, we can infer that there is not only at most one line but also exactly one line that passes through A and does not intersect a, and we can prove that the sum of the three interior angles of a triangle equals that of two right angles.

As our exposition hereafter will involve no axioms of congruence or order, we shall take the axiom of parallels in a sharper form which was denoted as axiom of parallels IV* in Hilbert's book and is denoted here simply as axiom of parallels IV, stated as follows.

IV. Let a be any line and A a point not on a. Then there exists *one and only one* line determined by a and A that passes through A and does not intersect a.

These two non-intersecting lines are said to be *parallel*, which is denoted by

$$a \parallel b.$$

It can be seen from Axiom IV that if two lines are both parallel to a third line, they are also parallel to each other.

H V Axioms of continuity

Here we do not need to introduce any new fundamental relation. There are two axioms:

V1 (Axiom of measure or Archimedes' axiom). If AB and CD are any two segments, then there exists a finite number of points $A_1, A_2, \ldots, A_n$ on the ray from A through B such that the segments $AA_1, A_1A_2, \ldots, A_{n-1}A_n$ are all congruent to the segment CD while B lies between A and A_n.

V2 (Axiom of line completeness). An extension of a set of points on a line with its order and the first congruence axiom as well as Archimedes' axiom (i.e., Axioms H I, H III, H III1, H V1) being preserved is impossible.

Among the above five groups of axioms, the fifth which consists of two axioms, i.e., the axiom of measure or Archimedes' axiom and the axiom of line completeness, has special significance for this book. Axioms in this group deal with properties of intuitive continuity, and the role, as stated in Hilbert's book, of the first is on preparation for the requirement of continuity and of the second together with other axioms is for completing the whole axiom system. In addition, Hilbert stated that the exposition of "Grundlagen der Geometrie" is principally based on Archimedes' axiom and generally does not assume the validity of the second axiom of line completeness. In fact, in the first edition of Hilbert's book the fifth group of axioms only consists of Archimedes' axiom as the axiom of line completeness was added only in later editions. The addition of this axiom, aimed purely at filling the "gap" between plane and space and at making the geometry discussed uniquely become the usual Euclidean geometry, is quite unnatural. Not only did the axiom of line completeness not play any role but the use of Archimedes' axiom was also avoided for all discussions when Hilbert established the important theory about geometry in his book. Therefore, the Euclidean geometry of Hilbert is essentially a non-Archimedean geometry. In the preface to the Russian edition of "Grundlagen der Geometrie," Rashevsky made a brilliant comment on this point.

It is very important to point out this fact for the mechanization of geometry to be explained in this book. First of all, axioms of continuity unavoidably involve concepts like the set of all points or infinite number of points on a line or in a plane. On the other hand, in the hypothesis and conclusion of an axiom or a theorem, it is always the case that only a finite number of points, lines, circles or other geometric objects are involved. Also, the process of proving a theorem is no more than a repeated application of the axioms and given theorems to this finite number of points, lines or circles in a finite number of constructive steps, so that the proof can proceed from hypothesis to conclusion. The so-called mechanization of theorem proving does mean that the constructive steps of proof for some classes of theorems can be given step by step according to a certain mechanical procedure, which ensures either getting to the conclusion from the hypothesis or reporting the impossibility of eliciting the conclusion after a finite number of steps. This finite number of mechanical steps can be easily realized on computer and is what mechanical proving aims to do. Since electronic computers can only deal with a finite number of things, the finiteness of stating and proving theorems becomes a prerequisite to using computers. Hilbert's development of geometry by avoiding axioms of continuity as exhibited in "Grundlagen der Geometrie" removes the obstacle for the mechanization and the use of computers in proving geometric theorems and thus provides a necessary precondition for its success.

Actually, Hilbert himself already gave a concrete result on the mechanization of geometry theorem proving (see Chap. 3 for details). Hilbert started from his axiom system without assuming axioms of continuity and arrived at a certain level of concrete mechanization. His method is principally that of introducing the number system determined by geometry on the basis of axioms. This process is called the *algebraization* of geometry. According to the correspondence relation between points on a line and numbers in a system, coordinates may be introduced. Geometric theorems can then be transformed into theorems about algebraic expressions, of which the problem of mechanical proving can be described much more clearly and is usually easily solved. Those algebraic expressions are either polynomial equations or polynomial inequalities in the coordinates, of which the latter reflect some relations of order in geometry. However, except for those in elementary plane geometry, theorems which really deal with relations of order are generally rather rare. In addition, our mechanical method of theorem proving is very efficient in case we are dealing with those theorems whose algebraic form corresponds to polynomial equations. Whereas in the case of dealing with inequality relations, the problem becomes much more complicated (with respect to the complexity and feasibility of computation). Therefore, in theory there are reasons to avoid using axioms of order to establish more general ordinary geometry. For this purpose, in later sections of this chapter we shall introduce an axiom system with a slight difference from Hilbert's which avoids using not only the axioms of continuity but also the axioms of order. Based on this axiom system, we shall set up many kinds of ordinary geometries in a broad sense. This and the next chapter will explain how to get to coordinatization from axiomatization while the remaining chapters will be focused on the mechanization problem of theorem proving in these geometries.

1.2 The axiom of infinity and Desargues' axioms

In the (plane) geometry considered below, we shall still take points and lines as fundamental objects and that a point lies on a line as a fundamental relation. We assume Hilbert's axioms of incidence H I and axiom of parallels H IV but do not introduce any concept nor assume any axiom of order. In this geometry, words like segment and ray do not have any meaning and thus angle, triangle, and polygon cannot be defined in the usual way. Even so, we can still define the concepts of triangles and parallelograms in another way as follows. Remember that, when parallel lines are mentioned, they are always meant to be neither intersecting nor the same.

Definition 1. Three points A, B, C with a fixed order, if distinct from each other and not lying on the same line, constitute a *triangle* ABC, denoted as $\triangle ABC$. In this case, the points A, B, C are called the *vertices* and the connecting lines AB, AC, BC the *sides* of this triangle. For two triangles ABC and $A'B'C'$, we say, according to the order of their vertices, that A, A'; B, B' and C, C' are the three pairs of *corresponding* vertices and $AB, A'B'$; $AC, A'C'$; $BC, B'C'$ are the three pairs of *corresponding* sides (cf. Fig. 1.1).

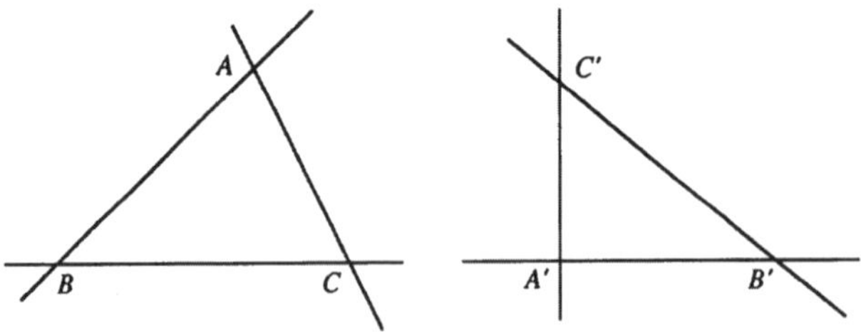

Fig. 1.1

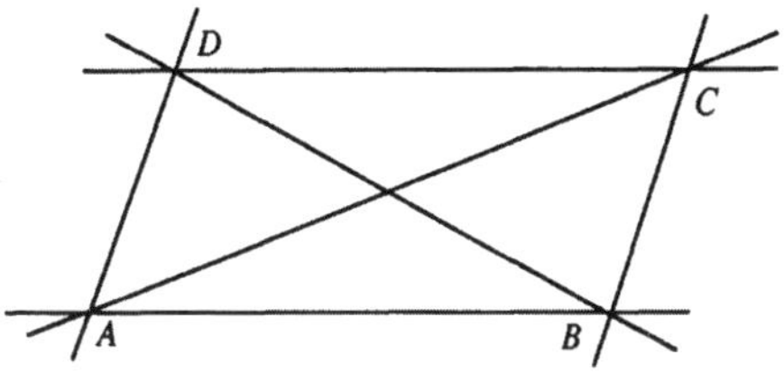

Fig. 1.2

Definition 2. Four points A, B, C, D with a fixed order constitute a *parallelogram*, denoted as $\square ABCD$; if these points are distinct from each other, any three of them are not collinear and the connecting line AB of A, B and the connecting line CD of C, D are parallel, and so are the connecting line AD of A, D and the connecting line BC of B, C, i.e., $AB \parallel CD$, $AD \parallel BC$. In this case the points A, B, C, D are called the *vertices* and, A, C and B, D two pairs of *opposite vertices* of $\square ABCD$. We say that AB, BC, CD, AD are four *sides* of $\square ABCD$, in which AB, CD are a pair of *opposite sides*, so are AD, BC. Furthermore, we call the connecting line AC of A, C and the connecting line BD of B, D the *diagonals* of $\square ABCD$ (cf. Fig. 1.2).

Starting from two points A, B one can construct a parallelogram as follows. By Axiom H I, there is another point, say C, not on the line AB. By Axiom H IV, we may draw through B a parallel to AC and through C a parallel to AB. By Axiom H IV again, these two parallels are not parallel but necessarily meet at a point D; then $ABDC$ is a parallelogram.

Instead of the axioms of order, we introduce the following.

Axiom of infinity I. Let A_0, A_1 be any two points on a line l and construct an arbitrary $\square A_0A_1BC$.[1] Through B draw a line $BA_2 \parallel CA_1$, meeting l at point A_2, and then a line $BA_3 \parallel CA_2$, meeting l at point A_3, and so on analogously. Similarly, through C draw a line $CA_{-1} \parallel BA_0$, meeting l at point A_{-1}, and then a line $CA_{-2} \parallel BA_{-1}$, meeting l at point A_{-2}, and so on analogously. Then, in the infinite series

$$\ldots, A_{-2}, A_{-1}, A_0, A_1, A_2, \ldots,$$

no two points are the same (cf. Fig. 1.3).

This Axiom I is sometimes denoted as *Axiom* D_∞ and the part $A_2 \neq A_0$ is equivalent to saying that the diagonals of $\square A_0A_1BC$ must meet each other, which is called *Fano's axiom* in the literature.

Axiom I ensures that the diagonals of a parallelogram must intersect and

1 Throughout the book we use a symbolized expression typically of the form "draw through A, $AB \parallel a$, meeting b at C." This expression may be restated in full length as "draw through the point A a parallel AB to the line a. This parallel meets the line b at a point C." [Transl.]

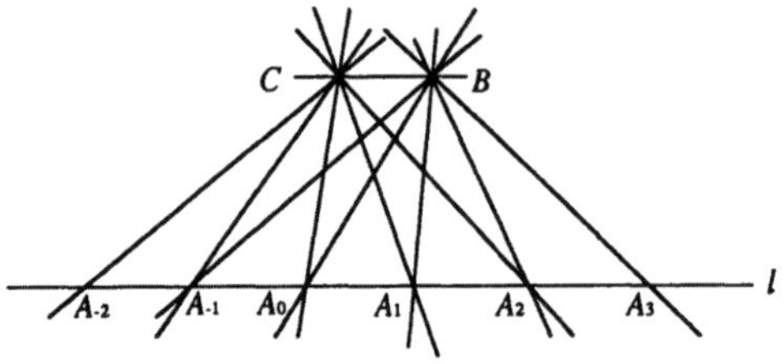

Fig. 1.3

that there are infinitely many points on a line and thus in a plane. From this we also know that there are infinitely many distinct lines through an arbitrary point.

Besides the above axiom of infinity I, we introduce two axioms associated with the name of Desargues which play a decisive role in establishing the algebraization and then the mechanization of geometry. These two axioms are stated as follows.

Desargues' axiom D_1. If the three pairs of the corresponding sides of two triangles ABC and $A'B'C'$ are all parallel to each other, i.e.,

$$AB \parallel A'B', \quad AC \parallel A'C', \quad BC \parallel B'C',$$

then the three lines AA', BB', CC' joining the corresponding vertices of these two triangles are either parallel to each other or concurrent.

Desargues' axiom D_2. If two pairs of the corresponding sides of two triangles ABC and $A'B'C'$ are parallel to each other, say

$$AB \parallel A'B', \quad AC \parallel A'C',$$

and the three lines joining the corresponding vertices are distinct yet either concurrent or parallel to each other, then the third pair of the corresponding sides are also parallel to each other, i.e.,

$$BC \parallel B'C'$$

(Fig. 1.4).

Evidently, Desargues' two axioms D_1 and D_2 are not independent of each other, i.e., under the assumption of the Axioms H I, H IV, and I (or D_∞) D_1 and D_2 can be deduced from each other.

A direct corollary of Desargues' two axioms is that one may introduce the concepts of *midpoints* and *symmetric points* as follows.

As shown in Fig. 1.5, let A, B be two arbitrary points on a line l. Construct $\square ABDE$ arbitrarily as before. Through D draw a parallel to BE, meeting l at a point C. By the axiom of infinity I, C and A are distinct. By applying Desargues' axiom, we may show that point C is independent of the construction of $\square ABDE$ as follows.

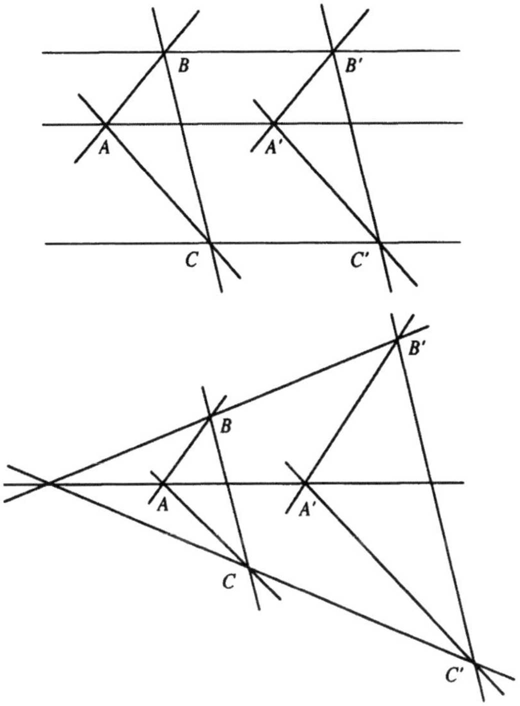

Fig. 1.4

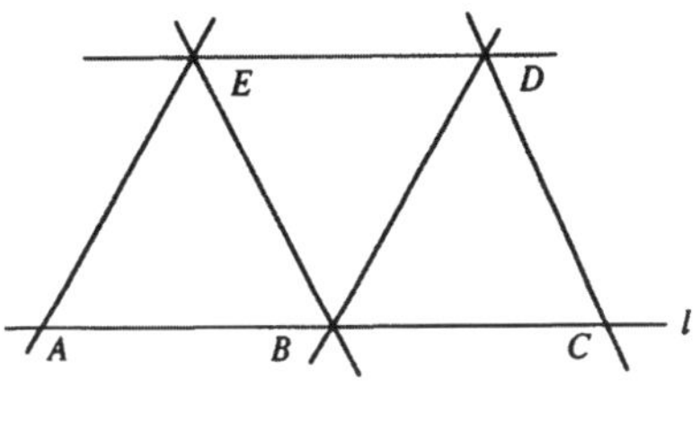

Fig. 1.5

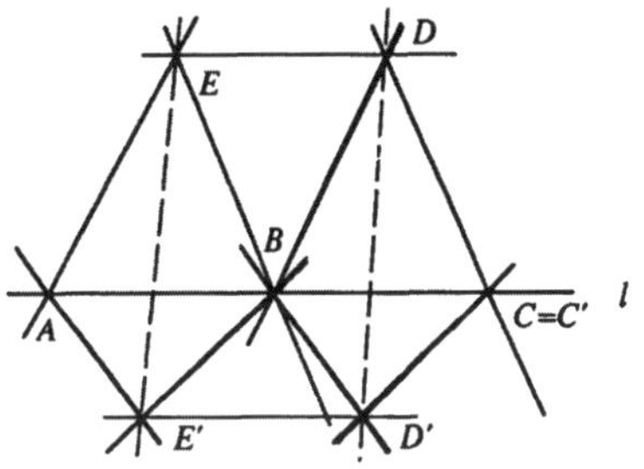

Fig. 1.6

Let $ABD'E'$ be another parallelogram and draw $D'C' \parallel E'B$, meeting l at point C'. We need to prove $C' = C$. Let us discuss some different cases respectively.

Firstly, suppose that the line AE' is distinct from AE, the three points E, B, E' do not lie on the same line and EE' is not parallel to l (cf. Fig. 1.6). In this case $DE, D'E', l$ are pairwise distinct but parallel to each other. As A, E, E' do not lie on the same line, they form a triangle. By Axiom H IV, B, D, D' are not collinear and thus also constitute a triangle. Applying Desargues' axiom D_2 to $\triangle AEE'$ and $\triangle BDD'$, we have $EE' \parallel DD'$. Moreover, since B, E, E' are not collinear and thus form a triangle, by Axiom H IV, C, D, D' are not

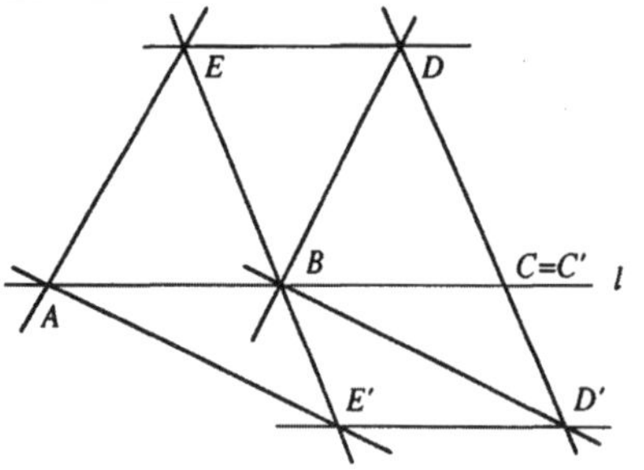

Fig. 1.7

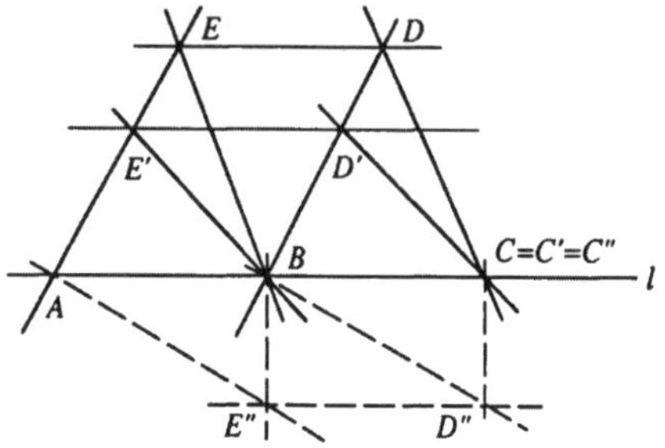

Fig. 1.8

collinear and also form a triangle. Applying Desargues' axiom D_2 once more to $\triangle BEE'$ and $\triangle CDD'$ yields $D'C \parallel E'B$. Then by Axiom H IV, the two lines $D'C$ and $D'C'$ are the same, so C and C' coincide.

Secondly, suppose that the line AE' is still distinct from AE while E, B, E' are collinear (cf. Fig. 1.7). As before, we still have $DD' \parallel EE'$. Thus, the lines $D'C'$, DD' and DC are the same, which implies that $C' = C$.

Finally, suppose that the lines AE' and AE are the same (Fig. 1.8), or they are distinct but $EE' \parallel l$ (Fig. 1.9). In either case one can always construct a line through A distinct from AE, AE' and l, and choose a point E'' distinct from A on this line such that both EE'', $E'E''$ are not parallel to l. Construct in addition $\square ABD''E''$ and $D''C'' \parallel E''B$, meeting l at point C''. By applying the case we have proved before to $\square ABDE$ and $\square ABD''E''$, it follows that $C'' = C$. Applying that case again to $\square ABD'E'$ and $\square ABD''E''$ yields $C'' = C'$. Hence, $C' = C$.

We have taken great pains to write down the explicit proof above, for which the reason will be explained in Sect. 3.1.

From the above proof, we see that the following definition is proper.

Definition 3. For two arbitrary points $A \neq B$ on a line l, let us construct $\square ABDE$ and $DC \parallel EB$, meeting l at C. Then point C is independent of the construction of $\square ABDE$. We say that C is the *symmetric point* of A with

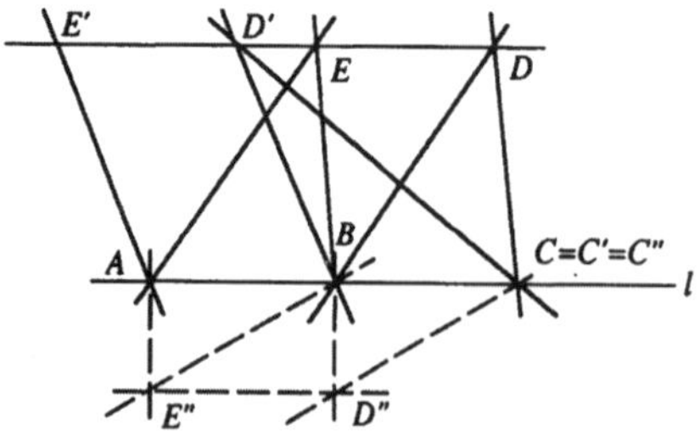

Fig. 1.9

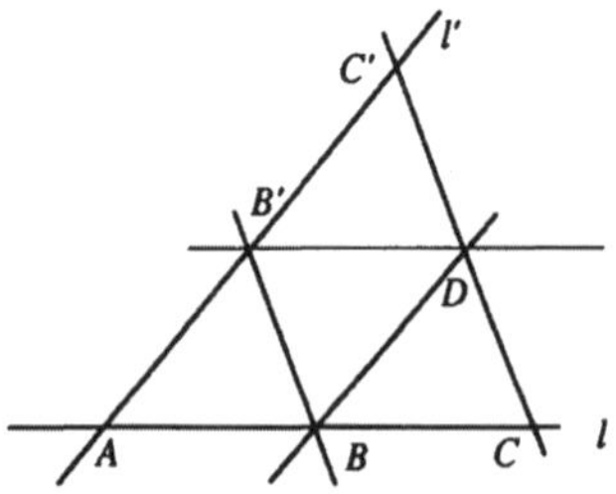

Fig. 1.10

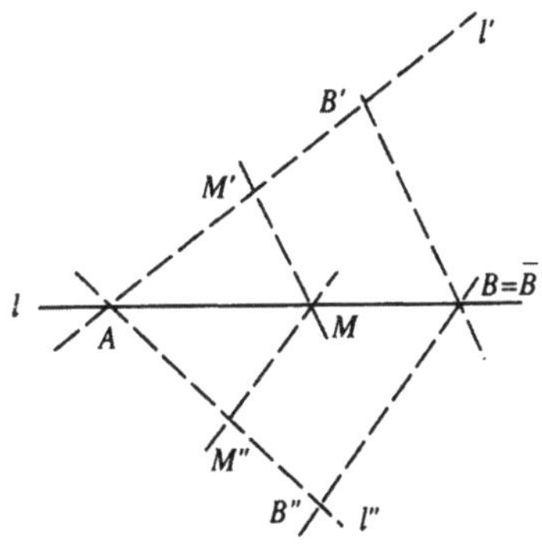

Fig. 1.11

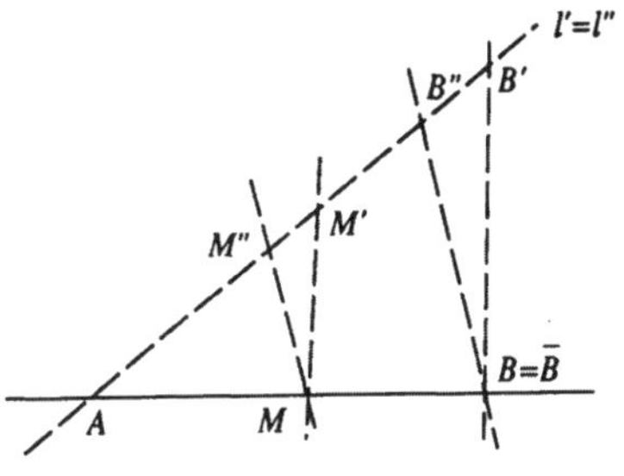

Fig. 1.12

respect to B. In addition, any point is said to be a *symmetric point* of the point with respect to itself.

By this definition, it is easy to prove that if C is the symmetric point of A with respect to B, then A is also the symmetric point of C with respect to B. One may also prove that if two lines l, l' meet at a point A, and two points B, B' distinct from A lie on l, l', respectively, while C, C' are the symmetric points of A with respect to B, B', respectively, then CC' is parallel to BB'.

The proof is self-evident from Fig. 1.10, in which $ABDB'$ is a parallelogram.

We now introduce the concept of a *midpoint* of two points as follows.

Let A and B be any two points on a line l as in Fig. 1.11. Draw through A a line l' distinct from l and take thereon a point M', distinct from A. Construct the symmetric point B' of A with respect to M' and the parallel of BB' through M', meeting l at M. Let us prove that M is independent of the choice of l' and M' as follows.

Note first that the symmetric point of A with respect to M is B. To see this, let the symmetric point of A with respect to M be $\bar{B}$; we have $B'\bar{B} \parallel M'M$ from the above. Then the two lines $B'\bar{B}$ and $B'B$ are the same. Hence, $\bar{B} = B$, i.e., the symmetric point of A with respect to M is B.

To prove that M is independent of the choice of (l', M'), we may assume that the line l'' is distinct from l and that M'' is a point distinct from A on l'' (l'' may be the same as l') (Figs. 1.11 and 1.12). Construct the symmetric point B'' of A with respect to M''. Then $B''B \parallel M''M$ from the above, for the symmetric point of A with respect to M is B. Hence, (l'', M'') and (l', M') lead to the same point M.

Based on the above proof, the following definition is proper.

Definition 4. Let A, B be two points on a line l. Draw through A a line l' distinct from l and take thereon a point M' distinct from A. Construct the symmetric point B' of A with respect to M' and draw through M' a line $M'M \parallel B'B$, meeting l at M. Then M is independent of the choice of l', M', and is called

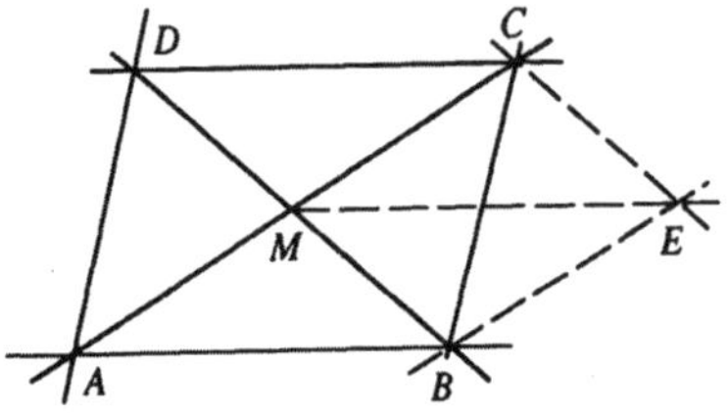

Fig. 1.13

the *midpoint* of A and B. Define the midpoint of two coincident points to be the point itself.

By this definition, it is easy to show that the midpoint of A, B is also the midpoint of B, A and that, for a $\triangle ABC$, if the midpoints of A, B and A, C are M and N, respectively, then MN is parallel to AB.

Theorem 1. The two diagonals of a parallelogram must intersect and bisect one another. Namely, the point of intersection of the two diagonals is the common midpoint of the two pairs of opposite vertices.

Proof. As in Fig. 1.13, let $\square ABCD$ be a parallelogram. By Axiom I, its diagonals AC, BD intersect, say at a point M. Draw through B, C the parallels of AC, BD, respectively, and let their point of intersection be E. Application of Desargues' axiom D_1 to $\triangle AMD$ and $\triangle BEC$ yields $ME \parallel AB \parallel CD$. As both $AMEB$ and $MCEB$ are parallelograms, C is the symmetric point of A with respect to M, i.e., M is the midpoint of AC. Similarly, M is the midpoint of BD. Therefore, AC, BD bisect one another and meet at point M. This completes the proof. □

Apparently, the inverse of the above theorem holds as well. In other words, we have the following.

Theorem 2. Suppose four points A, B, C, D are pairwise distinct and any three of them are not collinear. If the midpoint of A, C is the same as the midpoint of B, D, then $ABCD$ is a parallelogram.

From the results on midpoints and symmetric points, one knows that the points A_n $(n = \pm 1, \pm 2, \ldots)$ determined by A_0, A_1 in the axiom of infinity I (or D_∞) may also be determined in the following way.

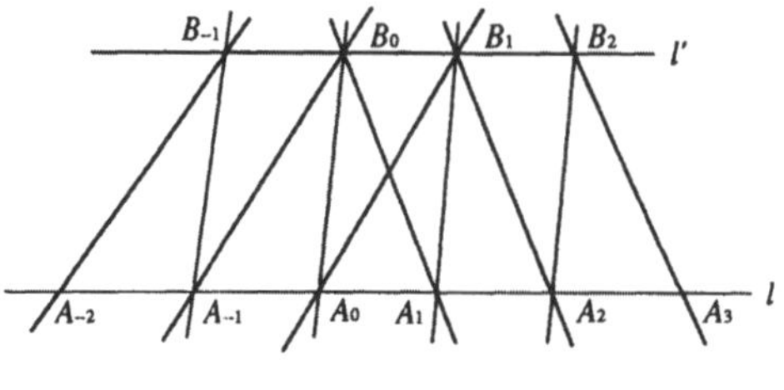

Fig. 1.14

As in Fig. 1.14, draw $l' \parallel l$ and, $A_0B_0 \parallel A_1B_1$, meeting l' at points B_0 and B_1, respectively. Draw through B_1, $B_1A_2 \parallel B_0A_1$, meeting l at A_2 and through A_2, $A_2B_2 \parallel A_1B_1$, meeting l' at B_2. Draw further, through B_2, $B_2A_3 \parallel B_1A_2$, meeting l at A_3. By analogy, we get successively the points $A_2, A_3, A_4, \ldots$. Draw, on the other hand, through B_0, $B_0A_{-1} \parallel B_1A_0$, meeting l at A_{-1} and through A_{-1}, $A_{-1}B_{-1} \parallel A_0B_0$, meeting l' at B_{-1}. Construct then $B_{-1}A_{-2}$ through B_{-1}, meeting l at A_{-2}. Analogously, we get the points $A_{-1}, A_{-2}, A_{-3}, \ldots$.

It is easy to see that all these points A_n are the same as those constructed in Axiom I. One may also consider all distinct points obtained from the present construction as Axiom I, then introduce the concepts of symmetric points and midpoints and finally prove that the points A_n in the original axiom are the same as those constructed in the present way. In one word, under the assumption of Axioms H I, H IV and D_1, D_2, two different statements about the axiom of infinity I are equivalent to each other.

1.3 Rational points in a Desarguesian plane

We have introduced the axiom of infinity D_∞ and Desargues' axioms D_1, D_2 in the last section. From them, the concepts of midpoints and symmetric points are derived (under the assumption of H I, H IV). Looking back at Fig. 1.3 about the axiom of infinity D_∞ at the beginning of Sect. 1.2, it is seen that A_1 is the midpoint of A_0 and A_2, A_2 is the midpoint of A_1 and A_3, and A_0 is the midpoint of A_{-1} and A_1, etc. In this section we further explore this kind of relation.

Definition 1. Let the Axioms D_∞, D_1 and D_2 altogether be denoted as D. Then the set of points and lines which satisfy Hilbert's axioms of incidence H I and axiom of parallels H IV and the Axioms D is said to constitute a *Desarguesian plane*, and the corresponding geometry is called a *Desarguesian (plane) geometry*.

In Hilbert's original axiom system, the congruence relation is an undefined fundamental relation. In Desarguesian geometry, neither is congruence introduced as a fundamental concept nor is any axiom of congruence assumed, but the *congruence* relation and a certain *addition* may still be defined on an arbitrary line such that the series (see Fig. 1.3 for the axiom of infinity I or D_∞)

$$\ldots, A_{-n}, \ldots, A_{-1}, A_0, A_1, A_2, \ldots, A_n, \ldots$$

has some isomorphism relation to the integer sequence. Observe the following.

First of all, any two points A, B with a fixed order, whether they are the same or not, are called a *pair of points*, denoted by (AB). In case A, B are distinct, we define the *midpoint* of the pair (AB) to be the midpoint of A and B and, in case A, B are the same, the *midpoint* of the pair (AB) is defined to be A. Now let us define the concept of congruence on a line.

Definition 2. Let four points A, B, C, D, not necessarily distinct, lie on a line l. If the pairs (AD) and (BC) have the same midpoint, we say that the pair (AB)

is congruent to pair (CD), denoted by

$$(AB) \equiv (CD).$$

From some simple properties of midpoints and the above definition, it is clearly seen that in case $(AB) \equiv (CD)$ we also have

$$(AC) \equiv (BD).$$

Let A, B, C be three, not necessarily distinct, fixed points on a line l. The point D such that $(AB) \equiv (CD)$ will be uniquely determined by A, B, C. This point can also be constructed in the following way.

Obviously, we have $D = C$ when $B = A$, and $D = B$ when $C = A$. Thus, we may assume that both B and C are distinct from A. Since in this case $B \neq A$, we may construct $\square ABB'A'$. Now through B' draw $B'D \parallel A'C$, meeting l at D. Then D is the point desired such that $(AB) \equiv (CD)$. This can be shown as follows.

Suppose first that $C = B$. By definition, C is the midpoint of A, D and of (BC). Hence $(AB) \equiv (CD)$ and point D is what we wanted (cf. Fig. 1.15).

Suppose next that $C \neq B$ and M is the midpoint of B, C. If $M = A$, then $B'D \parallel A'C$ and it is the same as $B'A$, so D is the same as A and the midpoint of (AD) is also M. Therefore $(AB) \equiv (CD)$ and point D is what we wanted (cf. Fig. 1.16).

Suppose finally that $C \neq B$ and the midpoint M of B, C does not coincide with A. Construct, as in Fig. 1.17, the diagonals $A'D$ and $B'C$ of $\square A'B'DC$, meeting at point N. By Theorem 1 in the last section, N is the midpoint of both B', C and A', D. From $\triangle BCB'$ one knows $MN \parallel BB'$ and $MN \parallel AA'$. From

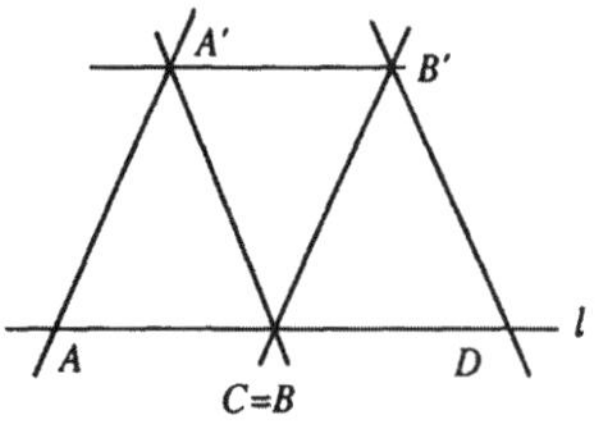

Fig. 1.15

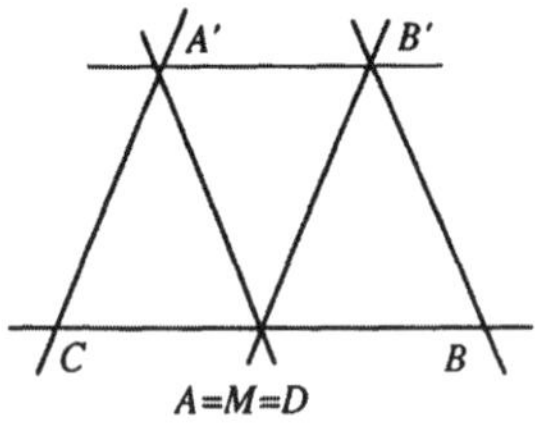

Fig. 1.16

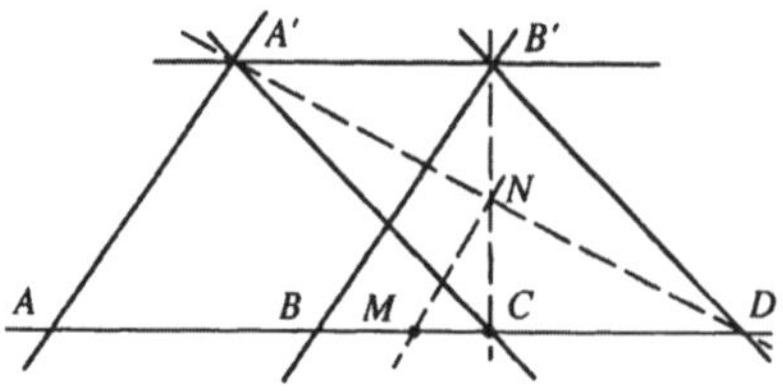

Fig. 1.17

$\triangle ADA'$, one knows that M is also the midpoint of A, D. So, by definition we have

$$(AB) \equiv (CD)$$

and thus point D is what we wanted.

The above proof is rather cumbersome and the interested reader may refer to Sect. 3.1.

From this proof and the definition itself, it is easy to see the reflexivity, symmetry, and transitivity of the congruence relation. In other words, we have the following.

Theorem. The congruence relation "$\equiv$" among pairs of points on a line is an equivalence relation.

Below we shall introduce the addition of pairs of points. For this purpose, we first prove the following.

Lemma. Let the points A, B, C, D, E, F lie on one and the same line l with

$$(AB) \equiv (DE), \quad (BC) \equiv (EF).$$

Then whether or not some of these points are the same, one always has

$$(AC) \equiv (DF).$$

Proof. See Fig. 1.18. If A, B, C are all distinct, one may construct $\square ABB'A'$ and $\square BCC'B'$. Then, by the construction of congruence $A'DEB'$ and $B'EFC'$ are both parallelograms, and so are $ACC'A'$ and $DFC'A'$. By the construction of congruence again, one obtains $(AC) \equiv (DF)$.

If some of A, B, C are the same, one may still show $(AC) \equiv (DF)$ by verifying every case. We omit the cumbersome verifications, even though they are necessary (cf. Sect. 3.1). □

According to the above lemma, the following definition is proper.

Definition 3. Fix a point A_0 on a line l and construct, for any two points R, S on l, a point T such that $(ST) \equiv (A_0R)$. We say that the pair (A_0T) is the *sum*

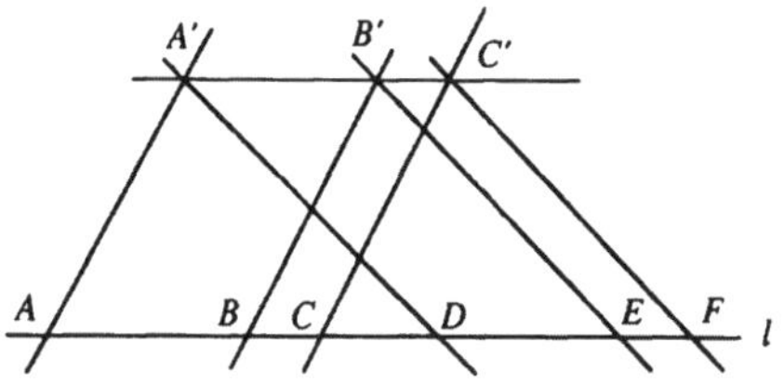

Fig. 1.18

of the pairs (A_0R) and (A_0S), which is denoted by

$$(A_0T) = (A_0R) + (A_0S).$$

For fixed A_0, the above expression is denoted simply as

$$T = R + S.$$

T is called the *sum* of R and S (for fixed A_0).

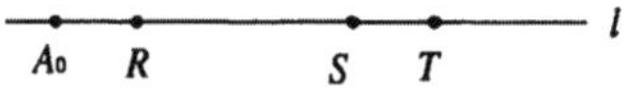

Fig. 1.19

By this definition, we have in particular (with respect to A_0)

$$R = A_0 + R.$$

At the beginning of the last section, we have constructed from two points, A_0, A_1, an infinite sequence of points

$$\ldots, A_{-n}, \ldots, A_{-1}, A_0, A_1, A_2, \ldots, A_n, \ldots$$

on a line l. It is easy to prove, according to the above definition of addition among pairs of points, that if (A_0A_n) corresponds to the integer n, then the addition among the pairs (A_0A_n) corresponds to the addition among the integers, i.e.,

$$(A_0A_n) + (A_0A_m) = (A_0A_{n+m}),$$

or

$$A_n + A_m = A_{n+m},$$

for fixed A_0, A_1.

Besides this, we may construct, for any rational number $r = p/q$ $(p, q$ are integers and q is positive), a point A_r on line l such that in the case that (A_0A_r) corresponds to r, the addition relation is still valid (r, s are any rational

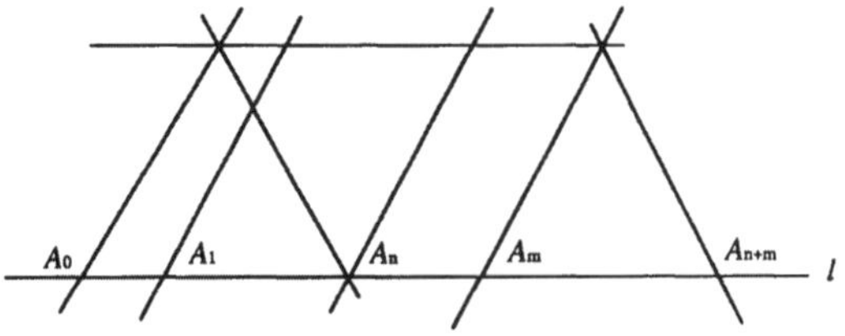

Fig. 1.20

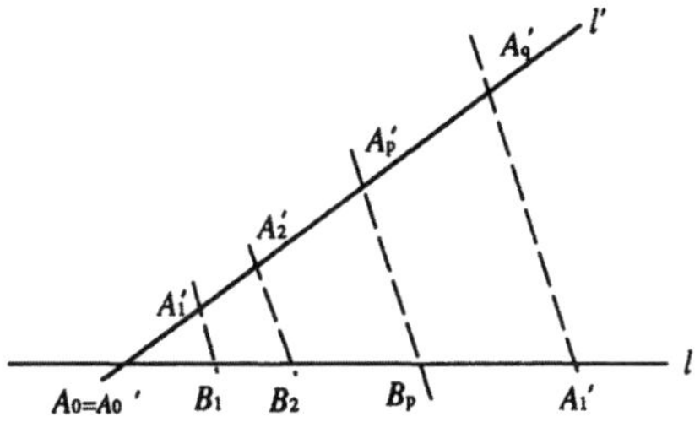

Fig. 1.21

numbers):

$$(A_0A_r) + (A_0A_s) = (A_0A_{r+s}),$$

or written simply as

$$A_r + A_s = A_{r+s}$$

for a fixed A_0.

Construct through A_0, as shown in Fig. 1.21, a line l' distinct from l, take thereon a point A'_1 distinct from A_0, and rewrite A_0 as A'_0. Starting from A'_0, A'_1, as in Fig. 1.3 for the axiom of infinity D_∞ in Sect. 1.2, construct the series

$$\ldots, A'_{-n}, \ldots, A'_{-1}, A'_0, A'_1, \ldots, A'_n, \ldots$$

and the connecting line $A_1A'_q$. Draw $A'_nB_n \parallel A'_qA_1$, meeting l at B_n. Then, it can be easily seen that B_p depends upon neither the choice of l' and A'_1 nor the expression $r = p/q$, so B_p can be denoted as

$$A_r = A_{p/q} = B_p.$$

Definition 4. In the case $r = p/q$ is a rational number, we say that the point $A_{p/q}$ *depends rationally upon* the points A_0, A_1.

According to this definition,

$$A_{p/q} \to p/q$$

induces a one-to-one correspondence between all the points depending rationally upon A_0, A_1 and all the rational numbers. Under this correspondence, the addition among A_r with respect to A_0, A_1 corresponds to that among the rational numbers. If we define the *product* of A_r and A_s as A_{rs} (still with respect to A_0, A_1), then the multiplication among these points also corresponds to that among the rational numbers and, the addition and multiplication among these points retain all operation rules corresponding to those among rational numbers, especially A_0 corresponding to 0 and A_1 to 1.

There is a greatness relation among the rational numbers which makes the set of rational numbers ordered, wherein the usual relations of order, addition, and multiplication are satisfied. Therefore, under the correspondence $A_{p/q} \to$

p/q, we can introduce the concept of order among those points which depend rationally upon A_0, A_1 according to the corresponding order among the rational numbers such that the usual relations of order are satisfied. In particular, the points which depend rationally upon A_0, A_1, except A_0, A_1 themselves, can be divided into two parts: one consists of those points which are *between* A_0, A_1, i.e., those points A_r which correspond to such rational numbers r that are greater than 0 and less than 1. They are called the *interior* points of the *rational segment* A_0A_1. The other part consists of those points A_r which are not between A_0, A_1, i.e., those points for which r is either less than 0 or greater than 1. They are called the *exterior* points of the rational segment A_0A_1. In one word, for any three points A_r, A_s, A_t which depend rationally upon A_0, A_1, we can define, according to the order of the rational numbers r, s, t, the order relation that one point lies *between* the other two. Such a betweenness relation satisfies the order axioms of Hilbert on the line, but is restricted only to the points which depend rationally upon A_0, A_1. It should be stressed that in Desarguesian geometry discussed in this section, three *arbitrary* points on a line cannot lead to the relations of order which satisfy Hilbert's axioms of order.

Starting from the rational dependence of points on a line for three points A, B, C, not collinear, in the plane it is also easy to define the points which depend rationally upon them. First, take an arbitrary point D which depends rationally upon B, C on the line BC, and then an arbitrary point E which depends rationally upon A, D on the line AD. So, point E is one which *depends* rationally upon A, B, C. It is easy to prove that all points, except for those lying on the three sides AB, AC and BC of the triangle ABC, which depend rationally upon A, B, C can be divided into the interior and the exterior two parts – called the *rational interior part* and *rational exterior part* – of triangle ABC which have those properties concerning the interior and exterior of triangles in usual geometry. Although for all points in a plane the corresponding Pasch's axiom may not necessarily be satisfied or may even become meaningless, we can still prove that all points which depend rationally upon A, B, C satisfy the so-called Pasch's axiom of the triangle ABC.

For a parallelogram, we may define *rational points* with respect to it and prove that the rational points which do not lie on the four sides of the parallelogram can be divided into the *rational interior* and the *rational exterior* parts which satisfy those properties concerning the interior and exterior of parallelograms derived from axioms of order in usual geometry.

The proofs of many of the above assertions are easy and are omitted here.

1.4 The Desarguesian number system and rational number subsystem

The points and lines satisfying the Axioms H I, H IV and D constitute a Desarguesian plane. In the preceding section, we have proved that after fixing two points on an arbitrary line in a Desarguesian plane, one can determine a correspondence between an infinite number of points and the rational number system such that the two fixed points correspond to 0 and 1 respectively. However, besides these points there are generally other points on the line. In the following sections we shall prove that among all points on a line, after having fixed two, one can introduce some operations such that they correspond to numbers in a certain system, called a Desarguesian *number system*, which has some special properties and contains all rational numbers. It will organically relate to and be uniquely determined by the Desarguesian plane. For this end, let us first define the Desarguesian number system as follows.

Definition. Let N be a set in which there are two binary operations *addition* and *multiplication* among its elements, satisfying the following Axioms 1–12. We call N a *Desarguesian number system* and the elements of N *numbers.*

The axioms are divided into three groups.

Group I Axioms of incidence (N 1–N 6)

N 1. There is a binary operation, called *addition* (+), such that for any two numbers $a, b \in N$, there is a definite number $c \in N$. Symbolically,

$$c = a + b \quad \text{or} \quad a + b = c.$$

N 2. For any two numbers $a, b \in N$, there always exists one and only one number $x \in N$ such that

$$a + x = b$$

and one and only one number $y \in N$ such that

$$y + a = b.$$

N 3. There exists a definite number 0, called "zero," such that for every $a \in N$ both

$$a + 0 = a \quad \text{and} \quad 0 + a = a.$$

N 4. There is a binary operation, called *multiplication* (·), such that for any two numbers $a, b \in N$, there is a definite number $d \in N$. Symbolically,

$$a \cdot b = d \quad \text{or} \quad d = a \cdot b$$

or by omitting the dot "·" as

$$ab = d \quad \text{or} \quad d = ab.$$

N 5. For any given numbers $a, b \in N$, where $a \neq 0$, there always exists one and only one number $x \in N$ such that

$$ax = b$$

and also one and only one number $y \in N$ such that

$$ya = b.$$

N 6. There exists a definite number 1, called "unit," such that for every $a \in N$ both

$$a \cdot 1 = a \quad \text{and} \quad 1 \cdot a = a.$$

Group II Axioms of operation (N 7–N 11)

Let a, b, c be any three numbers in N. Then the following rules of operation hold:

N 7. $a + (b + c) = (a + b) + c$.
N 8. $a + b = b + a$.
N 9. $a(bc) = (ab)c$.
N 10. $a(b + c) = ab + ac$.
N 11. $(a + b)c = ac + bc$.

Because of the Axioms N 7 and N 9, some parentheses can be omitted.

Group III Axiom of infinity (N 12)

N 12. For the zero "0" and unit "1," we consider them as 0 and 1 of the usual natural numbers and define $2 = 1 + 1$, $3 = 2 + 1$, etc. Then these numbers are pairwise distinct. In other words, the usual natural numbers altogether can be considered as a subset of N.

The above Axioms N 1–N 12 are not independent of each other (see Hilbert 1900, for instance).

Proposition 1. Axiom N 3 can be deduced from Axioms N 1, N 2 and N 7.

Proof. Take any fixed number $a \in N$. By Axioms N 1, N 2, there is an $x \in N$ such that

$$a + x = a.$$

For any $b \in N$ set

$$a + b = e.$$

Then

$$(a + x) + b = e.$$

By Axiom N 7, we have

$$a + (x + b) = e,$$

and by the uniqueness in Axiom N 2

$$x + b = b.$$

Similarly, for an arbitrary $c \in N$, the above x satisfies

$$c + x = c.$$

Therefore, this x can be regarded as 0 in Axiom N 3. □

Proposition 2. Axiom N 6 can be deduced from Axioms N 4, N 5, N 9.

Proof. Similar to the proof of Proposition 1 above. □

Proposition 3. Axiom N 8 can be deduced from the axioms of incidence N 1–N 6 and the axioms of operation N 7, N 10, N 11.

Proof. For any $a, b \in N$, by applying first Axiom N 10 and then Axiom N 11 we have

$$(a+b)(1+1) = (a+b)1 + (a+b)1 = (a+b) + (a+b).$$

And by applying first Axiom N 11 and then Axiom N 10, we have

$$(a+b)(1+1) = a(1+1) + b(1+1) = (a+a) + (b+b).$$

Hence,

$$(a+b) + (a+b) = (a+a) + (b+b)$$

or, by Axiom N 7,

$$a + (b+a) + b = a + (a+b) + b.$$

According to Axioms N 2, N 7, we have therefore

$$b + a = a + b$$

which is Axiom N 8. □

Some simple facts below can also be easily proved.

Proposition 4. According to Axioms N 2, N 3, N 8, for an arbitrary $a \in N$ there exists one and only one number $x \in N$ such that

$$x + a = a + x = 0.$$

This number is denoted as $-a$ and is called the *negative* of a.

The concept of negative numbers observes the usual rules such as $-(-a) = a$. But we should note that in a Desarguesian number system, there is no concept of order nor of absolute values. Any number necessarily has its negative, but the number itself does not have positiveness or negativeness.

Proposition 5. For any number $a \in N$, we have

$$0a = a0 = 0.$$

Proof. By Axioms N 11 and N 3, we have

$$0a + 0a = (0 + 0)a = 0a.$$

Hence, from Axiom N 2 it follows that $0a = 0$. Similarly, we have $a0 = 0$. □

Proposition 6. For any number $a \neq 0$, by Axiom N 5 there are numbers $x, y \in N$ such that

$$xa = 1, \quad ay = 1.$$

They are called the *left inverse element* and *right inverse element* of a. Furthermore,

$$x = y,$$

i.e., the left inverse element and the right inverse element of a are equal, simply called the *inverse element* of a and denoted by a^{-1}.

Proof. By Axioms N 9 and N 6, we have

$$(ax)a = a(xa) = a \cdot 1 = a.$$

Since

$$(ay)a = 1 \cdot a = a,$$

one obtains from Axiom N 5 that $ax = ay$. Applying Axiom N 5 again yields $x = y$. □

Proposition 7. Starting from the natural numbers, one may define, by Axioms N 2 and N 5, any positive or negative rational number p/q (where p is an integer and q a positive integer) such that it satisfies the usual rational operation laws and such that for any rational number r and any number $a \in N$,

$$ra = ar.$$

From the last proposition, one knows that in a Desarguesian number system the concept of rational numbers may be introduced, while the multiplication of a rational number and any other number is commutative. In general, the multiplication of any two numbers is not necessarily commutative. Furthermore, we can introduce a greatness relation among the rational numbers in a Desarguesian number system, satisfying the usual properties. But, generally speaking, there is no greatness relation between two arbitrary numbers. In order to meet our later requirements, we shall extend the concept of number systems and introduce some supplementary axioms below. Whether or not a number system is assumed to satisfy these axioms will depend on the concrete situation.

Commutative axiom of multiplication (N 13)

N 13. For any two numbers $a, b \in N$,

$$ab = ba$$

always holds.

Axioms of order (N 14–N 17)

N 14. Let a, b be any two distinct numbers in N. Then one and only one of them is always *greater than* the other. Let the former be a; then the latter b is said to be *less than* a. Symbolically,

$$a > b \quad \text{and} \quad b < a.$$

For any number a, $a > a$ does not hold.

N 15. If $a > b$ and $b > c$, then

$$a > c.$$

N 16. If $a > b$, then

$$a + c > b + c$$

for any c.

N 17. If $a > b$ and $c > 0$, then

$$ac > bc.$$

Axioms of continuity (N 18–N 20)

N 18 (Archimedes' axiom). Let $a > 0$ and $b > 0$ be any two numbers. Then it is always possible to add a to itself a sufficient number of times so that the resulting sum is greater than b. Symbolically,

$$\underbrace{a + a + \cdots + a}_{n} > b.$$

In other words, there is a natural number n such that $na > b$.

N 19 (Rolle's axiom). Let

$$f(x) = c_0 x^n + c_1 x^{n-1} + \cdots + c_n$$

be a polynomial with coefficients $c_0, c_1, \ldots, c_n$ in N. If, for any two numbers $a, b \in N$ with $a < b$, $f(a) \cdot f(b) < 0$, i.e., both $f(a)$ and $f(b)$ are not zero, one of them is greater than 0 and the other is less than 0 (or said to have *different signs*), then there must be a number $\xi \in \boldsymbol{N}$ greater than a and less than b such that $f(\xi) = 0$.

N 20 (Axiom of completeness). Let $\boldsymbol{N}$ satisfy Axioms N 1–N 18. Then it is impossible to adjoin new objects (called *new numbers*) to $\boldsymbol{N}$ so that in the extended new number system the original addition, multiplication, and greatness relations are preserved while the Axioms N 1–N 18 still hold; or briefly, under the assumption of the Axioms N 1–N 18 and by preserving all relations, $\boldsymbol{N}$ admits no extension.

The above axioms are listed and classified more or less according to Hilbert's "Grundlagen der Geometrie." Only some slight modifications are made on a few axioms in order to match the special need of this book. Following the original terminology of Hilbert, a set of numbers which satisfies some of the Axioms N 1–N 20 will be called a *complex number system*. In particular, a complex number system satisfying the Axioms N 1–N 12 has already been called a *Desarguesian number system*, which is the most fundamental complex number system for the algebraization and mechanization of geometry. In modern terminology a complex number system satisfying the Axioms N 1–N 13, i.e., a Desarguesian number system satisfying the commutative axiom of multiplication, is called a *field*, whereas a Desarguesian number system is called a skew field, or sometimes a *sfield* or a *division ring*. Axiom N 12 is not completely necessary for these concepts, but it restricts the field or sfield considered in the case of *characteristic* 0. A field satisfying the axioms of order N 13–N 17 is now commonly called an *ordered field*. If, furthermore, it satisfies Rolle's axiom N 19, then it is called a *real closed field*; if it satisfies all the Axioms N 1–N 20, then it must be isomorphic to the usual real field. Generally speaking, the construction of fields and sfields may be rather complicated. The real and complex number fields and the four-element sfield are only a few very simple and classical examples, but they are representative in a certain sense.

The above newly added axioms are also not necessarily independent of each other. For example, the axiom of infinity N 12 and Rolle's axiom N 19 can both be deduced from the other Axioms N 1–N 11, N 13–N 18, N 20. In addition, Hilbert proved the following proposition.

Proposition 8. If a complex number system $\boldsymbol{N}$ satisfies the Axioms N 1–N 16 and Archimedes' axiom N 18, then it must also satisfy the commutative axiom of multiplication N 17 and thus is a field.

Proof. See Hilbert (1899: theorem 59). □

The above axioms about the complex number system all have a certain geometric background. In the following sections and the next chapter, we shall explain how to determine a Desarguesian number system from a Desarguesian geometry, the geometric fact corresponding to the commutative axiom of multiplication and the corresponding relations between geometry and complex number system and between axioms of order and axioms of continuity etc. Based on Hilbert's concept of "numbers" (see Hilbert 1900), we shall recall a *field* a *number field*, a *sfield* a *number sfield* and their elements *numbers* in this book.

1.5 The Desarguesian number system on a line

The points and lines satisfying the geometric axioms H I, H IV and D constitute a Desarguesian plane and the corresponding geometry will be called a Desarguesian (plane) geometry. A complex number system satisfying the Axioms N 1–N 12 in Sect. 1.4 will be called a Desarguesian number system. In this and the next section, we shall prove that every Desarguesian plane has a definite Desarguesian number system in correspondence. For this purpose let us first consider the problem of introducing a Desarguesian number system on a line.

Let l be a line in a Desarguesian plane and O, I be two points chosen arbitrarily thereon. We shall construct a Desarguesian number system $N = N(l, O, I)$ such that the points on l have a one-to-one correspondence with the numbers in N, while O corresponds to 0 and I to 1. In this case line l will be said to be the *base line* of the number system N.

To do so, we consider the totality of points on l as a set N. Points will be denoted by capital Roman letters, and by the corresponding lowercase letters when they are considered as elements of N (except for 0 and 1). We shall introduce addition and multiplication in N and prove that N constitutes a Desarguesian number system under the addition and multiplication, while O, I correspond exactly to 0 and 1.

Let A, B be two arbitrary points on l, denoted as elements a, b of N.

First introduce addition, i.e., define $a + b \in N$, as follows.

Let $B \neq O$ as in Fig. 1.22. In this case, one may draw through O a line l' distinct from l. Take a point A' distinct from O on l' and construct a parallelogram $OBQA'$. Through Q draw $QC \parallel A'A$, meeting l at C. Then the corresponding element c of C in N is defined to be $a + b$.

As shown in Fig. 1.23, in the case $B = O$, i.e., $b = 0$, one can directly define C to be A. Then the corresponding element $c = a$ is defined as $a+b = a+0$. If we suppose that $Q = A'$ and consider $OBQA'$ as a *degenerate* parallelogram, then the direct definition here may be put in a nutshell as a *degenerate case* of the above definition.

Now we need to prove the propriety of the definition of addition, i.e., to prove that the constructed point C is independent of the choice of l' and A'. In

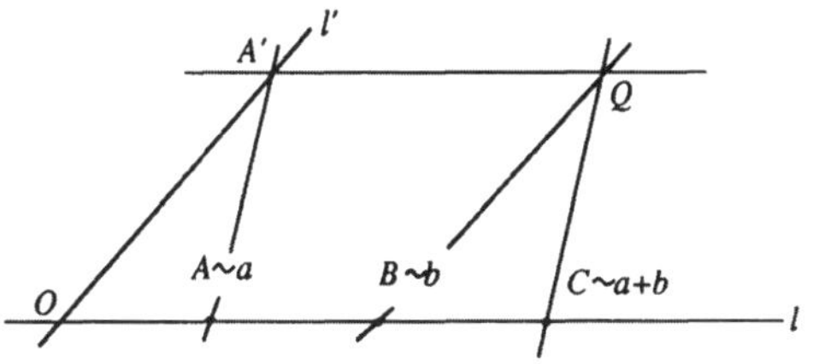

Fig. 1.22

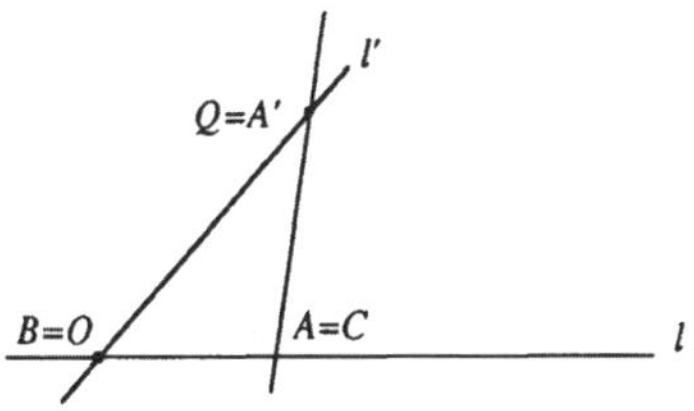

Fig. 1.23

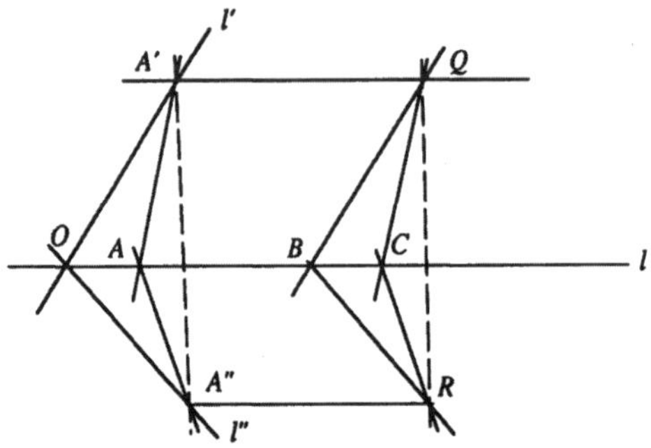

Fig. 1.24

case $B = O$, $C = A$ is obviously independent of the choice of l' and A', so we need only consider the case $B \neq O$.

As in Fig. 1.24, let us draw through O a line l'' distinct from l, take thereon an arbitrary point A'' and construct further a parallelogram $OBRA''$. We only need to prove $RC \parallel AA''$.

The proof requires repeated use of Desargues' axioms. Note first that in the statement of Desargues' axioms, triangles and parallel lines are all defined according to those in Sects. 1.1 and 1.2. Hence, while speaking about a triangle, we mean neither that the three vertices are collinear nor that two of them are coincident. Also, parallel lines must be distinct lines. Similarly, parallelograms are defined according to Definition 2 in Sect. 1.2.

Suppose first that l'' is distinct from l', the points A', A, A'' are not collinear and $A'A''$ is not parallel to l. In this case $OA'A''$, $AA'A''$, BQR, CQR are all triangles. Apply now Desargues' axiom D_2 to $\triangle OA'A''$ and $\triangle BQR$. Since the connecting lines of the corresponding vertices OB, $A'Q$, $A''R$ are parallel to each other and so are two pairs of the corresponding sides of the triangles, i.e., $OA' \parallel BQ$, $OA'' \parallel BR$, we have $A'A'' \parallel QR$. Apply Desargues' axiom D_2 again to $\triangle AA'A''$ and $\triangle CQR$. Now all the connecting lines of the corresponding vertices are parallel to each other as well, and so are two pairs of the corresponding sides, i.e., $AA' \parallel CQ$, $A'A'' \parallel QR$. Hence $AA'' \parallel CR$. This completes our proof.

In the case l'' and l' are the same, or they are not, but A', A, A'' are collinear, or $A'A'' \parallel l$, we are unable to apply Desargues' axioms as above and need to deal with each case in another way. The process is rather tedious, but completely necessary. The reason can still be seen in Sect. 3.1. As the process of dealing with each case is similar to some proofs in Sect. 1.2, it is omitted here.

It may be seen from the definition of $a + b$ that, if O corresponds to the element 0 of N, then $a+0 = a$, $0+a = a$, i.e., 0 plays the role of the element 0 of the Desarguesian number system, see Axiom N 3 in Sect. 1.4.

Secondly, let us define $a \cdot b$ or $ab \in N$.

In the definition of $a + b$, the point I plays no role, but in the definition of ab, it is indispensable.

Define now ab as follows. As in Fig. 1.25, draw through O a line l' distinct from l and take thereon a point I' distinct from O. For l' and I' we determine a point D on l as ab in the following way: Suppose first that $A \neq O$, $B \neq I$. In

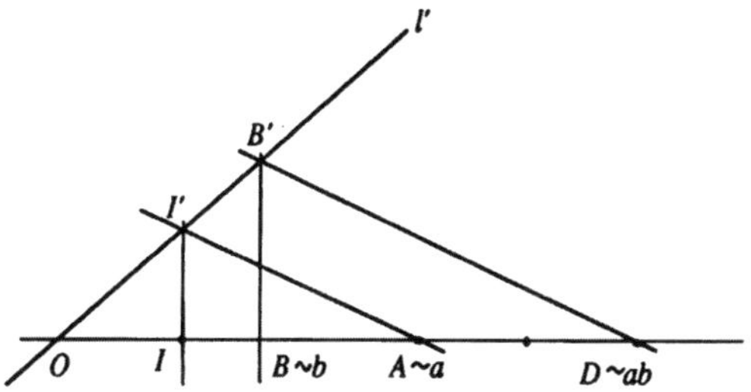

Fig. 1.25

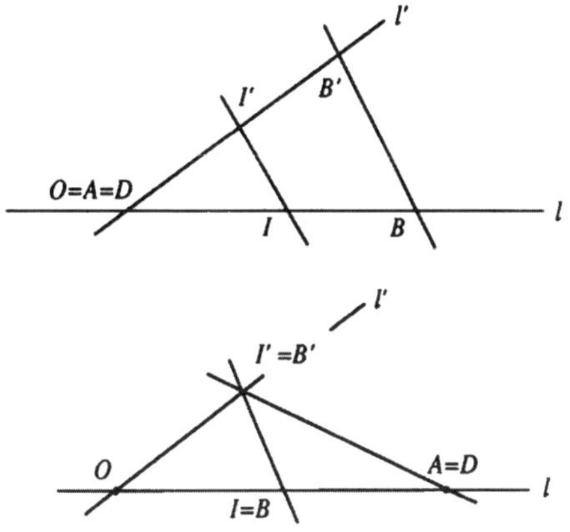

Fig. 1.26

this case let us draw through B, $BB' \parallel II'$, meeting l' at B'. Since B' does not lie on the line $I'A$, we may draw $B'D \parallel I'A$, meeting l at D. The corresponding element d of the point D in N is then defined as ab, i.e., $d = ab$. If $A = O$ or $B = I$, then the above construction is no longer possible. But we may define $D = A$, i.e., $0 \cdot b = 0, a \cdot 1 = a$. If we consider two coincident lines as two *degenerate* parallel lines, then in case $A = O$ and $B = I$, the definition of ab can be put in a nutshell as a *degenerate* case of the definition for $A \neq O, B \neq I$ (see Fig. 1.26).

Now we need to prove the propriety of the above definition of multiplication, i.e., the constructed point D is independent of the choice of l' and I'. In case $A = O$ or $B = I$, the proof is quite obvious, so the following proof is restricted to the case $A \neq O$ and $B \neq I$.

As in Fig. 1.27, let us draw through O a line l'' distinct from l, l' and take an arbitrary point I'' thereon. Suppose first that I'' lies neither on the line $I'I$ nor on $I'A$. Draw through B, $BB'' \parallel II''$, meeting l'' at B''. We want to prove that $B''D \parallel I''A$.

Under the above conditions, we can apply Desargues' axiom D_2 to $\triangle II'I''$ and $\triangle BB'B''$ to get $I'I'' \parallel B'B''$ and to $\triangle AI'I''$ and $\triangle DB'B''$ to get $B''D \parallel I''A$. This is actually what we need. In case the conditions are not satisfied, we should verify each case individually for the same reasons as before. In any case it is easy to show that $B''D \parallel I''A$. Hence, the construction of the point D is

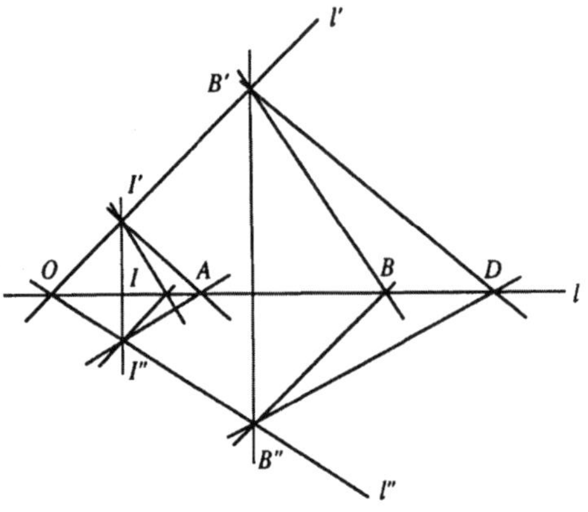

Fig. 1.27

always independent of the choice of l'' and I'', which ensures the propriety of the definition of multiplication.

By definition, we immediately have $1 \cdot a = a$ and $a \cdot 1 = a$. Therefore, the role of the element 1 corresponding to I is equivalent to that of the element 1 in Axiom N 6 of the complex number system, which thus causes no confusion in notation.

When we have defined the elements 0 and 1 as well as the operations of addition and multiplication in N, we can prove by repeated use of Desargues' axioms that N satisfies all the Axioms N 1–N 12 in a Desarguesian number system. However, the proofs are cumbersome and not always easy. They can all be found in Hilbert's "Grundlagen der Geometrie." It should be noted that Hilbert's proofs were only given for the *generic* cases, whereas the *degenerate* cases also need to be considered (cf. Sect. 3.1). Thus, the complete proofs are actually much more cumbersome than the original ones. We shall omit most of the proofs and give only some additional explanation for the inverse elements and a few special problems. As for the proofs in Axioms N 1–N 12, one may refer to Hilbert (1899).

Direct proof for the existence of inverse elements. Let l, O, I be as before and A be a point on l, not identical to O or I, corresponding to the number $a \neq 0, 1$ (cf. Fig. 1.28). Draw through O a line l' distinct from l and take a point $I' \neq O$ thereon. Construct moreover, through A, $AA' \parallel II'$, meeting l' at A' and, through I', $I'X \parallel A'I$, meeting l at X, which corresponds to a number x in N. By the definition of multiplication, $xa = 1$, i.e., x is the left inverse element of a. Construct then, through I, $IY' \parallel AI'$, meeting l' at Y' and, through Y', $Y'Y \parallel I'I$, meeting l at Y, which corresponds to a number y in N. By the definition of multiplication, $ay = 1$, i.e., y is the right inverse element of a. Let us prove $x = y$, i.e., the left inverse element is equal to the right inverse element, as follows.

Consider first the case $IA' \parallel I'A$. Now $I'X$ coincides with $I'A$, so does X with A. As IY' also coincides with IA', so does Y' with A'. Hence, $Y'Y$ coincides with $A'A$ and so does Y with A and thus with X, that is, $x = y$.

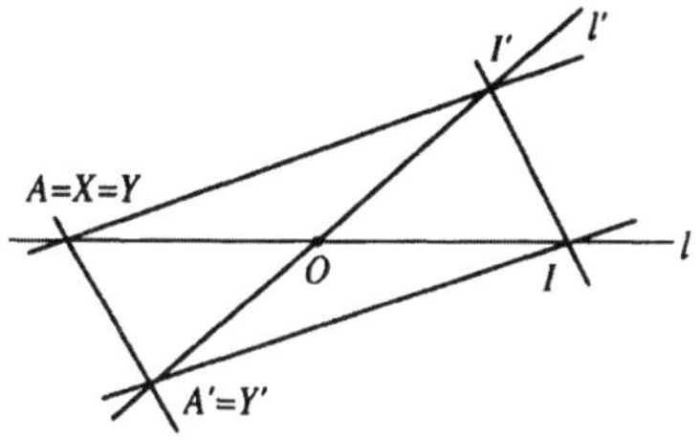

Fig. 1.28

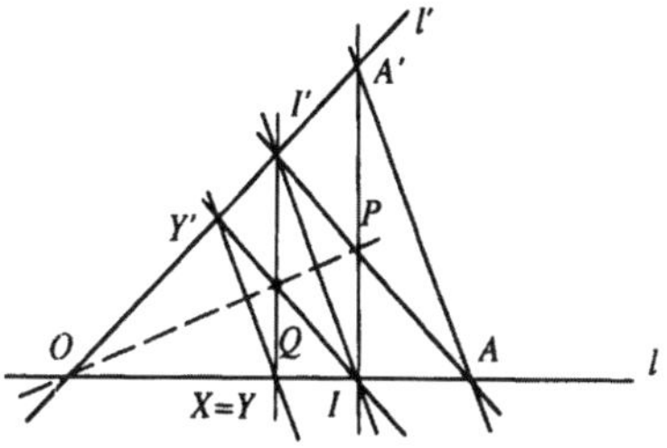

Fig. 1.29

Now suppose that IA' is not parallel to $I'A$, so they meet at a point, say P, as in Fig. 1.29. Then IY' and $I'X$ are not parallel either, so they meet at a point, say Q. Applying Desargues' axiom D_1 to $\triangle II'Q$ and $\triangle AA'P$, we see that the line PQ passes through O. Applying Desargues' axiom D_2 again to $\triangle PII'$ and $\triangle QXY'$, we have $II' \parallel XY'$. Hence, the line XY' coincides with YY' and so does X with Y, i.e., $x = y$. This is what we wanted to prove.

We have assumed $A \neq O, I$ above. If $A \neq O$ but $A = I$, then the inverse element of $a = 1$ is clearly 1 itself. □

Concept of midpoints

Let the line l and the points O, I be as before and let A, B be any two points on l corresponding to two numbers a and b in N. Let M be the midpoint of the pair (AB), corresponding to a number m. Then

$$m = \frac{1}{2}(a + b).$$

To prove this, suppose that A, B, O, M are all pairwise distinct (in the case some of them are the same, one can verify directly). In this case we may construct $\square AMCD$ as shown in Fig. 1.30. Since B is the symmetric point of A with respect to M, we have $CB \parallel DM$ by definition. Now draw through O, $l' \parallel AD \parallel MC$. Let CD meet l' at a point E and draw through D, $DF \parallel EB$, meeting l at F. According to the definition of addition, F corresponds to the number $b + a$ in N, or by Axiom N 8, i.e., $a + b$. Draw then, through C, $CF' \parallel EM$, meeting l at F'. By definition, F' corresponds to the number $2m$.

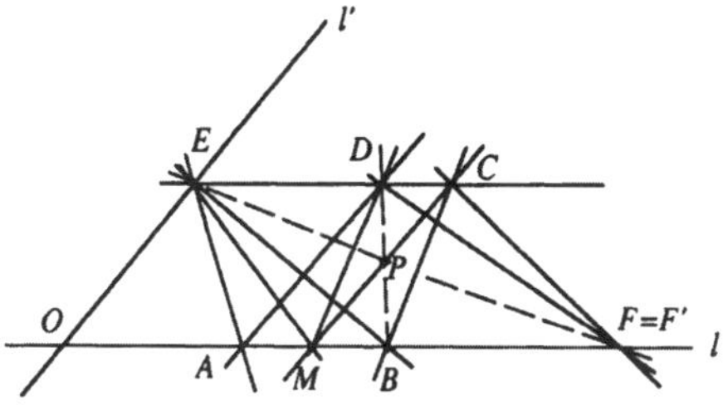

Fig. 1.30

By Theorem 1 in Sect. 1.2, the diagonals MC and BD of $\square MBCD$ bisect the pairs (BD) and (MC) at their common midpoint, say P. Similarly, the diagonals EF and BD of $\square BFDE$ also bisect one another, in particular, EF passes through P. For $\square MF'CE$, one sees that EF' passes through P as well. Therefore EF coincides with EF' and so does F with F'. Thus $2m = a + b$.

Concept of congruence

Let four points A, B, C, D on l correspond to a, b, c, d in N, respectively. As in Sect. 1.3, we define the congruence

$$(AB) \equiv (CD)$$

of two pairs of points, which says equivalently that the pairs (BC) and (AD) have a common midpoint M. By the concept of midpoints, we have

$$2m = b + c = a + d.$$

This implies that a condition for $(AB) = (CD)$ is

$$b - a = d - c.$$

1.6 The Desarguesian number system associated with a Desarguesian plane

Choose an arbitrary line l in a Desarguesian plane and two distinct points O, I on l. Taking O and I as 0 and 1, we may define, according to Sect. 1.5, a Desarguesian number system $N = N(l, O, I)$. This section aims to prove that this Desarguesian number system is actually independent of the choice of l, O, I and is determined by the Desarguesian plane. In other words, we have the following:

Theorem 1. Take two points $O \neq I$ on a line l as 0 and 1. The Desarguesian number system N defined according to Sect. 1.5 is independent of the choice of l, O, I.

Proof. Take another line l' in the plane and two distinct points O' and I' on l' as $0'$ and $1'$ so as to define another Desarguesian number system $N' = N(l', O', I')$. The theorem means that N is isomorphic to N', i.e., there is a one-to-one correspondence

$$F: \quad N \to N'$$

such that under F, 0 corresponds to $0'$ and 1 to $1'$ while the addition and multiplication of the number systems are preserved. That is, for arbitrary $a, b \in N$, we have

$$F(a + b) = F(a) + F(b),$$
$$F(a \cdot b) = F(a) \cdot F(b).$$

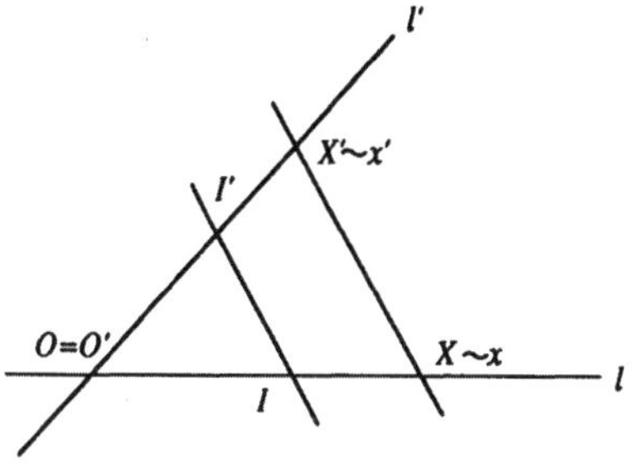

Fig. 1.31

To prove this, we shall discuss several separate cases.

Case 1. l and l' meet at a point $O = O'$ (cf. Fig. 1.31).

For an arbitrary point X on l, let us draw $XX' \parallel II'$, meeting l' at X' when $X \neq I$, and take $X' = I'$ when $X = I$. If X corresponds to $x \in N$, then the corresponding number of X' in N' will be denoted by x'. According to

$$F(x) = x',$$

we define a correspondence relation

$$F\colon \quad N \to N'.$$

In particular,

$$F(0) = 0', \quad F(1) = 1',$$

i.e., 0 and 1 in N'. Thus, the theorem is equivalent to saying that for arbitrary $a, b \in N$,

$$(a + b)' = a' + b'$$

and

$$(ab)' = a'b'.$$

We first prove $(a + b)' = a' + b'$ as follows.

Let A, B on l correspond to $a, b \in N$ and A', B' on l' to $a', b' \in N'$. In the case $A = B$ and the case $A = O$ or $B = O$, $(a+b)' = a'+b'$ clearly holds. So we suppose hereafter that A, B, O are distinct from each other (see Figs. 1.32 and 1.33). In this case $AA' \parallel II'$, $BB' \parallel II'$. Now construct $\square OBQB'$ and, through Q, $QC \parallel AB'$, $QC' \parallel A'B$, meeting l, l' at C, C' respectively. By the definition of addition in N, N', we know that C, C' correspond respectively to $a + b \in N, a' + b' \in N'$.

Since the three pairs of the corresponding sides of $\triangle OAA'$ and $\triangle QB'B$ are parallel to each other, by Desargues' axiom D_1 the lines OQ, AB', $A'B$ are also parallel to each other or are concurrent at a point P. In the former case, $C = C' = O = O'$. Hence, $a' + b' = 0' = (a + b)'$ and the conclusion is proved. In the latter case, the connecting lines of the corresponding vertices of

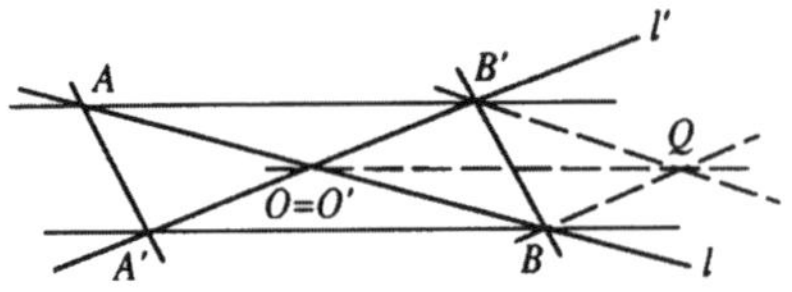

Fig. 1.32

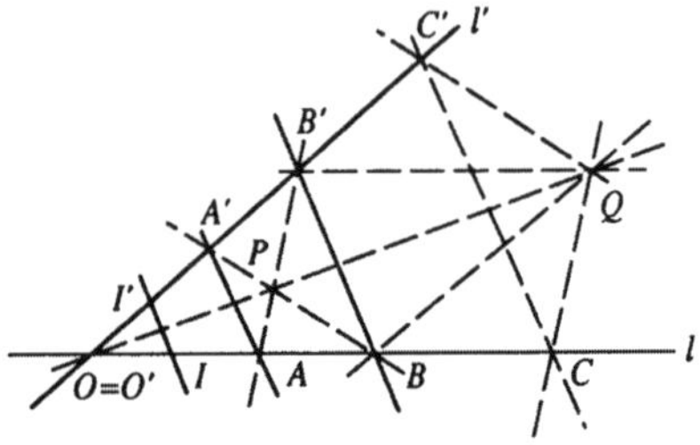

Fig. 1.33

$\triangle QCC'$ and $\triangle PAA'$ are concurrent at point O, and $QC \parallel PA$, $QC' \parallel PA'$, so $CC' \parallel AA'$. Therefore $CC' \parallel II'$ and $(a+b)' = a' + b'$ still holds.

Next, we prove that $(ab)' = a'b'$.

Symbolically, A, A', B, B' are as before. In case $A = B$ or one of A, B is the same as O or I, it is easy to verify $(ab)' = a'b'$. In what follows, we suppose that A, B, O, I are all pairwise distinct.

Draw $B'D \parallel I'A$, meeting l at D, and $BD' \parallel IA'$, meeting l' at D'. Then D corresponds to $ab \in N$ and D' to $a'b' \in N'$. Let us first assume that IA' is not parallel to $I'A$; they meet at a point, say P (see Fig. 1.34). In this case, BD' is not parallel to $B'D$ either. Let them meet at a point Q. Application of Desargues' axiom D_1 to $\triangle II'P$ and $\triangle BB'Q$ implies that PQ passes through point O. Applying Desargues' axiom D_2 to $\triangle PAA'$ and $\triangle QDD'$, one sees that $DD' \parallel AA'$. Hence, $DD' \parallel II'$ and $(ab)' = a'b'$ holds.

Consider now $IA' \parallel I'A$. The above proof is not applicable and we need to

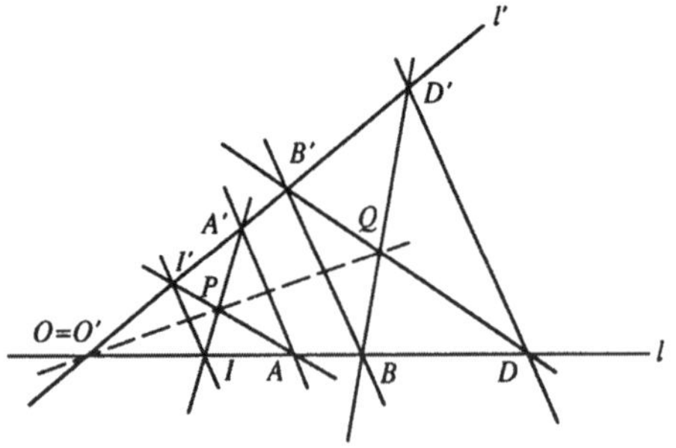

Fig. 1.34

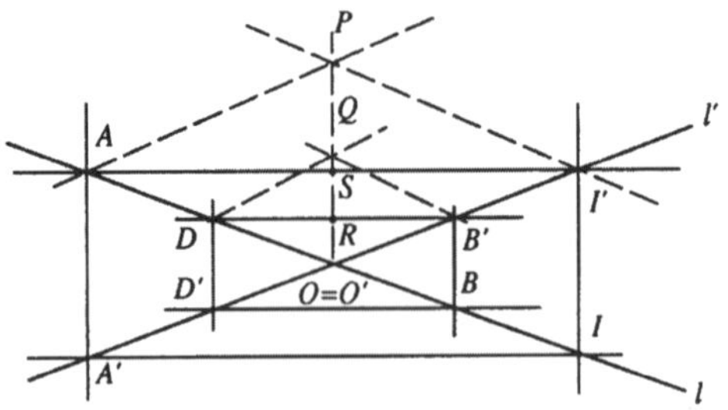

Fig. 1.35

find another proof. To do so, we draw through A a parallel to l', through I' a parallel to l and let them meet at a point P (see Fig. 1.35). Draw again, through B', $B'Q \parallel l$, meeting OP at Q. Applying Desargues' axiom D_2 to $\triangle PAI'$ and $\triangle QDB'$ yields $PA \parallel QD$. Hence, $QD \parallel l'$ and $OB'QD$ is a parallelogram. Applying Theorem 1 in Sect. 1.2 to $\square OB'QD$, $\square OI'PA$ and $\square AA'II'$, we know that the intersection points R, S, O of the corresponding diagonals are the midpoints of B', D; A, I'; and A, I respectively. For $\triangle AII'$, the connecting line of the midpoints of two sides is parallel to II', i.e., $OS \parallel II'$. From this we see that R is the midpoint of $B'D$ in $\triangle DBB'$ and $OR \parallel BB'$. Hence, O is the midpoint of B, D. Considering l' instead of l, we should infer that O is also the midpoint of B', D'. From $\triangle B'DD'$ it follows that $DD' \parallel OR$. Therefore, $DD' \parallel BB' \parallel II'$, i.e., $(ab)' = a'b'$ still holds.

This direct geometric proof is too tortuous. In fact, we can give a simple proof by using properties of the given number system as follows. By Proposition 4 in Sect. 1.4 and the concept of midpoints in Sect. 1.5, we have $a = -1$, $a' = -1'$ and thus D corresponds to $ab = (-1)b = -b$ and D' to $a'b' = (-1')b' = -b'$. By using the result that the operation of addition remains under the correspondence F, one knows

$$(-b)' + b' = ((-b) + b)' = 0'.$$

Hence,

$$(-b)' = -b'$$

or

$$(ab)' = a'b'.$$

This is what we needed to prove.

Similarly, for proving $(a + b)' = a' + b'$, in the case $AB' \parallel A'B$ the direct geometric proof can also be transformed into a simple proof by using properties of the number system. In comparing these two kinds of proofs, i.e., the direct geometric proof starting from geometric axioms and the algebraic proof based on properties of the number system, one may observe that for the former the introduction of auxiliary lines is not simple and the proof steps follow no rules, whereas for the latter the deduction rules can be more easily sought from the calculations, which thus provides the possibility for mechanical proving. Here we have given only some rather primitive examples.

Case 2. $l \parallel l'$ and $II' \parallel OO'$.

Through a point $X \neq O$ on l, draw $XX' \parallel OO'$, meeting l' at X'. If X and X' correspond to $x \in N$ and $x' \in N'$ respectively, then we define

$$F\colon \quad N \to N'$$

as

$$F(x) = x'$$

and set

$$F(0) = 0'.$$

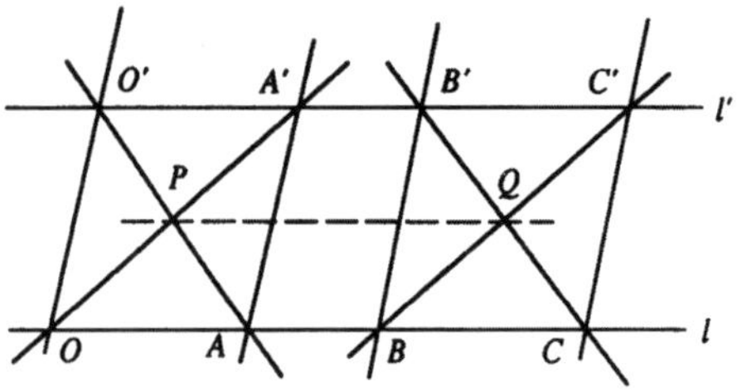

Fig. 1.36

Then under F, 0 and 1 in N correspond to $0'$ and $1'$ in N' respectively, i.e.,

$$F(0) = 0', \quad F(1) = 1'.$$

Now take arbitrary A, B on l, corresponding to a, b, and A', B' on l', corresponding to $a', b' \in N'$ respectively. Then AA', BB' are both parallel to OO' or II'. We want to prove

$$(a+b)' = a' + b',$$
$$(ab)' = a'b'.$$

First, we prove that $(a+b)' = a' + b'$. The case in which A or B is equal to O is quite evident, so we may assume that both A and B are not equal to O.

As in Fig. 1.36, draw $B'C \parallel O'A$, meeting l at C. Since $OBB'O'$ is a parallelogram, by the definition of addition C corresponds to $a + b \in N$. Similarly, draw through B, $BC' \parallel OA'$, meeting l' at C'. Then C' corresponds to $a' + b' \in N'$. Let the intersection point of the diagonals of $\square OAA'O'$ be P. Since BC' and $B'C$ are parallel to OA' and $O'A$ respectively, they must meet at a point, say Q. Applying Desargues' axiom D_1 to $\triangle POO'$ and $\triangle QBB'$ yields $PQ \parallel l \parallel l'$, and D_2 to $\triangle PAA'$ and $\triangle QCC'$ yields $CC' \parallel AA'$. Thus $CC' \parallel OO'$, i.e., $(a+b)' = a' + b'$.

Next we prove that $(ab)' = a'b'$.

If B is equal to I or O and thus B' is equal to I' or O', the equality obviously holds. Let us suppose in what follows that $B \neq I$, $B \neq O$ and also that $A \neq I, O$.

As shown in Fig. 1.37, in this case BI' meets OO' at a point E'. Draw through I, $IE \parallel BE'$, meeting OO' at E. Applying Desargues' axiom D_2 to $\triangle IAE$ and $\triangle I'A'E'$ yields $AE \parallel A'E'$. By the definition of multiplication, we know that the intersection point D of $A'E'$ and l corresponds to $ab \in N$. Now draw through B', $B'F \parallel I'E'$, meeting OO' at F and through F, $FD' \parallel A'E'$, meeting l' at D'. By the definition of multiplication again, D' corresponds to $a'b' \in N'$. Applying Desargues' axiom D_1 to $\triangle BDE'$ and $\triangle B'D'F$ yields $DD' \parallel BB'$. Thus $DD' \parallel OO'$, i.e., $(ab)' = a'b'$ holds.

Case 3. l and l' are neither coincident nor parallel and their intersection point is not equal to O or O' (see Fig. 1.38). In this case we may draw through O, $l'' \parallel l'$ and take I'' on l'' such that $I'I'' \parallel O'O$. Set $O = O''$ and construct a

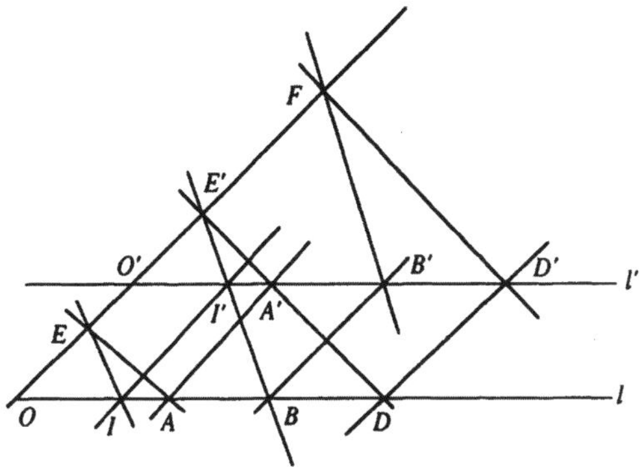

Fig. 1.37

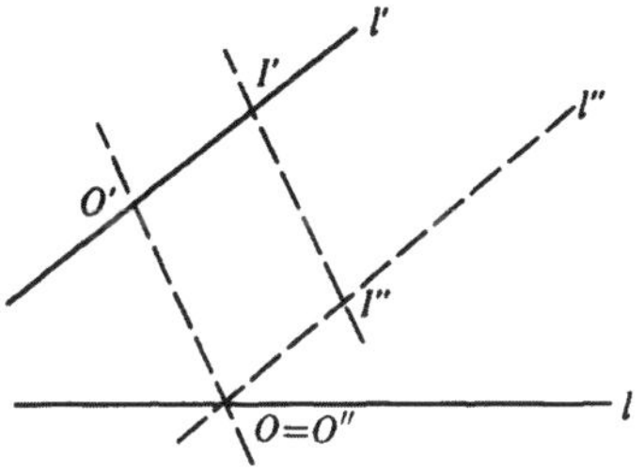

Fig. 1.38

Desarguesian number system N'' by taking O'' and I'' as 0 and 1 on l''. From Case 1, N is isomorphic to N'', denoted as $\bar{F}$: $N \to N''$, with $\bar{F}(0) = 0''$, $\bar{F}(1) = 1''$. From Case 2, there is an isomorphism $\bar{F}'$: $N'' \to N'$ such that $\bar{F}'(0'') = 0'$, $\bar{F}'(1'') = 1'$. This implies that under $\bar{F}'\bar{F}$, N is isomorphic to N' and such that 0, 1 and $0'$, $1'$ correspond to each other.

Apparently the general case may be put in a nutshell as one of the above three cases with the aid of at most a third line. This completes the proof of Theorem 1. □

Theorem 1 has shown that two Desarguesian number systems determined by any two lines l and l' are isomorphic to each other, i.e.,

$$F: \quad N(l, O, I) \approx N(l', O', I').$$

In addition to this, we shall further prove the following.

Theorem 2. The above isomorphism F can be determined in a unique manner, i.e., there exists a canonical isomorphism between any two Desarguesian number systems.

To prove this, we note first that in the proof of Theorem 1, two kinds of isomorphism correspondences have been used. They correspond to the two cases 1 and 2 in the proof and are denoted by F_I and F_{II} respectively. In the first kind of isomorphism correspondence F_I, l, l' are two distinct lines meeting at $O = O'$. In the second kind of isomorphism correspondence F_{II}, l, l' are two distinct parallel lines with $OO' \parallel II'$. Now suppose that there is a series of isomorphism correspondences between two Desarguesian number systems N and N':

$$N = N(l, O, I) \overset{F_1}{\underset{\approx}{\to}} N(l_1, O_1, I_1) \overset{F_2}{\underset{\approx}{\to}} N(l_2, O_2, I_2)$$

$$\underset{\approx}{\to} \cdots \overset{F_n}{\underset{\approx}{\to}} N(l_n, O_n, I_n) \overset{F'}{\underset{\approx}{\to}} N(l', O', I') = N'.$$

To prove the theorem, it suffices to prove that regardless of whether the intermediate isomorphisms are of the kind F_I or F_{II}, the final result $F'F_n \cdots F_2F_1$ is always the same.

Before proving Theorem 2, we state a few simple assertions below.

1. The order of action of two successive isomorphism correspondences of different kinds can be interchanged, i.e., $F_I F_{II} = F'_{II} F'_I$. We state it in detail as follows.

Let

$$F_{II}: \quad N(l_1, O_1, I_1) \approx N(l_2, O_2, I_2)$$

be an isomorphism of the second kind and

$$F_I: \quad N(l_2, O_2, I_2) \approx N(l_3, O_3, I_3)$$

be an isomorphism of the first kind. Draw through O_1, $l_4 \parallel l_3$ and, through I_3, $I_3I_4 \parallel O_1O_2 \parallel I_1I_2$, meeting l_4 at I_4 as in Fig. 1.39. Then we have an isomorphism of the first kind

$$F'_I: \quad N(l_1, O_1, I_1) \approx N(l_4, O_4, I_4),$$

in which $O_4 = O_1$, and an isomorphism of the second kind

$$F'_{II}: \quad N(l_4, O_4, I_4) \approx N(l_3, O_3, I_3).$$

We shall prove that $F_I F_{II}$ can be interchanged with $F'_{II} F'_I$.

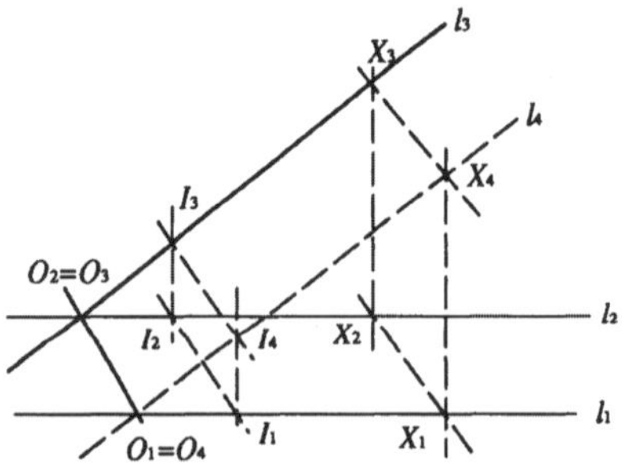

Fig. 1.39

If I_1, I_2, I_3 lie on the same line, then I_4 also lies on this line. In this case,

$$F_I F_{II} = F'_{II} F'_I: \quad N(l_1, O_1, I_1) \approx N(l_3, O_3, I_3)$$

is quite evident. In what follows we assume that I_1, I_2, I_3 are not collinear.

Applying Desargues' axiom D_2 to $\triangle O_1 I_1 I_4$ and $\triangle O_2 I_2 I_3$, one has $I_1 I_4 \parallel I_2 I_3$. Now draw through a point X_1 ($\neq O_1, I_1$) on line l_1, $X_1 X_2 \parallel O_1 O_2$, meeting l_2 at X_2, through X_2, $X_2 X_3 \parallel I_2 I_3$, meeting l_3 at X_3 and, through X_3, $X_3 X_4 \parallel I_3 I_4$, meeting l_4 at X_4. Applying Desargues' axiom D_2 to $\triangle O_4 X_1 X_4$ and $\triangle O_3 X_2 X_3$ yields $X_1 X_4 \parallel X_2 X_3$, and they are parallel to $I_2 I_3$ and $I_1 I_4$. By definition, we have

$$F_{II}(X_1) = X_2, \quad F_I\ (X_2) = X_3,$$
$$F'_I\ (X_1) = X_4, \quad F'_{II}(X_4) = X_3.$$

Hence, for $X_1 \neq O_1, I_1$,

$$F_I F_{II}(X_1) = F'_{II} F'_I(X_1) = X_3.$$

In the case $X_1 = O_1, I_1$, the above formula is more obvious ($X_3 = O_1, I_1$), so

$$F_I F_{II} = F'_{II} F'_I.$$

Similarly, if there are two successive isomorphism correspondences $F_{II} F_I$ of different kinds, then there are F'_I and F'_{II} such that $F_{II} F_I = F'_I F'_{II}$. This proves assertion 1.

2. In case l_1 is distinct from l_3, the action of two successive isomorphism correspondences

$$F_I: \quad N(l_1, O_1, I_1) \approx N(l_2, O_2, I_2)$$

and

$$F'_I: \quad N(l_2, O_2, I_2) \approx N(l_3, O_3, I_3)$$

of the first kind can be merged into one isomorphism correspondence

$$F''_I = F'_I F_I: \quad N(l_1, O_1, I_1) \approx N(l_3, O_3, I_3)$$

of the first kind. The proof is quite easy and is thus omitted here (cf. Fig. 1.40).

3. In case l_1 is distinct from l_3, the action of two successive isomorphism correspondences

$$F_{II}: \quad N(l_1, O_1, I_1) \approx N(l_2, O_2, I_2)$$

and

$$F'_{II}: \quad N(l_2, O_2, I_2) \approx N(l_3, O_3, I_3)$$

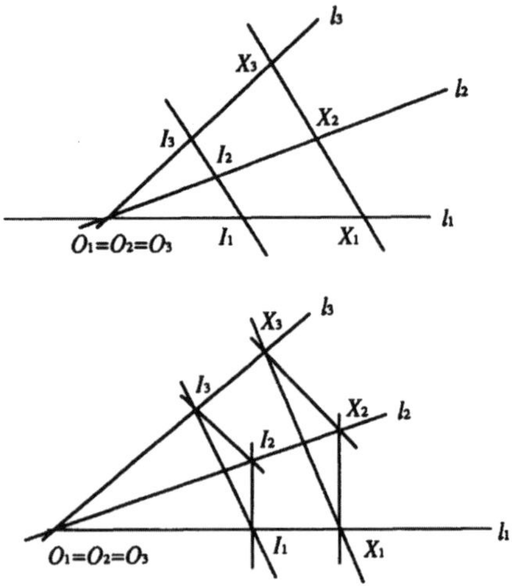

Fig. 1.40

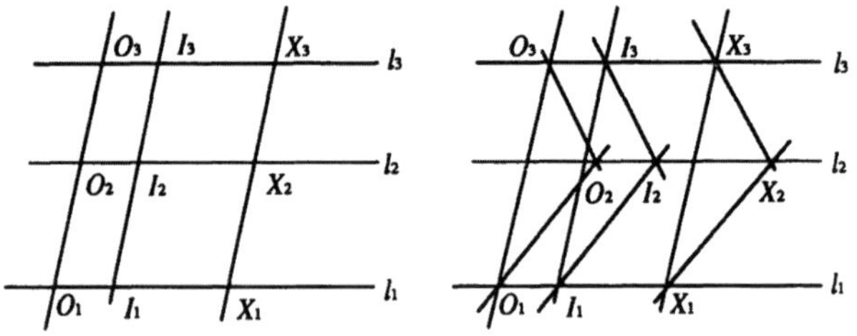

Fig. 1.41

of the second kind can be merged into one isomorphism correspondence

$$F''_{II} = F'_{II}F_{II}: \quad N(l_1, O_1, I_1) \approx N(l_3, O_3, I_3)$$

of the second kind.

For the proof, see Fig. 1.41.

4. Let

$$F_I: \quad N(l_1, O_1, I_1) \approx N(l_2, O_2, I_2),$$
$$F'_I: \quad N(l_2, O_2, I_2) \approx N(l_3, O_3, I_3)$$

be two successive isomorphism correspondences of the first kind, where $O_1 = O_2 = O_3$ and the beginning and the end base lines are the same, i.e., $l_1 = l_3$. Then the intermediate base line l_2 and I_2 can be arbitrarily alternated to another base line l_4 and I_4. That is, while introducing isomorphism correspondences

$$F''_I: \quad N(l_1, O_1, I_1) \approx N(l_4, O_4, I_4),$$
$$F'''_I: \quad N(l_4, O_4, I_4) \approx N(l_3, O_3, I_3),$$

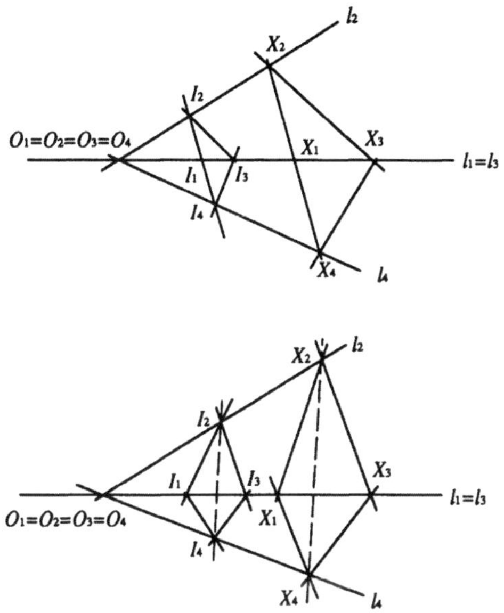

Fig. 1.42

where $O_4 = O_1$, one should have

$$F_I' F_I = F_I''' F_I''.$$

For the proof, see Fig. 1.42. If $X_1, \ldots, X_4$ lie on $l_1, \ldots, l_4$ respectively, and $X_1X_2 \parallel I_1I_2$, $X_2X_3 \parallel I_2I_3$, $X_1X_4 \parallel I_1I_4$, then application of Desargues' axiom yields $X_3X_4 \parallel I_3I_4$. Hence,

$$F_I' F_I(X_1) = F_I''' F_I''(X_1) = X_3,$$

i.e., $F_I' F_I$ is independent of the choice of l_2 and I_2.

We call $F_I' F_I$ an isomorphism correspondence of the third kind, denoted as F_{III} or so.

5. Let

$$F_{II}: \quad N(l_1, O_1, I_1) \approx N(l_2, O_2, I_2),$$

$$F_{II}': \quad N(l_2, O_2, I_2) \approx N(l_3, O_3, I_3),$$

where $l_1 \parallel l_2, l_2 \parallel l_3$ and $l_3 = l_1$, be two successive isomorphism correspondences of the second kind. Then the intermediate base line l_2 and O_2 can be arbitrarily alternated to another base line l_4 and O_4. In this case,

$$F_{II}' F_{II}: \quad N(l_1, O_1, I_1) \approx N(l_3, O_3, I_3),$$

which is called an isomorphism correspondence of the fourth kind, denoted as F_{IV}, or F_{IV}' etc.

The proof is analogous to 4, see Fig. 1.43.

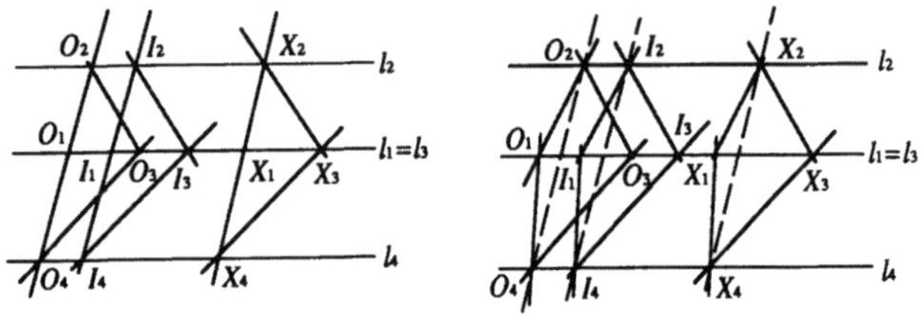

Fig. 1.43

6. The action of two successive isomorphism correspondences

$$F_{III}: \quad N(l_1, O_1, I_1) \approx N(l_1, O_2, I_2),$$

$$F'_{III}: \quad N(l_1, O_3, I_3) \approx N(l_1, O_4, I_4)$$

of the third kind, where $O_1 = O_2 = O_3 = O_4$, $I_2 = I_3$, can be merged into one isomorphism correspondence of the third kind.

According to 4, we can take two lines l_5, l_6 through O_1, which are both distinct from each other and distinct from l_1, and two points I_5 and I_6 on l_5 and l_6 respectively. Set $O_5 = O_6 = O_1$ such that

$$F_{III} = F_I^* F_I, \quad F'_{III} = F'^*_I F'_I,$$

where

$$F_I: \quad N(l_1, O_1, I_1) \approx N(l_5, O_5, I_5),$$

$$F_I^*: \quad N(l_5, O_5, I_5) \approx N(l_1, O_2, I_2),$$

$$F'_I: \quad N(l_1, O_3, I_3) \approx N(l_6, O_6, I_6),$$

$$F'^*_I: \quad N(l_6, O_6, I_6) \approx N(l_1, O_4, I_4)$$

are all isomorphism correspondences of the first kind. By 2, $F'_I F_I^*$ is an isomorphism correspondence of the first kind. Similarly, $(F'_I F_I^*)F_I$ is also an isomorphism correspondence of the first kind. Therefore, $F'_{III} F_{III} = F'^*_I((F'_I F_I^*)F_I)$ is an isomorphism correspondence of the third kind.

In the same way, we get:

7. The action of two successive isomorphism correspondences of the fourth kind can be merged into one isomorphism correspondence of the fourth kind.

Proof of Theorem 2. Let

$$F: \quad N(l, O, I) \approx N(l', O', I')$$

be a correspondence obtained from the action of a sequence of successive isomorphism correspondences of kinds I and II. According to 1, by interchanging the order of isomorphism correspondences of the first kind and of the second kind in this sequence one can make F become the composition of first the action of a sequence of successive isomorphism correspondences of the second kind and then the action of a sequence of successive isomorphism correspondences of the first kind. According to 2 and 6, one can merge the composition of a sequence of isomorphism correspondences of kind I into one isomorphism correspondence of kind I or of kind III. According to 3 and 7, one can merge in the same way the composition of a sequence of isomorphism correspondences of kind II into one isomorphism correspondence of kind II or of kind IV. Hence, the original F must be equivalent to one of the following eight kinds of isomorphism correspondences:

$$F_I, F_{II}, F_{III}, F_{IV}, F_I F_{II}, F_I F_{IV}, F_{III} F_{II}, F_{III} F_{IV}.$$

Denote the unit isomorphism by F_0 and suppose that the isomorphisms F_{III}, F_{IV} of the above eight kinds are both not unit. Then the original isomorphism correspondence F is equivalent to one of the following nine kinds of isomorphism correspondences whose geometric characters are shown together:

F_0	$l = l'$		$O = O'$	$I = I'$
F_I	$l \neq l'$	$l \nparallel l'$	$O = O'$	$(= l \wedge l')$
F_{II}	$l \neq l'$	$l \parallel l'$	$OO' \parallel II'$	
F_{III}	$l = l'$		$O = O'$	$I \neq I'$
F_{IV}	$l = l'$		$O \neq O'$	$(OI) \equiv (O'I')$
$F_I F_{II}$	$l \neq l'$	$l \nparallel l'$	$O \neq O'$	O' is not on l
$F_I F_{IV}$	$l \neq l'$	$l \nparallel l'$	$O \neq O'$	O' is on l
$F_{III} F_{II}$	$l \neq l'$	$l \parallel l'$	$OO' \nparallel II'$	
$F_{III} F_{IV}$	$l = l'$		$O \neq O'$	$(OI) \not\equiv (O'I')$.

From the above, we see that the nine different kinds of isomorphisms correspond to different geometric characters (cf. Fig. 1.44). According to the geometric characters of l, l', O, O' and I, I', one knows which of the nine kinds of isomorphism correspondences is the isomorphism correspondence F. Moreover, this correspondence is completely determined by l, l', O, O' and I, I'. Therefore, from the number system $N(l, O, I)$ to $N(l', O', I')$, no matter how the sequence of isomorphism correspondences of the kinds I and II proceeds, the isomorphism correspondence F finally obtained is completely determined by the two number systems. This proves Theorem 2. □

From Theorem 1 of this section, it is known that one can determine, after taking any two distinct points as 0 and 1, a Desarguesian number system, while number systems determined in this way are all isomorphic to each other. Furthermore, it is known from Theorem 2 that the isomorphism among the number systems can be uniquely determined. Thus, under this unique canonical isomorphism, we can identify the number systems to a single one, denoted by N. Then

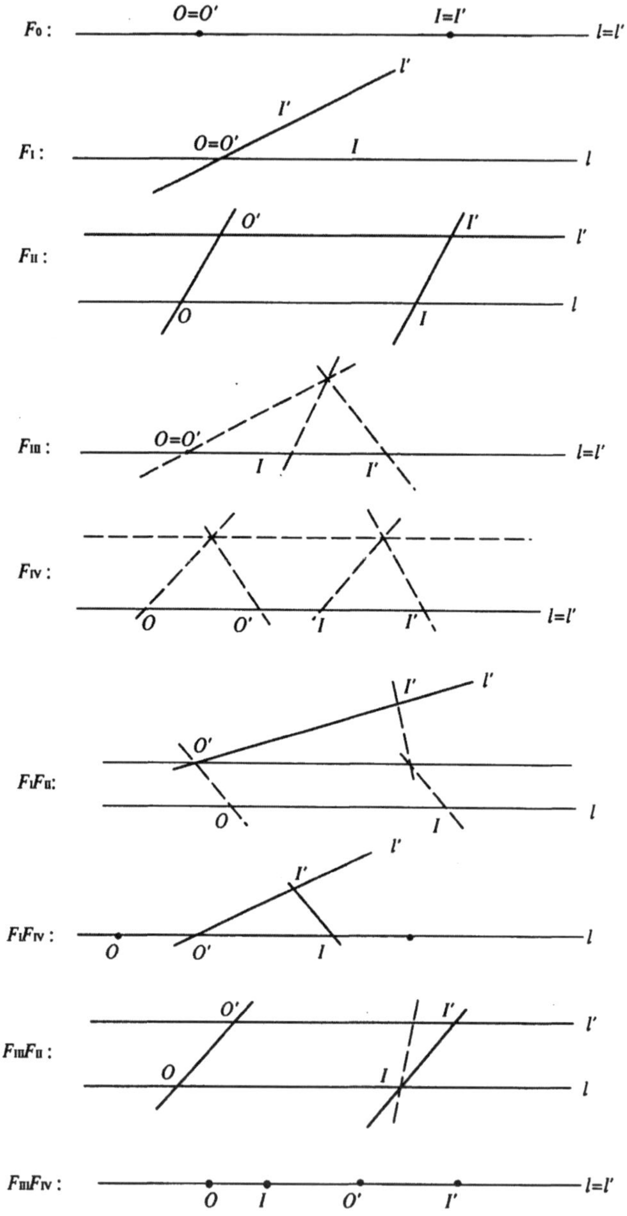

Fig. 1.44

there is a canonical isomorphism

$$F: \quad \boldsymbol{N} \approx N(l, O, I)$$

between $\boldsymbol{N}$ and any Desarguesian number system $N(l, O, I)$ on each line. Since $\boldsymbol{N}$ is a sfield of characteristic 0, we shall say $\boldsymbol{N}$ is a *Desarguesian* sfield associated with the original Desarguesian plane or Desarguesian geometry. In general, it is not a field, i.e., the commutative law of multiplication generally does not hold. Whether or not it holds is related to Pappus' axiom or what Hilbert called Pascal's axiom (for intersecting lines), see Sect. 2.1. The associated Desarguesian number sfield depends upon the considered geometry, but it sufficiently reflects the character of the original geometry. In fact, the original geometry can be recovered from its associated Desarguesian number sfield (see the book of Hilbert or other books on the foundations of geometry).

Remark. The content of Theorem 2 in this section is by no means simple. Actually, in projective geometry if we assume only the axioms of incidence, the axiom of infinity, and axioms equivalent to Desargues' axioms but no others, then the corresponding theorem 1 holds, while theorem 2 falls into fallacy. In detail, under the assumption of the above-mentioned axioms, while taking three distinct points as $0, 1, \infty$ on a line, one can uniquely determine a Desarguesian number sfield (except for ∞) and all number sfields determined in this way are isomorphic to each other. But these isomorphisms are non-canonical unless there is an additional assumption of other axioms such as the so-called Pappus' axiom. Only under the assumption of Pappus' axiom, can one prove the fundamental theorem of projective geometry, establish canonical isomorphism correspondences among various Desarguesian number sfields and prove the satisfaction of the commutative law of multiplication so that the sfield becomes a number field. The proofs are all non-trivial and are the most abstruse part of the foundations of projective geometry, for which the reader may refer to some general books in projective geometry or Sect. 6.2 in this book.

1.7 The coordinate system of Desarguesian plane geometry

According to Sects. 1.5 and 1.6, one can uniquely determine a Desarguesian number system N in a Desarguesian plane such that, after taking two points O and I on an arbitrary line, the points on this line will correspond one-to-one to the numbers in N with O to 0 and I to 1. From this we can introduce a plane coordinate system by the usual method of analytic geometry such that the points in the plane correspond one-to-one to the pairs of numbers in $N \times N$. We explain this in detail in what follows.

As shown in Fig. 1.45, we take two arbitrary intersecting lines l_1 and l_2 in

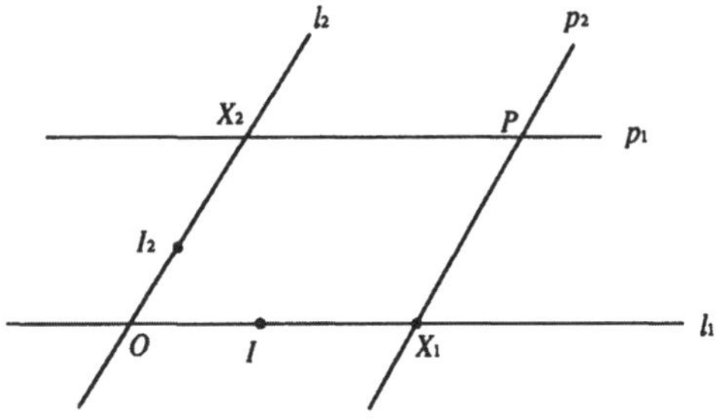

Fig. 1.45

the plane with O as their intersection point. Take two points I_1 and I_2 distinct from O on l_1 and l_2 respectively. Then, by mapping O, I_1 to 0, 1 we can set up a canonical one-to-one correspondence between the points on l_1 and the numbers in $\boldsymbol{N}$. Similarly, by mapping O, I_2 to 0, 1 there is a canonical one-to-one correspondence between the points on l_2 and the numbers in $\boldsymbol{N}$, too. As usual, O, l_1, l_2, I_1 and I_2 constitute a *coordinate system*, of which O is called the *origin*, l_1, l_2 are called the *first* and the *second axes* and I_1, I_2 the *units*.

Now, let P be a point in the plane. If P does not lie on line l_1, we draw through P a parallel p_1 to l_1. Otherwise, set $p_1 = l_1$. Similarly, in the case P does not lie on l_2, let p_2 be a line through P and parallel to l_2 and, otherwise, set $p_2 = l_2$. Suppose that p_2, p_1 meet l_1, l_2 at X_1, X_2 respectively, which correspond to two numbers x_1, x_2 in $\boldsymbol{N}$. Then

$$P \leftrightarrow (x_1, x_2)$$

clearly induces a one-to-one correspondence between points in the plane and pairs of numbers in $\boldsymbol{N} \times \boldsymbol{N}$, where x_1 and x_2 are called the *first* and the *second coordinates* of P respectively in the above coordinate system and (x_1, x_2) is called the *coordinate representation* of P, denoted as

$$P = (x_1, x_2).$$

After fixing a coordinate system, the one-to-one correspondence relation between points and coordinates makes it possible to express the properties and relations of a geometric configuration in terms of numbers and relations among numbers. This is what we call the *algebraization of geometry*. In the following, we take a few most fundamental relations in Desarguesian geometry as examples for illustration.

Example 1. Let $R = (z_1, z_2)$ be the midpoint of two points $P = (x_1, x_2)$ and $Q = (y_1, y_2)$. Express this geometric relation in terms of relations among numbers.

As in Fig. 1.46, let us draw through P, Q, R three lines p_2, q_2, r_2 respectively, such that each of them is parallel to l_2 if the corresponding point P, Q or R does not lie on l_2, or otherwise is just the line l_2 itself. Let p_2, q_2, r_2 meet l_1 at points X_1, Y_1, Z_1 respectively, which correspond to three numbers

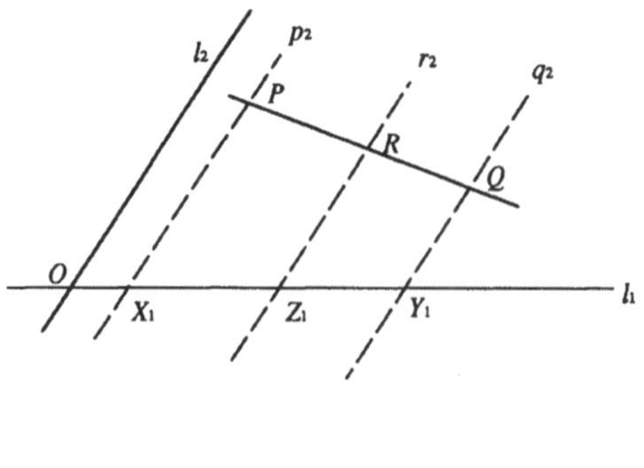

Fig. 1.46

x_1, y_1, z_1 in N. It is easy to prove that in each case Z_1 is always the midpoint of the pair (X_1, Y_1). According to the concept of midpoints in Sect. 1.5, one should have $2z_1 = x_1 + y_1$ and, similarly, $2z_2 = x_2 + y_2$. On the contrary, if there are relations

$$2z_1 = x_1 + y_1, \quad 2z_2 = x_2 + y_2$$

among the coordinates of the three points $P = (x_1, x_2)$, $Q = (y_1, y_2)$, $R = (z_1, z_2)$, then point R must be the midpoint of the pair (P, Q). Hence, the above two equations give an algebraic expression for the geometric relation that "R is the midpoint of (P, Q)."

Why we explain this simple example in so much detail can be seen in Sect. 3.1.

Example 2. Let

$$P = (x_1, x_2), \quad Q = (y_1, y_2), \quad R = (z_1, z_2), \quad S = (u_1, u_2)$$

be four points with $P \neq Q$, $R \neq S$. Find an algebraic relation corresponding to the geometric relation that the two lines PQ and RS are parallel or coincident.

To do so, let us first prove two lemmas below.

Lemma 1. Let $P = (x_1, x_2)$, $Q = (y_1, y_2)$ with $P \neq Q$. Take a point A_1 on the first axis l_1 corresponding to the number $y_1 - x_1$ in N, and a point A_2 on the second axis l_2 corresponding to the number $x_2 - y_2$ in N. Then A_1A_2 and PQ are either parallel or coincident.

Proof. We only consider the case in which P, Q do not lie on l_1, l_2 and PQ neither passes through the origin O nor is parallel to l_1, l_2. As in Fig. 1.47, draw, through P, $PX_1 \parallel l_2$, $PX_2 \parallel l_1$, meeting l_1, l_2 at X_1, X_2 and, through Q, $QY_1 \parallel l_2$, $QY_2 \parallel l_1$, meeting l_1, l_2 at Y_1, Y_2 respectively. Then X_1, Y_1 correspond to x_1, x_2 in N and X_2, Y_2 to x_2, y_2 in N respectively. Since $(y_1 - x_1) + x_1 = y_1$, by the definition of addition on l_1 we have $PY_1 \parallel X_2A_1$. Similarly, we have

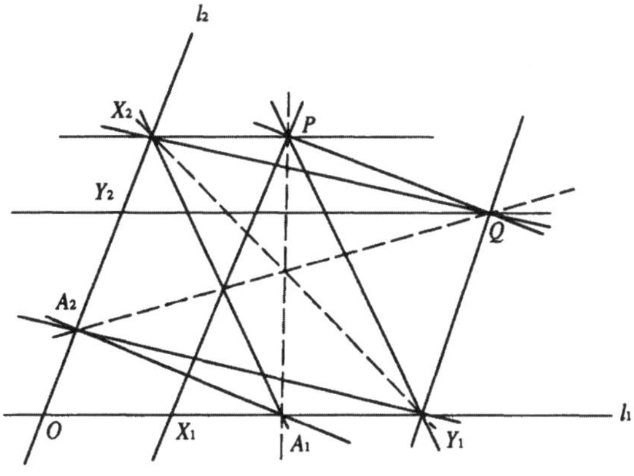

Fig. 1.47

$QX_2 \parallel Y_1A_2$ as $(x_2 - y_2) + y_2 = x_2$ on l_2. Since $PX_2A_1Y_1$ is a parallelogram, by Theorem 1 in Sect. 1.2 the two diagonals PA_1 and X_2Y_1 bisect one another. Similarly, the diagonals QA_2 and X_2Y_1 of $\square QX_2A_2Y_1$ also bisect one another. Hence, the three lines PA_1, QA_2, X_2Y_1 pass through one common point, i.e., the common midpoint of the pairs (PA_1), (QA_2), (X_2Y_1). Applying Desargues' axiom D_2 to $\triangle PQX_2$ and $\triangle A_1A_2Y_1$ yields $A_1A_2 \parallel PQ$, which is what we wanted to prove.

For the other cases, the above proof is not applicable and further verification one by one is required. The reason can be seen again in Sect. 3.1. The proofs are easy yet cumbersome and are all omitted here. □

Lemma 2. Let A, B be two points on the axis l_1 corresponding to two numbers a, b in N, and C, D two points on the axis l_2 corresponding to two numbers c, d in N. Suppose A, C are both distinct from O, so that a^{-1}, c^{-1} can be defined. Then a necessary and sufficient condition for the two lines AC and BD to be parallel or coincident is

$$a^{-1}b = c^{-1}d.$$

Proof. We suppose that AC is parallel to or coincident with BD as to prove $a^{-1}b = c^{-1}d$, from which the inverse can be easily obtained.

Below we shall assume that the points A, B are distinct from I_1, the points C, D are distinct from I_2, and $AC \parallel BD$ (and thus not coincident), while both are not parallel to I_1I_2.

Under these conditions, we may draw, through I_1, $I_1\bar{A} \parallel I_2A$, meeting l_2 at $\bar{A}$ and, through I_2, $I_2\bar{C} \parallel I_1C$, meeting l_1 at $\bar{C}$ (see Fig. 1.48). By the definition of multiplication, $\bar{A}, \bar{C}$ correspond respectively to a^{-1}, c^{-1} in N. Now draw through B, $BF \parallel I_1\bar{A}$, meeting l_2 at F and through D, $DE \parallel I_2\bar{C}$, meeting l_1 at E. Again, by the definition of multiplication, E, F correspond respectively to $c^{-1}d, a^{-1}b$ in N. We need to prove that $a^{-1}b = c^{-1}d$ or $EF \parallel I_1I_2$.

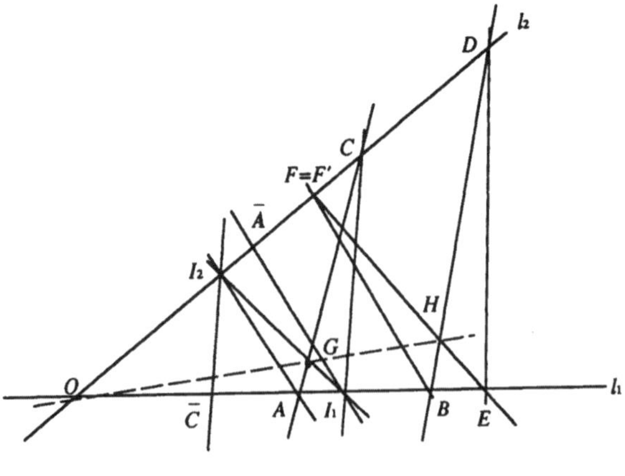

Fig. 1.48

To do so, let us draw through E, $EF' \parallel I_1I_2$, meeting l_2 at F'. Then both F' and E correspond to $c^{-1}d$. Since AC is not parallel to I_1I_2, it meets I_1I_2 at a point, say G. Since BD, EF' are parallel to AC, I_1I_2 respectively, they also meet at a point, say H. Applying Desargues' axiom D_1 to ΔGI_1C and ΔHED, it follows that HG passes through O. Applying Desargues' axiom D_2 to ΔGAI_2 and $\Delta HBF'$ yields $BF' \parallel AI_2$. Because $BF \parallel AI_2$, BF coincides with BF', as does F with F'. That is,

$$a^{-1}b = c^{-1}d,$$

and the proof is complete.

It is also not difficult to check all other cases one by one. We omit the verifications. □

Remark. The condition in Lemma 2, in the case that B, D are also distinct from O, can be rewritten as

$$b^{-1}a = d^{-1}c.$$

But it cannot be rewritten as $ba^{-1} = dc^{-1}$, since the commutative law of multiplication generally does not hold (cf. Sect. 2.1).

Returning to Example 2, we have the following.

Theorem 1. Let

$$P = (x_1, x_2), \quad Q = (y_1, y_2), \quad R = (z_1, z_2), \quad S = (u_1, u_2)$$

be four points with $P \neq Q$, $R \neq S$. Suppose both PQ, RS are neither parallel to or coincident with l_1 nor parallel to or coincident with l_2. Then a necessary and sufficient condition for the two lines PQ and RS to be parallel or coincident is

$$(y_1 - x_1)^{-1}(u_1 - z_1) = (y_2 - x_2)^{-1}(u_2 - z_2).$$

Proof. Take two points A, B on l_1, corresponding to $y_1 - x_1, u_1 - z_1$ in N, and two points C, D on l_2, corresponding to $x_2 - y_2, z_2 - u_2$ in N. Since PQ is not parallel to nor coincident with l_2, $x_1 \neq y_1$ and thus A is distinct from O. Similarly, B, C, D are all distinct from O. By Lemma 1, AC and PQ are either parallel or coincident, and so are BD and RS. Therefore, a necessary and sufficient condition for PQ and RS to be parallel or coincident is that AC and BD are parallel or coincident. By Lemma 2, a necessary and sufficient condition for the latter is

$$(y_1 - x_1)^{-1}(u_1 - z_1) = (x_2 - y_2)^{-1}(z_2 - u_2).$$

This is the formula we wanted to prove. Apparently, the condition may also be rewritten as

$$(u_1 - z_1)^{-1}(y_1 - x_1) = (z_2 - u_2)^{-1}(x_2 - y_2).$$

□

Example 3. We call any necessary and sufficient condition between x_1 and x_2 that the point (x_1, x_2) in the figure satisfies the *equation* of this figure. Derive the equation of a line and a necessary and sufficient condition for a point to lie on a line.

Let the line be L and the point be $P = (x_1, x_2)$. If L is parallel to or coincident with l_2 (or l_1), then the equation of L is clearly of the form

$$x_1 = c \quad (\text{or } x_2 = c),$$

in which c is a constant corresponding to the intersection point of L and l_1 (or l_2) (see Fig. 1.49).

We assume that L is not parallel to l_1, l_2 but meets l_2 at C, which corresponds to a number c in N (cf. Fig. 1.50). Consider first the case in which L does not pass through I_1. Then we can draw through I_1, I_1A parallel to L, meeting l_2 at A, which corresponds to a number $a \neq 0$ in N. The numbers c and a are all determined by the line L. Now suppose $P = (x_1, x_2)$ lies on L but not on l_1 nor on l_2. Draw through P, $p_1 \parallel l_1, p_2 \parallel l_2$, meeting l_2, l_1 at X_2, X_1 respectively. Furthermore draw, through X_1, $X_1Z \parallel L$, meeting l_2 at Z. Then X_1, X_2 correspond respectively to x_1, x_2 in N. By the definition of multiplication, Z corresponds to ax_1. Moreover, by the definition of addition,

$$ax_1 + x_2 = c.$$

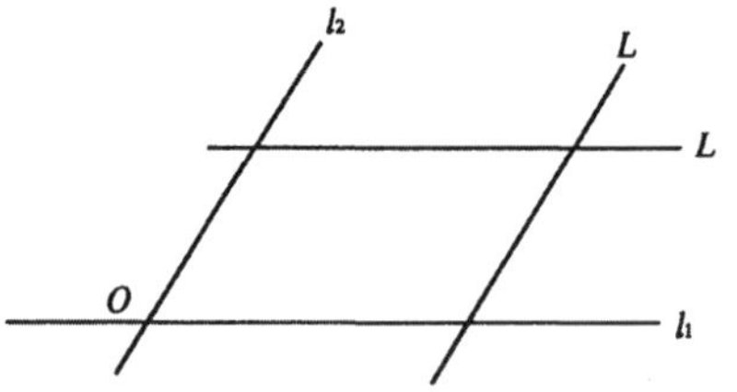

Fig. 1.49

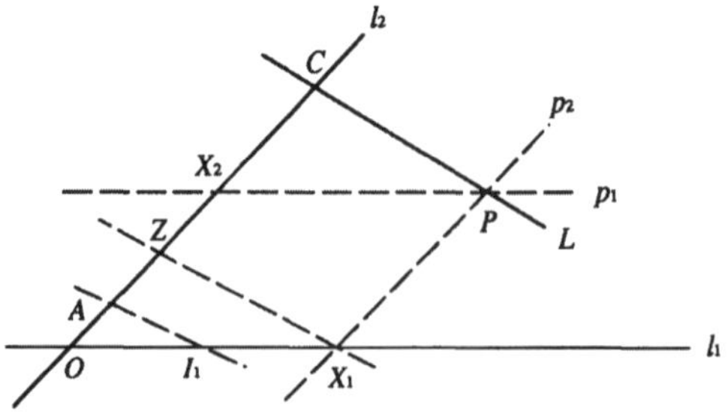

Fig. 1.50

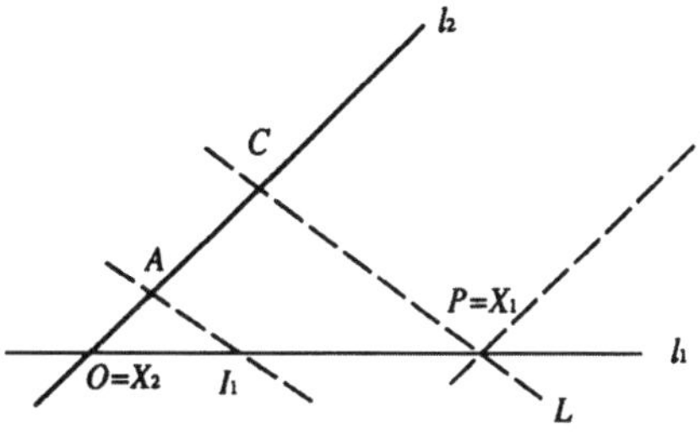

Fig. 1.51

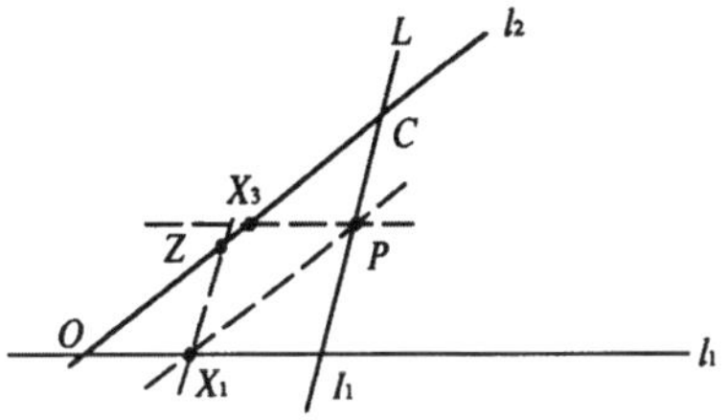

Fig. 1.52

If P lies on both L and l_1, then $x_2 = 0$ (cf. Fig. 1.51). As C corresponds to $c = ax_1$, x_1, x_2 still satisfy the above equation. If P is on both L and l_2, i.e., $P = C$, then $x_1 = 0, x_2 = c$ and x_1, x_2 also satisfy the above equation.

Finally, suppose that L is not parallel to l_1, l_2 but passes through I_1 (cf. Fig. 1.52). Let C be the intersection point of L and l_2, which corresponds to a number $c \neq 0$ in N. Then by an analogy to the proof before, an arbitrary point $P = (x_1, x_2)$ on L satisfies the equation

$$cx_1 + x_2 = c.$$

On the contrary, every point $P = (x_1, x_2)$ which satisfies one of the following equations lies on a definite line:

$$x_1 = c,$$
$$x_2 = c,$$
$$ax_1 + x_2 = c \quad (a \neq 0).$$

Summing up all the cases, we have:

Theorem 2. The equation of a line in the plane is

$$a_1x_1 + a_2x_2 = c,$$

where a_1 and a_2 are not simultaneously zero. If $a_2 \neq 0$, then the line meets l_2 and the point of intersection corresponds to the number $a_2^{-1}c$. If $a_1 \neq 0$, then the line meets l_1 and the point of intersection corresponds to the number $a_1^{-1}c$.

Remark. Since the multiplication in the associated Desarguesian number system of a Desarguesian plane in general is not commutative, the equation of a line above cannot be written as

$$x_1a_1 + x_2a_2 = c.$$

In fact, this equation does not even represent a line.

Example 4. Let $P=(x_1,x_2)$, $Q=(y_1,y_2)$, $R=(z_1,z_2)$ be three points. Find a necessary and sufficient condition for a line to pass through these three points, or equivalently for these three points to be *collinear*.

Clearly, a necessary and sufficient condition for the three points P, Q, R to be collinear is that either at least two of them coincide, or all three are distinct from each other while the lines PQ and PR coincide. Therefore, according to the lemmas of Example 2 the necessary and sufficient condition to be found is that *at least* one of the following relations holds:

1. $x_1=y_1=z_1$,
2. $x_2=y_2=z_2$,
3. $x_1=y_1$ and $x_2=y_2$,
4. $x_1=z_1$ and $x_2=z_2$,
5. $y_1=z_1$ and $y_2=z_2$,
6. $x_1\neq y_1, x_2\neq y_2$ and $(y_1-x_1)^{-1}(z_1-x_1)=(y_2-x_2)^{-1}(z_2-x_2)$,
7. $y_1\neq z_1, y_2\neq z_2$ and $(z_1-y_1)^{-1}(x_1-y_1)=(z_2-y_2)^{-1}(x_2-y_2)$,
8. $x_1\neq z_1, x_2\neq z_2$ and $(x_1-z_1)^{-1}(y_1-z_1)=(x_2-z_2)^{-1}(y_2-z_2)$.

2 Orthogonal geometry, metric geometry and ordinary geometry

2.1 The Pascalian axiom and commutative axiom of multiplication – (unordered) Pascalian geometry

In Desarguesian (plane) geometry which takes Hilbert's axioms of incidence H I, (sharper) axiom of parallels H IV, the axiom of infinity D_∞, and Desargues' axioms D as its basis, one can uniquely determine a Desarguesian number system N, called a geometry-associated Desarguesian number system, as has been exhibited in the previous sections. This number system is actually a *skew field* (of characteristic 0) and in general it does not satisfy the commutative axiom of multiplication N 13 of the complex number system. In order to let the commutative axiom of multiplication be satisfied, too, so that N becomes a *number field*, we must introduce other axioms in this geometry. One way, as shown in Hilbert's "Grundlagen der Geometrie," is to introduce the so-called Pascalian axiom. What Hilbert called the Pascalian axiom is actually a special case of the theorem commonly named after Pappus. It is also a special case of Pascal's theorem in usual projective geometry where the conic section degenerates into two lines. To distinguish the axiom considered by Hilbert from the general Pappus' and Pascal's theorems, we call it the *linear* Pascalian axiom, stated as follows.

Linear Pascalian axiom. Let two sets of points A, B, C and A', B', C' respectively on two distinct lines l and l' be pairwise distinct (see Fig. 2.1). If

$$BC' \parallel B'C, \quad AB' \parallel A'B,$$

then

$$AC' \parallel A'C.$$

In this l and l' are called two *base lines* of the axiom. When they intersect, the corresponding axiom is called the *Pascalian axiom for intersecting lines*. The Pascalian axiom given in Hilbert's book is the Pascalian axiom for intersecting lines. Below we shall call it *Axiom P*, denoted sometimes as Axiom

$$P\begin{bmatrix} A & B & C \\ A' & B' & C' \end{bmatrix} \quad \text{or} \quad P\begin{bmatrix} l \\ l' \end{bmatrix} \quad \text{or} \quad P\begin{bmatrix} l & / & A\,B\,C \\ l' & / & A'B'C' \end{bmatrix}.$$

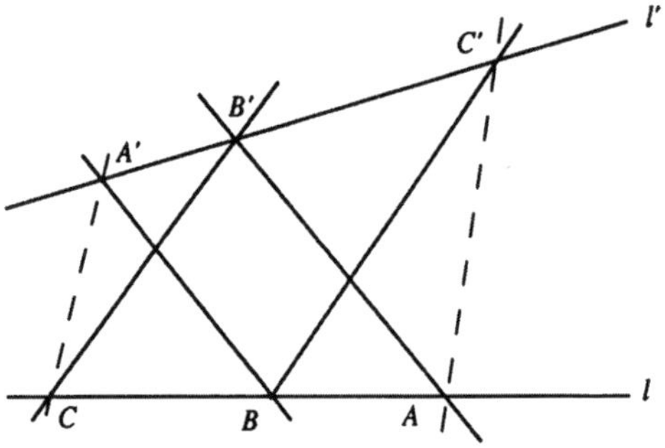

Fig. 2.1

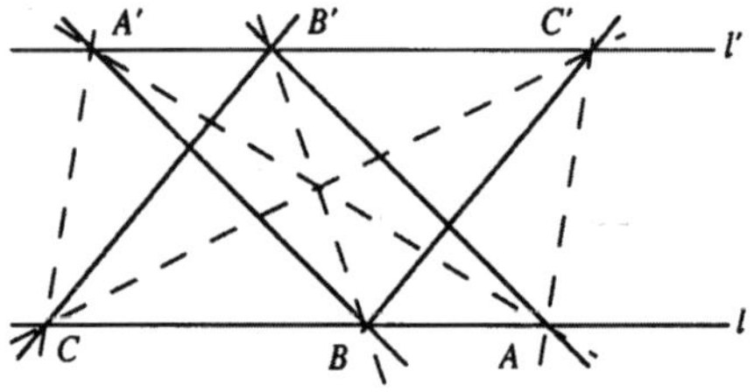

Fig. 2.2

The relation between this axiom and Desarguesian geometry is explained by the following points.

1. Linear Pascalian axiom with the base lines parallel to each other. In this case the axiom can be deduced from other axioms, so it in fact is a theorem in Desarguesian geometry. The proof proceeds as follows.

Since l and l' are parallel, the hypothesis implies that $CBC'B'$ is a parallelogram. By Theorem 1 in Sect. 1.2 the diagonals BB' and CC' bisect one another. Similarly, $ABA'B'$ is a parallelogram and thus AA' and BB' bisect one another. Thus AA' and CC' bisect one another too. Applying Theorem 2 in Sect. 1.2, one sees that $ACA'C'$ is also a parallelogram; therefore, $AC' \parallel A'C$. This is what we wanted to prove (cf. Fig. 2.2).

2. Pascalian axiom for intersecting lines P and the commutative axiom of multiplication N 13. The significance of the Pascalian axiom for intersecting lines P in Desarguesian geometry is that it provides a necessary and sufficient condition for the geometry-associated Desarguesian number system to satisfy the commutative axiom of multiplication N 13.

To explain this fact, we first assume that Axiom P holds. Take an arbitrary line l in the Desarguesian plane and two points O and I on l. Define a Desarguesian number system such that O corresponds to 0 and I to 1. Let A and B be two points on l corresponding respectively to a and b in $\boldsymbol{N}$. When $A = B$ or one of A and B is O or I, $ab = ba$ is apparent. So we assume that A, B, O, I are all pairwise distinct (see Fig. 2.3). Now, we may take any line l' through O but distinct from l and an arbitrary point I' distinct from O on l'. Draw $AB' \parallel II'$, $BA' \parallel II'$, meeting l' at B', A' respectively. Draw, moreover, through B', $B'C \parallel I'B$, meeting l at C. If we define a Desarguesian

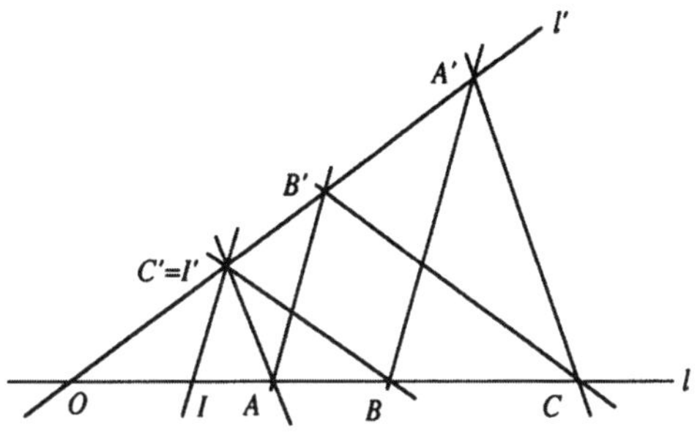

Fig. 2.3

number system $\boldsymbol{N}$ on l' such that O corresponds to 0 and I' to 1, then it follows from Sect. 1.6 that A' corresponds to b and B' to a in $\boldsymbol{N}$. By the definition of multiplication, one knows that C corresponds to ba. Application of the linear Pascalian axiom

$$P\begin{bmatrix} l & / & A\ B\ C \\ l' & / & A'B'I' \end{bmatrix}$$

yields $A'C \parallel I'A$. Therefore, the definition of multiplication implies that C also corresponds to ab. This means $ab = ba$, i.e., the commutative axiom of multiplication can be deduced from Axiom P. It is also easy to prove, on the other hand, that if the commutative axiom of multiplication of the geometry-associated Desarguesian number system $\boldsymbol{N}$ holds, then so does the Pascalian axiom for intersecting lines P. Therefore, the geometric axiom P is equivalent to the Axiom N 13 of number systems.

From the above proof, one also sees that if the Pascalian axiom holds with respect to a certain pair of intersecting base lines l, l' in Desarguesian geometry, i.e.,

$$P\begin{bmatrix} l \\ l' \end{bmatrix}$$

holds, then the commutative axiom of multiplication holds. Thus the Pascalian axiom holds with respect to an arbitrary pair of intersecting base lines, too.

3. Pascalian axiom for intersecting lines and the theory of proportion. In his "Grundlagen der Geometrie," Hilbert derived the Pascalian axiom as a theorem by axioms of congruence H III, and based on it he established the theory of proportion. However, in the establishment of this theory, axioms of congruence had to be used. In the supplement to a revised edition of Hilbert's book, Bernays simplified the theory of proportion but still made use of axioms of congruence. We shall point out below that in Desarguesian geometry, the theory of proportion can be completely set up under the assumption of the Pascalian axiom for intersecting lines P alone, without the use of axioms of congruence (and axioms of order). However, if we do not assume Axiom P, it is impossible to establish such a theory of proportion in Desarguesian geometry.

To demonstrate this recall the theorem that was proved in Sect. 1.7:

Let two distinct lines l and l' meet at a point O. Take two arbitrary points I, I' distinct from O on l, l' respectively, and define a Desarguesian number system N such that O corresponds to 0 and I, I' to 1. Let A, B be two points distinct from O on l, corresponding to two numbers a, b in N and C, D be two points distinct from O on l', corresponding to two numbers c, d in N. Then a necessary and sufficient condition for AC and BD to be parallel or coincident is

$$a^{-1}b = c^{-1}d$$

or

$$b^{-1}a = d^{-1}c.$$

Note first that in Desarguesian geometry, if we do not assume the Pascalian axiom for intersecting lines, then the above necessary and sufficient condition in general cannot be written as

$$ba^{-1} = dc^{-1}$$

or

$$ab^{-1} = cd^{-1}.$$

To prove this, we suppose that A and B are distinct from O, I (in the contrary case the condition may naturally be written in the above form) (see Fig. 2.4). For the sake of simplicity, let us consider the case $D = I'$, i.e., $d = 1$. Then, from a theorem proved before one sees that $AC \parallel BD$ implies

$$a^{-1}b = c^{-1} \quad \text{or} \quad b^{-1}a = c.$$

We need to prove that without the assumption of Axiom P, $AC \parallel BD$ does not imply

$$ba^{-1} = c^{-1} \quad \text{or} \quad ab^{-1} = c.$$

Since in the case $a \neq 0$ and $b \neq 0$, a^{-1} and b^{-1} may be arbitrary non-zero numbers, this conclusion can be deduced directly from 2 above. Let us write down the proof as follows.

Through I draw $IE \parallel BD$, meeting l' at E. Then, according to Sect. 1.5, E corresponds to the inverse element b^{-1} of b. Through E draw $EF \parallel AD$, meeting l at F. Then F corresponds to the number ab^{-1} in N. If the Pascalian

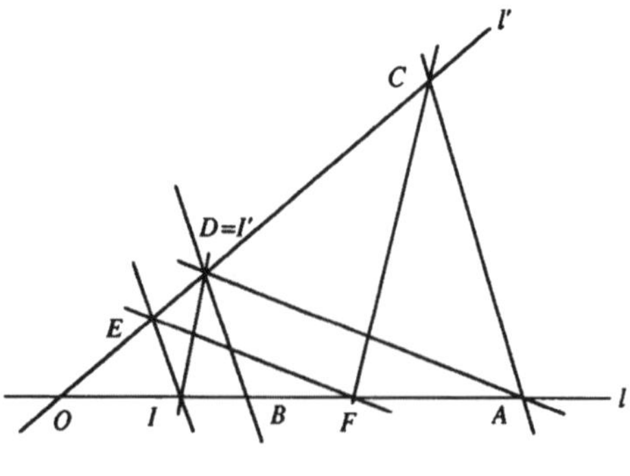

Fig. 2.4

axiom P holds, then its application to

$$P\begin{bmatrix} l & / & FAI \\ l' & / & DEC \end{bmatrix}$$

yields $CF \parallel DI$ and, therefore, $ab^{-1} = c$. If, on the contrary, the Pascalian axiom P does not hold, e.g.,

$$P\begin{bmatrix} l & / & FAI \\ l' & / & DEC \end{bmatrix}$$

is false, we may take $I' = D$ and draw $I'B \parallel IE$, meeting l at B. Thus CF is not parallel to ID. If we assume that A, B correspond to a, b and C to c, then F corresponds to ab^{-1}, while if CF is not parallel to ID then $ab^{-1} \neq c$. It follows that in the general case from $AC \parallel BD$ one can deduce $a^{-1}b = c^{-1}d$ but not $ba^{-1} = dc^{-1}$. Hence, it is impossible to set up a proper theory of proportion.

If we assume the Pascalian axiom for intersecting lines, then by 2 above the commutative axiom of multiplication holds. Thus, the condition $a^{-1}b = c^{-1}d$ for $AC \parallel BD$ can be written not only as

$$ba^{-1} \doteq dc^{-1}$$

but also as

$$ad = bc$$

which can be defined furthermore as

$$a : b = c : d.$$

Based on this, the establishment of the theory of proportion is easy.

4. From the Pascalian axiom for intersecting lines, one can deduce Desargues' axioms D_1, D_2 (under the assumption of other axioms). This is the well-known Hessenberg theorem (cf. Hessenberg 1905a).

Theorem (Hessenberg). In a geometry which satisfies the axioms of incidence H I, the (sharper) axiom of parallels H IV, the axiom of infinity D_∞, and the Pascalian axiom for intersecting lines P, Desargues' axioms D_1, D_2 become theorems.

Proof. We need only prove Desargues' axiom D_2. It is sufficient to prove only the case stated as follows, from which the other cases can be deduced.

Let the vertices of the two triangles ABC and $A'B'C'$ be distinct from each other and suppose that the connecting lines AA', BB', CC' of the corresponding vertices are distinct from each other and pass through one point O. Suppose

$$AB \parallel A'B', \quad AC \parallel A'C'.$$

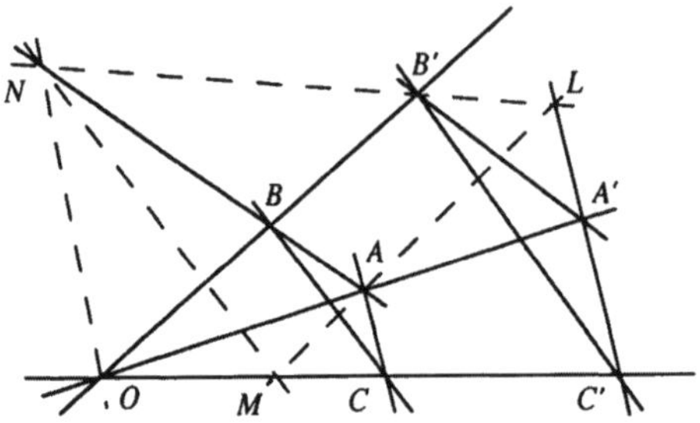

Fig. 2.5

We need to prove

$$BC \parallel B'C'.$$

Let us consider the following four different cases:

(1) AB is not parallel to OC and neither is AC to OB.
(2) AB is not parallel to OC but AC is parallel to OB.
(3) AC is not parallel to OB but AB is parallel to OC.
(4) AB is parallel to OC and so is AC to OB.

Case 1. See Fig. 2.5. Draw through A a parallel to BB': this parallel should meet $A'C'$ at a point, say L, and meet CC' at a point, say M. Assume for the time being that LB' is neither parallel to AA' nor to CC'. Let the intersection point of LB' and AB be N. Since LB' intersects AA', application of the Pascalian axiom for intersecting lines

$$P\begin{bmatrix} L & B' & N \\ O & A & A' \end{bmatrix}$$

yields $LA' \parallel NO$. Therefore, $AC \parallel NO$. Since AB also intersects CC', application of the Pascalian axiom for intersecting lines

$$P\begin{bmatrix} N & A & B \\ C & O & M \end{bmatrix}$$

yields $NM \parallel BC$. Applying once more the Pascalian axiom for intersecting lines

$$P\begin{bmatrix} N & L & B' \\ C' & O & M \end{bmatrix}$$

one obtains $NM \parallel B'C'$. Hence, $BC \parallel B'C'$. □

The above assumption can be treated in the following way (see Fig. 2.6). Draw through B two parallels to AA' and CC', meeting AC at P and Q respectively. Similarly, draw through B' two parallels to AA' and CC', meeting $A'C'$ at P' and Q' respectively. Draw through P, Q, P', Q' parallels to BB', meet-

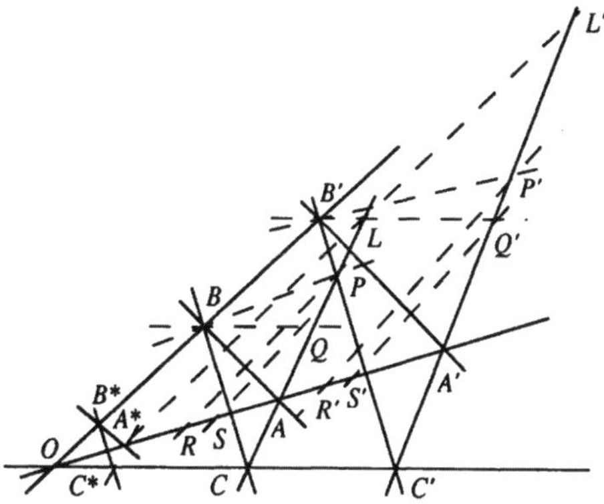

Fig. 2.6

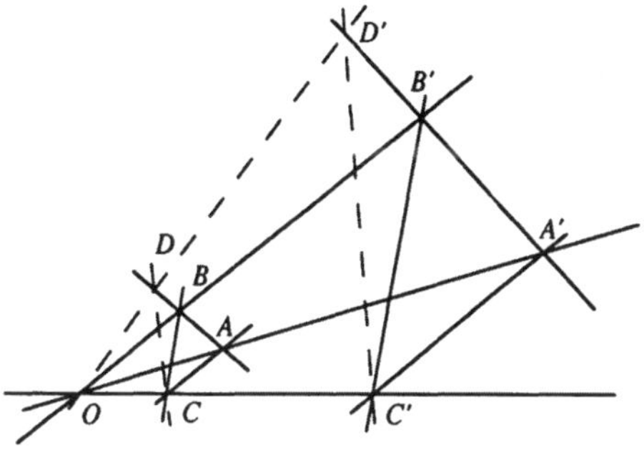

Fig. 2.7

ing AA' at R, S, R', S' respectively. Take an arbitrary point A^* on AA', distinct from O and any of R, S, R', S'. Draw through A^* a parallel to AB, meeting BB' at B^* and a parallel to AC, meeting CC' at C^*. Draw next through A^* a parallel to BB', meeting AC and $A'C'$ at L and L' respectively. Then L is distinct from P, Q and thus LB is neither parallel to AA' nor to CC'. Therefore, $\triangle A^*B^*C^*$ and $\triangle ABC$ satisfy the hypotheses of Case 1 and the assumed condition. By the previous proof, we have $B^*C^* \parallel BC$. Similarly, for $\triangle A^*B^*C^*$ and $\triangle A'B'C'$, we have $B^*C^* \parallel B'C'$ and thus $BC \parallel B'C'$.

Case 2. See Fig. 2.7. In this case, $AC \parallel BB'$, but AB is not parallel to CC'. Take a point D on AB such that CD is not parallel to BB'. Extend the line OD to meet $A'B'$ at D'. Then we may apply Case 1 to $\triangle ACD$ and $\triangle A'C'D'$ to get $CD \parallel C'D'$. Also, $\triangle DBC$ and $\triangle D'B'C'$ satisfy Case 1, so $BC \parallel B'C'$.

Case 3. This is similar to Case 2.

Case 4. In this case, $AB \parallel CC'$ and $AC \parallel BB'$. As in Case 2, let us draw a line distinct from AA', BB', CC' through O so as to intersect $AB, A'B'$ at D, D' respectively, whereas CD is not parallel to BB'. Then $\triangle ACD$ and $\triangle A'C'D'$ are as in Case 3, i.e., $CD \parallel C'D'$. Similarly, $\triangle DBC$ and $\triangle D'B'C'$ are as in Case 2 or Case 3; therefore, $BC \parallel B'C'$.

This completes the proof of Hessenberg's theorem. The reason why we are so patient to write down the detailed proof in all different cases is still as mentioned before (also see Sect. 3.1). □

5. Pascalian geometry

Definition. If a (plane) geometry satisfies the following axioms:

Hilbert's axioms of incidence H I;
Hilbert's (sharper) axiom of parallels H IV;
Axiom of infinity D_∞;
Pascalian axiom for intersecting lines P,

it is called *Pascalian* geometry.

In Pascalian geometry, by Hessenberg's theorem Desargues' axioms D_1, D_2 are naturally satisfied and thus, according to Sects. 1.5 and 1.6, there is an associated Desarguesian number system. According to 2 above, this system satisfies the commutative axiom of multiplication N 13. Thus, it is not only a number skew field but also a number field (of characteristic 0 due to the axiom of infinity). We shall call this field the *number field associated* with Pascalian geometry. In this geometry, after having fixed a coordinate system, the algebraic expressions of geometric relations become very simple because of the commutativity of multiplication. Consider, for instance, the lemma of Example 2 in Sect. 1.7. Then for four points $P = (x_1, x_2)$, $Q = (y_1, y_2)$, $R = (z_1, z_2)$ and $S = (u_1, u_2)$, so long as $P \neq Q$, $R \neq S$, a necessary and sufficient condition for PQ to be parallel to or coincident with RS becomes

$$(u_1 - z_1)(y_2 - x_2) = (y_1 - x_1)(u_2 - z_2).$$

Here we no longer need the restriction that PQ, RS are not parallel to or coincident with l_1, l_2.

2.2 Orthogonal axioms and (unordered) orthogonal geometry

The usual Euclidean geometry, called *ordinary geometry* in this book, is mainly the study of metric properties of geometric configurations. In an axiom system, especially in Hilbert's, the metric properties are expressed by means of the undefined fundamental concept *congruence* and some related congruence axioms. The introduction of these congruence axioms in the third group by Hilbert is not independent of the other groups of axioms, but uses the axioms of order. However, for those reasons stated in the preface, we shall establish geometry avoiding the use of the concept and axioms of order. Consequently, so-called segments, rays, angles, etc. all become meaningless, not to mention the length of segments or the size of angles. In order to establish, without assuming the axioms of order, an ordinary geometry similar to Euclidean geometry but including even wider metric geometries, whilst preserving the axioms of incidence

H I, the axiom of parallels H IV as well as the axiom of infinity and Desargues' axioms D, we shall introduce a few additional concepts and related axioms to replace partially the concepts and axioms of congruence. This and the next section are devoted to the exposition of such concepts.

The relation of orthogonality is one of the clearest and most natural of metric properties. Many ancient cultures discovered figures such as Kou-Ku triangles and rectangles. The concept of orthogonality may be even more fundamental than that of parallelism. For example, the parallel relation of two lines can be described by means of the relation that both of the lines are perpendicular to a third one. In ancient Chinese geometry, the relation of orthogonality pervaded the whole system through rectangles, Kou-Ku triangles and various other constructs. In contrast, the concept of parallelism was rare. In the modern study of geometric axioms, many authors have discarded the axiom of parallels but preserved the idea of orthogonality in order to unify all kinds of Euclidean and non-Euclidean geometries, cf. Bachmann (1959a, b) and the literature therein. The reader may also refer to Greenberg (1973). Those books all consider related problems from the viewpoint of axiomatization. Even so, from the viewpoint of mechanization, the orthogonal properties should also occupy a special position. Since the concepts of incidence and parallelism are essentially linear, this section focuses on *orthogonality*, which may be considered as the simplest non-linear one.

Now we introduce perpendicularity as an undefined fundamental concept: A line l is *perpendicular* to a line l', denoted as $l \perp l'$. This orthogonal concept satisfies the following axioms.

Orthogonal axioms O

O 1. If $l \perp l'$, then $l' \perp l$.

O 2. Given a point O and an arbitrary line l, there exists one and only one line $l' \perp l$ through O.

O 3. If two lines l' and l'' are both perpendicular to a third line l, i.e., $l' \perp l$ and $l'' \perp l$, then $l' \parallel l''$.

Our geometry will be assumed to satisfy the axioms of incidence H I, the axiom of parallels H IV, the axiom of infinity, and Desargues' axioms D as well as the above three orthogonal axioms.

In this geometry, a line l may be perpendicular to itself, i.e., $l \perp l$. Such a line is called an *isotropic line*. From Axioms O 2, O 3 and the axiom of parallels one knows that a line which is parallel to an isotropic line should also be an isotropic line, i.e., it is self-perpendicular. Moreover, any two parallel isotropic lines are perpendicular to each other. Therefore, if every line is isotropic then the concepts of orthogonality and of parallelism or congruence are identical. In such a geometry, the concept of orthogonality will be superfluous. For this reason we shall add some axioms.

O 4. Through any point, there always exists a non-isotropic line.

In order to make our geometry to have richer metric properties and to approach the ordinary geometry as closely as possible, we further add the following.

O 5 (Axiom of orthocenter). Let the vertices A, B, C of $\triangle ABC$ be non-collinear. Drop from A, B, C three perpendiculars $l_A \perp BC$, $l_B \perp AC$, $l_C \perp AB$ respectively. Then l_A, l_B, l_C are concurrent.

The three perpendiculars in this axiom dropped from the three vertices to the three sides of the triangle will be called the *altitudes* of the triangle. The common point of intersection of the three altitudes will be called the *orthocenter*.

Now, let A, B, C be non-collinear and suppose that both AB and AC are isotropic lines. Draw through B, C two parallels to AC, AB respectively. Then by the axiom of parallels, these two lines meet each other, say at A'. By the orthogonal axioms O 2 and O 3, CA' is also an isotropic line and it is both parallel and perpendicular to AB. Similarly, BA' is also an isotropic line and is both parallel and perpendicular to AC. By the axiom of orthocenter O 5, A' is the orthocenter of $\triangle ABC$ and $AA' \perp BC$. If BC is also an isotropic line, then $AA' \parallel BC$, too. Since $ABA'C$ is a parallelogram, this contradicts the axiom of infinity. Therefore, the three sides of a triangle whose vertices are not collinear cannot all be isotropic lines.

From this, one may also show that Axiom O 4 can actually be deduced from Axiom O 5 and the other axioms.

We call the geometry which satisfies the following axioms an *unordered orthogonal geometry* or, simply, an *orthogonal geometry:*

Hilbert's axioms of incidence H I;
Hilbert's (sharper) axiom of parallels H IV;
Axiom of infinity and Desargues' axioms D;
Orthogonal axioms O 1–O 5.

In this orthogonal geometry, the Pascalian axiom for intersecting lines is a theorem. A direct result is that the associated Desarguesian number system introduced according to Desargues' axioms is not only a sfield but also a field in which multiplication is commutative. This fact was found by Schur (1903). To prove it, we shall first prove the Pascalian axiom in the case that the intersecting lines are two distinct orthogonal lines.

Orthogonal Pascalian theorem. Let A, B, C and A', B', C' be two sets of points respectively on two distinct orthogonal lines l and l' that are distinct from each other and distinct from the intersection point O of l and l'. If

$$AB' \parallel A'B, \quad BC' \parallel B'C,$$

then

$$AC' \parallel A'C.$$

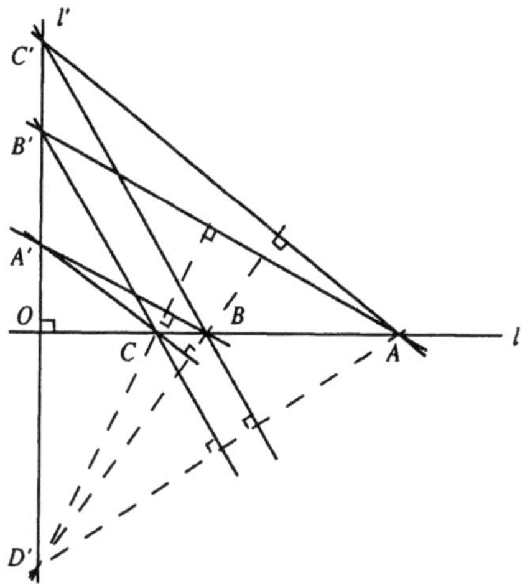

Fig. 2.8

Proof. See Fig. 2.8. First of all, both l and l' are non-isotropic lines. By Axiom O 2, construct a line through B perpendicular to $A'C$; then, according to Axiom O 3 and the axiom of parallels, this line intersects l' at a point, say D'.

Applying now the axiom of the orthocenter to $\triangle A'BD'$, we have $D'C \perp A'B$.

Since $AB' \parallel A'B$, Axiom O 3 and the axiom of parallels imply that $D'C \perp AB'$.

Applying the axiom of the orthocenter to $\triangle AB'D'$ yields $B'C \perp AD'$.

By $BC' \parallel B'C$, we have $BC' \perp AD'$.

Applying the axiom of the orthocenter to $\triangle C'AD'$ yields $AC' \perp BD'$.

Since AC' and $A'C$ are both perpendicular to BD', Axiom O 3 implies that $AC' \parallel A'C$. This completes the proof. □

According to Schur's theorem and its proof above, Pascal's theorem for intersecting lines holds for two orthogonal lines in orthogonal geometry. Since we have assumed Desargues' theorem as an axiom, according to Sect. 2.1 the multiplication in the Desarguesian number system associated with this geometry is commutative. Again, from 2 in Sect. 2.1 one knows that the general Pascalian theorem for intersecting lines still holds. This theorem can also be proved by a direct use of the orthogonal axioms, without reference to Sect. 2.1. The following proof originates from Guse (cf. Bachmann 1959a: p. 208).

We designate the intersection point of the two lines l, l' as O and two sets of points on these lines as A_1, A_2, A_3 and A_1', A_2', A_3' respectively. The hypothesis consists of

$$A_2A_3' \parallel A_3A_2', \quad A_1A_2' \parallel A_2A_1'.$$

We want to prove

$$A_1A_3' \parallel A_3A_1'.$$

For this purpose, we assume that A_1, A_2, A_3 are all distinct from each other and distinct from O, and so are A_1', A_2', A_3'. Suppose the two lines l and l' are

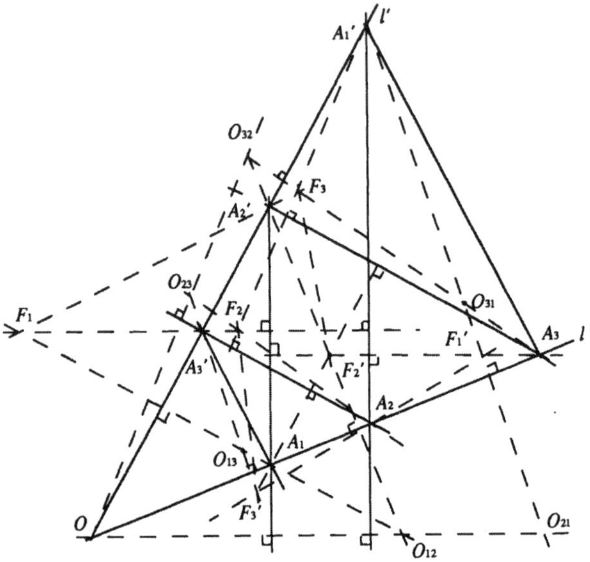

Fig. 2.9

not perpendicular nor isotropic and $A_iA'_j$ is not perpendicular to either l or l'. When all lines are perpendicular to each other, the proof below needs to be slightly modified.

Take an arbitrary permutation ijk of 123 and

let F_i be the orthocenter of $\triangle A_iA'_jA'_k$,
let F'_i be the orthocenter of $\triangle A'_iA_jA_k$,
let O_{ij} be the orthocenter of $\triangle OA_iA'_j$.

From Fig. 2.9, one sees that F_2F_3 and $O_{23}O_{32}$ are perpendicular to $A_2A'_3$ and $A_2A'_3 \parallel A_3A'_2$, so $F_2F_3 \parallel O_{23}O_{32}$. As F_2O_{23} and F_3O_{32} are perpendicular to l', we have $F_2O_{23} \parallel F_3O_{32}$. Similarly, $F'_2F'_3 \parallel O_{23}O_{32}$ and $F'_2O_{32} \parallel F'_3O_{23}$. Applying Desargues' theorem to $\triangle F_2F'_3O_{23}$ and $\triangle F_3F'_2O_{32}$ yields $F_2F'_3 \parallel F_3F'_2$. Therefore $F_2F_3F'_2F'_3$ is a parallelogram. According to Sect. 1.2, the diagonals $F_2F'_2$ and $F_3F'_3$ bisect one another.

In the same way, one can infer from $A_1A'_2 \parallel A_2A'_1$ that $F_1F_2F'_1F'_2$ is also a parallelogram and, therefore, the diagonals $F_1F'_1$ and $F_2F'_2$ bisect one another.

It follows from the above that $F_1F'_1$ and $F_3F'_3$ bisect one another. According to Sect. 1.2, $F_1F_3F'_1F'_3$ is a parallelogram. Now using Desargues' theorem for $\triangle F_1F'_3O_{13}$ and $\triangle F_3F'_1O_{31}$, one obtains $O_{13}O_{31} \parallel F_1F_3 \parallel F'_1F'_3$.

Assume that F_1F_3 does not pass through A'_2 so that $A'_2F_1F_3$ is a triangle. As A'_2F_1 and OO_{13} are both perpendicular to $A_1A'_3$, we have $A'_2F_1 \parallel OO_{13}$. Also A'_2F_3 and OO_{31} are both perpendicular to $A_3A'_1$, so $A'_2F_3 \parallel OO_{31}$. Using Desargues' theorem for $\triangle A'_2F_1F_3$ and $\triangle OO_{13}O_{31}$, one obtains $A'_2O \parallel F_1O_{13} \parallel F_3O_{31}$. But this contradicts the fact that $l' \perp F_1O_{13}$ and l' is a non-isotropic line. Hence, F_1F_3 passes through A'_2, i.e., $F_1F_3 \perp A_1A'_3$, $F_1F_3 \perp A_3A'_1$. Hence $A_1A'_3 \parallel A_3A'_1$.

Although we have used the axiom of the orthocenter to prove the Pascalian axiom for intersecting lines and, according to Sect. 2.1, one can deduce Desargues' axioms from the Pascalian axiom for intersecting lines by applying the axioms of incidence, parallels, infinity, etc., we still cannot readily omit Desargues' axioms from the axiom system of orthogonal geometry. The proofs of Schur and Guse both use Desargues' theorems and theorems about midpoints. Since our purpose is not to study the logical relations and independence among axioms, the problem whether Desargues' axioms can be deduced from other axioms of orthogonal geometry so as to be regarded as theorems will not be discussed further.

Since, on the other hand, there are concepts of both orthogonality and parallelism as well as midpoints, we may define, for any pair (AB) of two distinct points, the line through the midpoint of the pair (AB) and perpendicular to the line AB as the *perpendicular bisector* of (AB). Clearly, if AB is an isotropic line, then its perpendicular bisector is AB itself.

For $\triangle ABC$, the perpendicular bisectors of the pairs (AB), (AC), (BC) will be simply said to be the perpendicular bisectors of the three sides. As in ordinary geometry, the theorem on the circumcenter is still valid in orthogonal geometry. Namely, we have the following result.

Theorem 1. The perpendicular bisectors of the three sides of any triangle are concurrent.

Proof. As in Fig. 2.10, let the midpoints of the pairs $(BC), (AC), (AB)$ be A', B', C' respectively. According to Sect. 1.2, $B'C' \parallel BC$, $A'C' \parallel AC$, $A'B' \parallel AB$. By the orthogonal axioms, the perpendicular bisector of BC is perpendicular to $B'C'$, i.e., it is the altitude of $\triangle A'B'C'$ on side $B'C'$. Similarly, the perpendicular bisectors of AC, AB are the altitudes of $\triangle A'B'C'$ respectively on the sides $A'C'$, $A'B'$. From the axiom of the orthocenter, we see that the three perpendicular bisectors are concurrent. The theorem is then proved. □

As in ordinary geometry, we call the common intersection point of the three perpendicular bisectors of the three sides of a triangle the *circumcenter* of the triangle. Since the concepts of distance and a circle do not exist in orthogonal geometry, the circumcenter here does not have the properties that the circumcenter of a triangle has in ordinary geometry.

There are concepts of incenter and excenters of a triangle in ordinary geometry, but they cannot be extended directly in orthogonal geometry. In orthogonal geometry we do not assume any axiom of order nor do we have any concept of order and, thus, segments and angles have no meaning in this geometry. Moreover, we do not assume any axiom nor any concept of congruence in orthogonal geometry, so we cannot even have a concept of bisectors of an angle. Even so, we can still introduce the concept of *symmetric axes* in this geometry (instead partially of the concept of bisectors of angles in ordinary geometry), as we do have concepts of orthogonality and midpoints. Let us state this as follows.

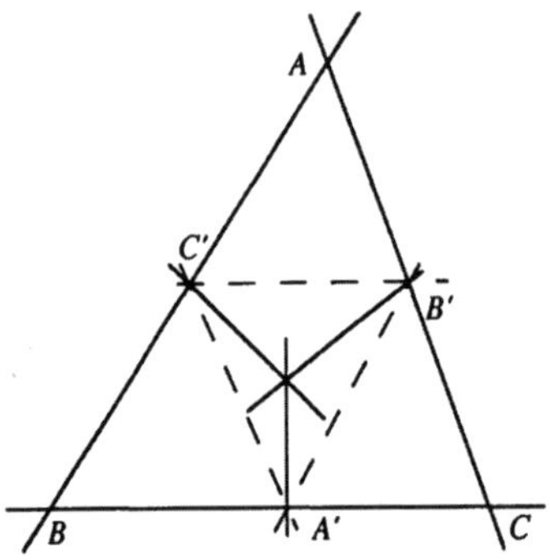

Fig. 2.10

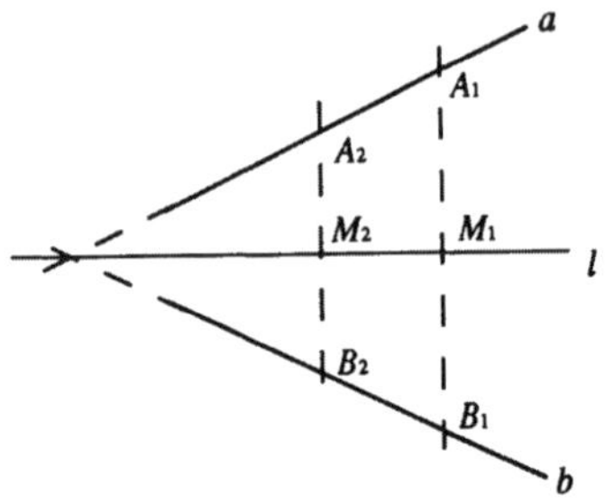

Fig. 2.11

Let the perpendicular bisector of the pair (AB) of two points A, B be l. We call A the *symmetric point* of B with respect to l or l the *symmetric axis* of (AB). Any point A on l is said to be a *symmetric point* of itself with respect to l. Obviously, this definition has meaning only if l is a non-isotropic line.

We may construct the symmetric point of any point A with respect to (a non-isotropic line) l in the same way as in ordinary geometry. By the orthogonal axiom O 2, there is a line through A and perpendicular to l. Since l is non-isotropic, this line must meet l at a point, say M. According to Sect. 1.2, we construct the symmetric point B of A with respect to M. Then B is the symmetric point of A with respect to l.

We show below some simple properties about symmetry.

Property 1. Let l be a non-isotropic line. If the points $A_1, A_2, \ldots$ are collinear, then the symmetric points $B_1, B_2, \ldots$ of $A_1, A_2, \ldots$ with respect to l are also collinear.

Proof. As in Fig. 2.11, construct perpendiculars to l through $A_1, A_2, \ldots$. Since l is non-isotropic, these perpendiculars meet l at points, $M_1, M_2, \ldots$, say. Then the symmetric points $B_1, B_2, \ldots$ of $A_1, A_2, \ldots$ with respect to l are those of $A_1, A_2, \ldots$ with respect to $M_1, M_2, \ldots$. According to Sect. 1.2, $B_1, B_2, \ldots$ are collinear. □

Let a and b be the lines determined by $A_1, A_2, \ldots$ and by $B_1, B_2, \ldots$ respectively. We also know that, if $a \parallel l$, then $b \parallel l$. If a intersects l at O, then b also intersects l at the same point O. Because of this property, we call b the *symmetric line* of a with respect to l, or l the *symmetric axis* of a and b.

Property 2. Let l be a non-isotropic line. If the lines a_1 and a_2 are parallel, then the symmetric lines b_1 and b_2 of a_1 and a_2 with respect to l are also parallel.

Proof. If b_1 and b_2 intersect, say at B, rather than being parallel, then a_1 and a_2 intersect at the symmetric point of B. This contradicts the hypothesis. □

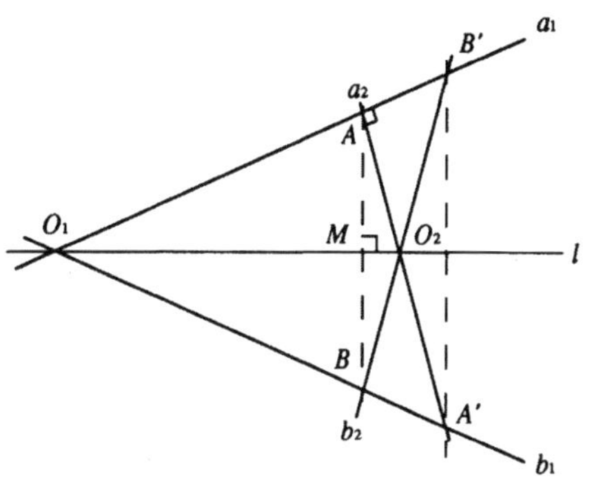

Fig. 2.12

If we replace parallelism in Property 2 by perpendicularity, then the proof becomes much more complicated, see details below.

Property 3. Let l be a non-isotropic line. If the lines a_1 and a_2 are perpendicular to each other, then the symmetric lines b_1 and b_2 of a_1 and a_2 with respect to l are also perpendicular to each other.

Proof. If one of a_1 and a_2, say a_1, is parallel to l, then b_1 is also parallel to l and a_2, b_2 are the same, both perpendicular to l. Hence $b_1 \perp b_2$. As shown in Fig. 2.12, suppose that both a_1 and a_2 are not parallel to l but meet l at O_1 and O_2 respectively. When a_1 or a_2 is isotropic, a_1 is coincident with a_2 and thus b_1 and b_2 should also be isotropic and coincident, therefore $b_1 \perp b_2$. Hence, in what follows we assume that a_1, a_2, b_1, b_2 are all non-isotropic lines.

Under this assumption, a_1 and a_2 are distinct and meet at a point, say A. Suppose first A is distinct from O_1 and O_2. Let the symmetric point of A with respect to l be B; then B is the intersection point of b_1 and b_2. In addition, $AB \perp l$, so the intersection point M of AB and l is the midpoint of the pair (AB).

Now, if $AO_2 \parallel O_1B$, then by Property 2 we have $BO_2 \parallel O_1A$, for the symmetric lines of AO_2 and O_1B with respect to l are BO_2 and O_1A respectively. Hence, $a_1 \perp a_2$ or $O_1A \perp O_2A$ implies that $O_1B \perp O_2B$ or $b_1 \perp b_2$.

If AO_2 is not parallel to O_1B, then we may assume that AO_2 (i.e., a_2) meets b_1 at A'. Similarly, b_2 should also meet a_1 at a point, say B'. Since the symmetric lines of a_2 and b_1 with respect to l are b_2 and a_1 respectively, the symmetric point of the intersection point A' of a_2 and b_1 with respect to l is the intersection point B' of b_2 and a_1. In particular, we have $A'B' \perp l$. Applying the axiom of the orthocenter to $\triangle O_1A'B'$, one sees that O_2 is the orthocenter of the triangle and thus $B'O_2 \perp O_1A'$, i.e., $b_1 \perp b_2$.

Suppose finally the intersection point A of a_1 and a_2 on l coincides with O_1, O_2. Then the intersection point B of l_1 and l_2 is the same as point A. As in Fig. 2.13, take a point A' distinct from A on a_1 and draw $a_2' \parallel a_2$ through A'. Let the symmetric point and line of A' and a_2' with respect to l be B' and b_2' respectively. Then $a_2' \parallel a_2$, $a_2 \perp a_1$ imply $a_2' \perp a_1$. Using the case we have

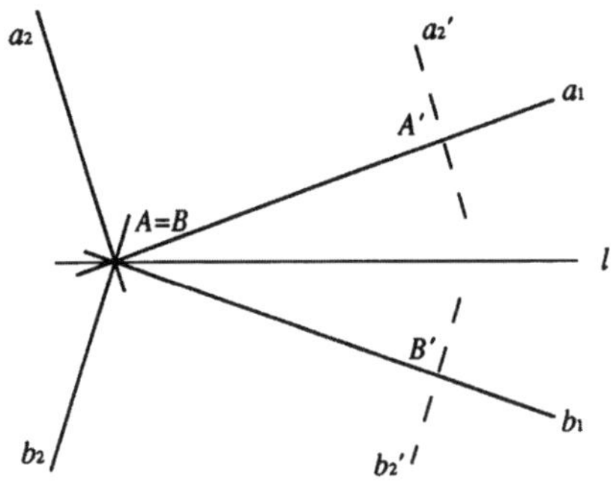

Fig. 2.13

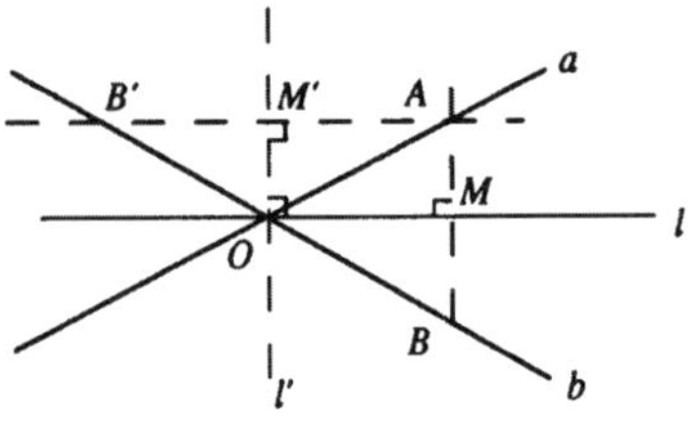

Fig. 2.14

proved already, it follows that $b_2' \perp b_1$ and thus $b_1 \perp b_2$. This completes the proof. □

Property 4. For any two non-isotropic parallel lines a and b, there is one and only one line l such that a and b are symmetric with respect to l.

Proof. If such a line l exists, then l should be parallel to a and b. Therefore, for an arbitrary point A on a, the symmetric point B of A, with respect to l, lies on b and $AB \perp l$, i.e., AB is perpendicular to a and b. Hence, we may take a point A on a and through it construct a perpendicular to a, meeting b at B. Thus the perpendicular bisector of AB is the unique symmetric axis l. □

For two intersecting lines, the symmetric axes are similar to, but not the same as, the bisectors of angles in ordinary geometry. In orthogonal geometry, there is no guarantee that two intersecting lines have a symmetric axis. If the symmetric axes of two intersecting lines exist, there are two of them and they are perpendicular to each other. In that case, they are analogous to the angular bisectors of two intersecting lines in ordinary geometry. In detail, we have the following.

Property 5. Two intersecting lines have either no symmetric axis or exactly two symmetric axes. In the latter case, the two axes are perpendicular to each other.

Proof. See Fig. 2.14. Let the lines a and b intersect at O and have l as their symmetric axis. Take a point A distinct from O on a and let the symmetric point of A with respect to l be B. Then B lies on b, $AB \perp l$ and the intersection point M of AB and l is the midpoint of the pair (AB).

Erect now the perpendicular l' to l at O and construct through A a perpendicular to l', meeting l' at M' and b at B'. As AB' and l are perpendicular to l', by the orthogonal axiom O 3 we have $AB' \parallel OM$. Also, as M is the midpoint of (AB), according to Sect. 1.2, O is the midpoint of the pair (BB'). Next, since AB and l' are perpendicular to l, we have $AB \parallel l'$. As O is the midpoint of (BB'), M' is the midpoint of the pair (AB'). This implies that l' is the perpendicular bisector of the pair (AB'). Thus, A and B' are symmetric with respect to l', and so are a and b. Hence a and b have two mutually perpendicular lines l and l' as their symmetric axes.

Suppose that a and b have another symmetric axis l'' different from l. Construct now A and B as before and the symmetric point B'' of A with respect to l''. Then l and l'' are the perpendicular bisectors of two sides AB and AB'' of $\triangle ABB''$, so their intersection point O is the circumcenter of $\triangle ABB''$. From the theorem about the circumcenter, one knows that the perpendicular bisector of (BB') passes through O, in particular O is the midpoint of (BB'). Similarly as before, we have $l'' \perp l$ while l'' coincides with l'. Thus there are exactly two symmetric axes of a and b, i.e., l and l'. The property is proved. □

Analogous to the theorem about incenter and excenters of a triangle in ordinary geometry, in orthogonal geometry we have the following.

Theorem 2. If the two lines AB, BC of $\triangle ABC$ have as their symmetric axes b, b' and the two lines AC, BC have as their symmetric axes c, c', then the two lines AB, AC also have two symmetric axes, say a, a'. Furthermore, all these symmetric axes, three-to-three, intersect at four distinct points respectively.

Proof. First of all, it is easy to know that the three sides of $\triangle ABC$ are non-isotropic.

Take a symmetric axis b of BA, BC and a symmetric axis c of CA, CB (see Fig. 2.15). Let us first show that b and c cannot be parallel. Assume that b and c are parallel. Let the midpoint of (BC) be M and construct through M a perpendicular to b and c. Let this perpendicular meet b, c at B', C' and BA, CA at P, Q. Note that BC is not perpendicular to b, c, otherwise BC would coincide with BA, CA. Hence B', C' are distinct from B, C. According to Chap. 1, M is the midpoint of $(B'C')$. Since B', C' are the midpoints of (MP) and (MQ), it is easy to show that M is the midpoint of (PQ). Therefore, $BA \parallel CA$ cannot be true.

From the above, b and c intersect at a point, say O. As in Fig. 2.16, through

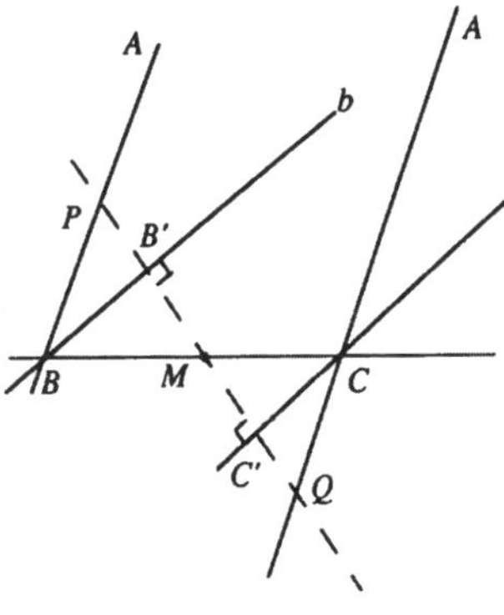

Fig. 2.15

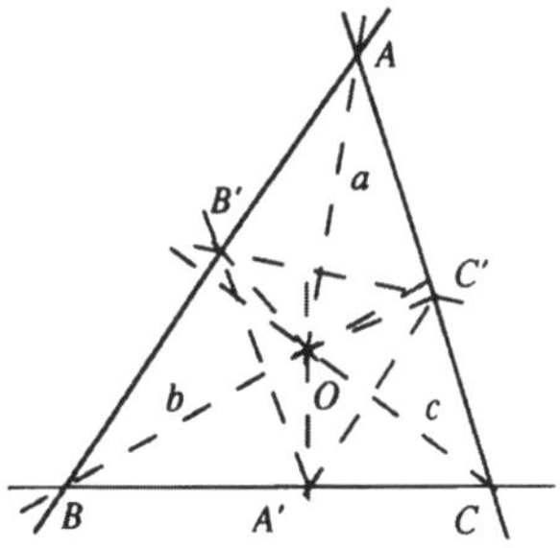

Fig. 2.16

O construct $OA' \perp BC$, $OB' \perp AB$, $OC' \perp AC$, and let these perpendiculars meet BC, AB, AC at A', B', C'. As the symmetric line of BA' with respect to b is BB', by Property 3 one knows that the symmetric line of OA' with respect to b is OB' and the symmetric point of A' with respect to b is B', so b is the perpendicular bisector of the pair $(A'B')$. Similarly, A' and C' are symmetric with respect to c, so c is the perpendicular bisector of the pair $(A'C')$. By the theorem about the circumcenter, O is the circumcenter of $\triangle A'B'C'$ and the perpendicular bisector, say a, of the pair $(B'C')$ also passes through O. Therefore B' and C' are symmetric with respect to a and the lines OB' and OC' are symmetric with respect to a, too. By Property 3 again, the lines AB' and AC' are symmetric with respect to a. Therefore, a passes through the point A and is a symmetric axis of AB and AC.

This implies that AB and AC have another symmetric axis, say a', different from a, and a, b', c' are concurrent as before. Similarly, b, c', a' and c, a', b' are respectively concurrent. This completes the proof. □

2.3 The orthogonal coordinate system of (unordered) orthogonal geometry

In Pascalian geometry or Desarguesian geometry, the choice of coordinate axes, except for being intersecting, may be arbitrary. In (unordered) orthogonal geometry, as there are orthogonal lines, we may take any two intersecting orthogonal and thus non-isotropic lines as the coordinate axes l_1 and l_2. The intersection point O is the origin and the unit points I_1 and I_2 can be arbitrarily chosen on l_1 and l_2, so long as they are distinct from O. Through a point P in the plane, construct two perpendiculars to l_1 and l_2, meeting them at X_1 and X_2 respectively. On l_1 and l_2 let X_1 and X_2 correspond respectively to x_1 and x_2 in the number system determined by mapping O to 0 and I_1, I_2 to 1. Then the coordinates of P are the pair (x_1, x_2) of numbers: $P = (x_1, x_2)$. Such a coordinate system determined by two intersecting orthogonal lines is called an *orthogonal coordinate system* and the coordinates of P are also called *orthogonal coordinates*. Since orthogonal geometry is a special kind of Pascalian geometry and the orthogonal coordinate system is a special kind of coordinate system, those algebraic expressions corresponding to the geometric relations such as parallelism, midpoint and equation of a line in Pascalian geometry are still valid. However, for concepts like perpendicularity, symmetry, etc. which appear in orthogonal geometry there are no corresponding ones in Pascalian geometry. The goal of this section is to set up the algebraic expressions for these geometric relations in an orthogonal coordinate system.

Below, the number system associated with orthogonal geometry will be denoted by $\boldsymbol{K}$. The number system defined by taking any line l and two distinct points O and I on it as 0 and 1 will be denoted by $N(l, O, I)$ as before. In view of Pascal's theorem in orthogonal geometry, those number systems are all number fields, i.e., the commutative law of multiplication holds, and there is a canonical isomorphism between $N(l, O, I)$ and $\boldsymbol{K}$. Therefore, we shall identify $N(l, O, I)$ and $\boldsymbol{K}$.

An orthogonal coordinate system determined by two intersecting orthogonal

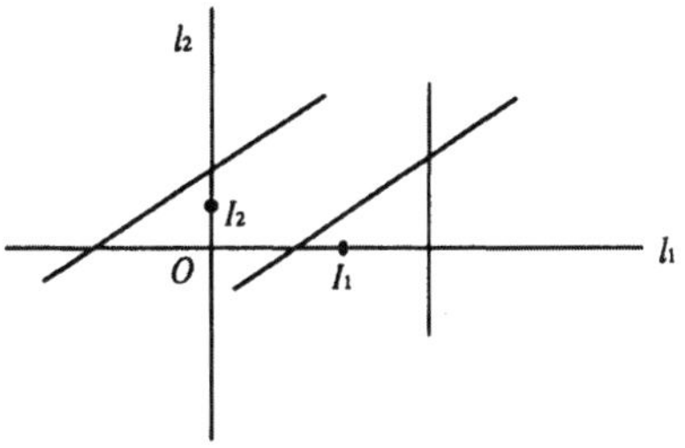

Fig. 2.17

lines l_1, l_2 and two points I_1, I_2 respectively on l_1, l_2 (distinct from the intersection point O of l_1, l_2) is denoted by $C(l_1, l_2, I_1, I_2)$ or simply by C. First of all, we introduce the concept of *slope* for a family of parallel lines with respect to the orthogonal coordinate system C.

According to Sect. 1.7, the equation of a line parallel to l_2 is always of the form

$$x_1 = c,$$

in which c is a constant in $\boldsymbol{K}$. And, the equation of a line not parallel to l_2 is of the form

$$x_2 = mx_1 + a,$$

in which a, m are also constants in $\boldsymbol{K}$. The equations of those lines that are not parallel to l_2 but parallel to each other may have different a but the same m. We call m the *slope* of those parallel lines. Especially, for those lines parallel to l_1, the slope will be $m = 0$. As for those parallel to l_2, we introduce a new symbol ∞ and define their slope to be ∞.

If the slope of a family of parallel lines is m, then those lines which are perpendicular to these are also parallel to each other and their slope is a function of m. Let this function be f. Then

$$f(0) = \infty, \quad f(\infty) = 0.$$

Below we shall derive the value of $f(m)$ for $m \neq 0$ and $m \neq \infty$.

See Fig. 2.18. To determine the function f, take J_2 as the symmetric point of I_2 with respect to O. Then the coordinates of J_2 are $(0, -1)$. Construct the orthocenter E of $\Delta I_1 I_2 J_2$ by the axiom of the orthocenter. Then E lies on the axis l_1. Let the coordinates of E be $(k, 0)$, where k is a constant in $\boldsymbol{K}$ determined by the orthogonal coordinate system $C(l_1, l_2, I_1, I_2)$. We say that k is the *orthogonal rate* of this coordinate system whose meaning is explained as follows.

As $I_1 = (1, 0)$, $I_2 = (0, 1)$, $J_2 = (0, -1)$, $E = (k, 0)$, the equation of the line $I_1 J_2$ is $x_2 = x_1 - 1$ which has slope 1. Similarly, the slopes of $I_1 I_2$, $I_2 E$, $J_2 E$ are $-1, -k^{-1}, k^{-1}$ respectively. Since $I_1 I_2 \perp J_2 E$ and $I_1 J_2 \perp I_2 E$, we have

$$f(1) = -k^{-1}, \quad f(-1) = k^{-1},$$

$$f(k^{-1}) = -1, \quad f(-k^{-1}) = 1.$$

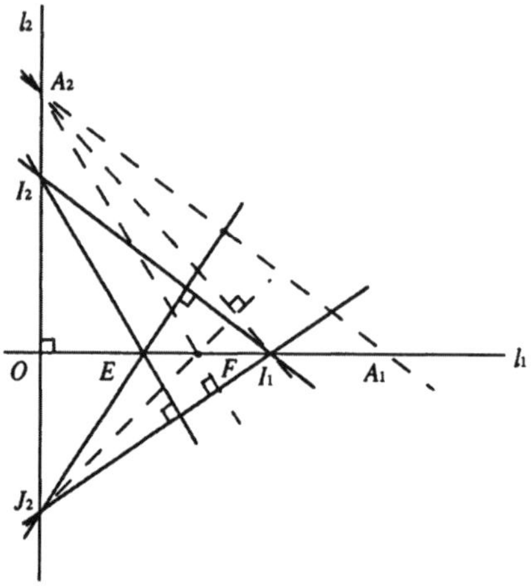

Fig. 2.18

From those formulas, one sees that if

$$m = +1,\ -1,\ k^{-1},\ \text{or}\ -k^{-1}$$

and

$$f(m) = m',$$

then

$$kmm' + 1 = 0.$$

Theorem 1. If the orthogonal rate of an orthogonal coordinate system is k, then for any two lines which are not parallel to an axis but perpendicular to each other, their slopes m and m' satisfy the relation

$$kmm' + 1 = 0$$

or

$$f(m) = -1/km.$$

Proof. Draw a line through I_1 with slope m ($m \neq 0, \infty$), meeting l_2 at A_2, and a line through A_2 parallel to I_1I_2, meeting the axis l_1 at A_1. Then the equation of I_1A_2 is $x_2 = mx_1 - m$ and the coordinates of A_2 are $(0, -m)$. In other words, the corresponding numbers of A_1 and A_2 in the number systems $N(l_1, O, I_1)$ and $N(l_2, O, I_2)$ are both $-m$. Construct the orthocenter F of $\triangle A_2I_1J_2$. Clearly F lies on the axis l_1. Since A_2F and I_2E are both perpendicular to I_1J_2, $A_2F \parallel I_2E$. As E and A_1 correspond respectively to the numbers k and $-m$ in the number system on l_1, F corresponds to $-km$ in the number system, i.e., $F = (-km, 0)$. Therefore, the slope of J_2F is $-1/km$. As $J_2F \perp I_1A_2$, we have $m' = -1/km$ and the theorem is proved. □

Let the equations of two perpendicular lines be expressed as

$$L_u:\quad u_1x_1+u_2x_2+u_3=0,$$
$$L_v:\quad v_1x_1+v_2x_2+v_3=0,$$

in which u_1, u_2 are not simultaneously 0, and neither are v_1, v_2. If none of u_1, u_2, v_1, v_2 is 0, then the slopes of L_u, L_v are $m_u=-u_1/u_2$, $m_v=-v_1/v_2$ respectively. Thus the condition for L_u, L_v to be perpendicular to each other becomes

$$ku_1v_1+u_2v_2=0.$$

If one of u_1, u_2, v_1, v_2, say u_2, is 0, then $m_u=\infty$. In this case, one should have $m_v=0$ or $v_1=0$, so the above condition still holds. The other cases are similar. That is, the above condition is satisfied in all cases, so Theorem 1 can be restated in the following form.

Theorem 1′. A necessary and sufficient condition for two lines L_u and L_v to be perpendicular to each other is

$$ku_1v_1+u_2v_2=0,$$

in which k is the orthogonal rate of the corresponding coordinate system.

In particular, the theorem remains true in the case of there being only one line. For instance, a necessary and sufficient condition for L_u to be isotropic is that L_u is perpendicular to itself and thus is

$$ku_1^2+u_2^2=0.$$

The orthogonal rate k depends apparently on the choice of coordinate systems. We shall investigate the relation between the orthogonal rates k and k^* determined by two different orthogonal coordinate systems $C=C(l_1, l_2, I_1, I_2)$ and $C^*=C(l_1^*, l_2^*, I_1^*, I_2^*)$. Consider first the case $l_1^*=l_1$, $l_2^*=l_2$ but I_1^*, I_2^* are not necessarily the same as I_1, I_2. We denote the number systems on l_1, l_2 simply by $N_1=N(l_1, O, I_1)$, $N_1^*=N(l_1, O, I_1^*)$, $N_2=N(l_2, O, I_2)$, $N_2^*=N(l_2, O, I_2^*)$. Construct as before the symmetric point J_2 of I_2 with respect to O and the orthocenter E of $\Delta I_1I_2J_2$ (see Fig. 2.19). Then the coordinates of E in C are $(k, 0)$, i.e., E corresponds to the number k in N_1. Let the coordinates of I_1^* and I_2^* in C be $(a_1, 0)$ and $(0, a_2)$, i.e., they correspond to the numbers a_1 and a_2 respectively in N_1 and N_2. Then the coordinates of I_1, I_2, J_2 in C^* are $(a_1^{-1}, 0)^*$, $(0, a_2^{-1})^*$, $(0, -a_2^{-1})^*$ respectively, or correspond to the numbers $a_1^{-1}, a_2^{-1}, -a_2^{-1}$ in N_1^*, N_2^*. Since E corresponds to the number ka_1^{-1} in N_1^*, its coordinates are $(ka_1^{-1}, 0)^*$ in C^*. Hence in C^* the slope of the line I_1J_2 is

$$m^*=a_2^{-1}/a_1^{-1}=a_1/a_2$$

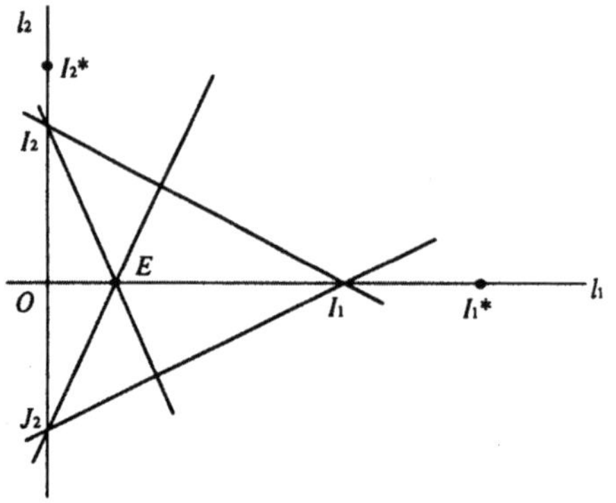

Fig. 2.19

and the slope of I_2E is

$$m^{*'} = -a_2^{-1}/(ka_1^{-1}) = -a_1/(ka_2).$$

As $I_2E \perp I_1J_2$, by the previous theorem we have

$$k^*m^*m^{*'} + 1 = 0,$$

where k^* is the corresponding orthogonal rate of C^*. This implies that

$$k^* = k \cdot (a_2/a_1)^2.$$

Therefore the orthogonal rate k^* differs from k by the square of a non-zero factor a_2/a_1, where a_2/a_1 is a number in $\boldsymbol{K}$.

We thus have the following result.

Theorem 2. The orthogonal rates of any two orthogonal coordinate systems are different by a non-zero square factor in $\boldsymbol{K}$.

From this theorem, one knows that orthogonal geometry determines a set of numbers $\{ka^2 \mid a \neq 0, a \in \boldsymbol{K}\}$ which are different from each other by non-zero square factors. We call this set of numbers the *orthogonal square set* of the geometry. A proof of the theorem follows.

Proof. If $l_1^* = l_1$ and $l_2^* = l_2$, the theorem has been proved before. Consider now the case in which O^* does not lie on either of l_1 and l_2, and $l_1^* \parallel l_1, l_2^* \parallel l_2$. As in Fig. 2.20, construct the connecting line OO^* and draw through I_1, I_2 two parallels to OO^*, meeting l_1^*, l_2^* at I_1^*, I_2^* respectively. Let us consider first the case in which the unit points in C^* are I_1^* and I_2^*. Construct the symmetric point J_2 of I_2 with respect to O, the symmetric point J_2^* of I_2^* with respect to O^*, and the orthocenters E and E^* of $\triangle I_1I_2J_2$ and $\triangle I_1^*I_2^*J_2^*$. Then the corresponding number k of E in the number system $N = N(l_1, O, I_1)$ is the

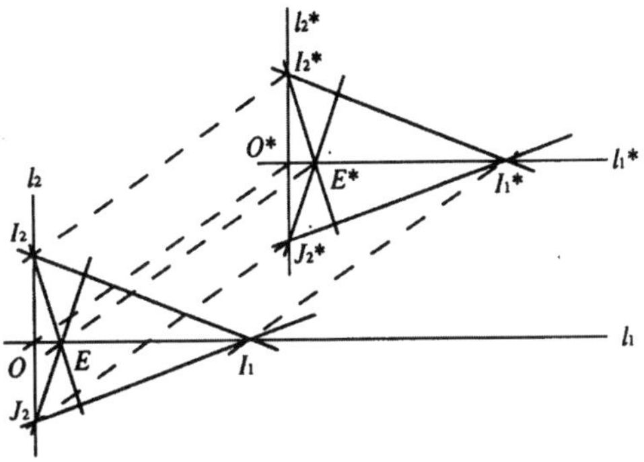

Fig. 2.20

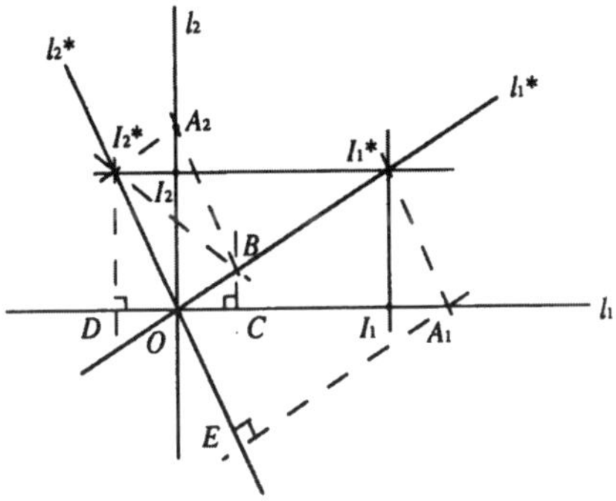

Fig. 2.21

orthogonal rate determined by the coordinate system $C(l_1, l_2, I_1, I_2)$. Similarly, E^* in $N^* = N(l_1^*, O^*, I_1^*)$ corresponds to the orthogonal rate k^* determined by $C^*(l_1^*, l_2^*, I_1^*, I_2^*)$. Applying Desargues' axiom to $\Delta I_1 O I_2$ and $\Delta I_1^* O^* I_2^*$, we have $I_1 I_2 \parallel I_1^* I_2^*$. Similarly, $I_1 J_2 \parallel I_1^* J_2^*$. Hence $J_2 E \parallel J_2^* E^*$ and $I_2 E \parallel I_2^* E^*$. Applying again Desargues' axiom to $\Delta E I_2 J_2$ and $\Delta E^* I_2^* J_2^*$, we have $EE^* \parallel I_2 I_2^* \parallel OO^*$. Thus E^* corresponds to the number k^* in N^*, i.e., the corresponding number of E in N is k and $k^* = k$. If we change the unit points to I_1^* and I_2^*, then by the previous proof the orthogonal rate differs from k^*, k by non-zero square factors. So the theorem holds in this case.

Consider now the case that O^* is the same as O but l_1^*, l_2^* are distinct from l_1, l_2. Take an arbitrary I_1 on l_1 as the unit point (see Fig. 2.21) and draw through I_1 a parallel to l_2, meeting l_1^* at I_1^*, considered as the unit point on l_1^*. Draw through I_1^* a parallel to l_1, meeting l_2, l_2^* at I_2, I_2^*, considered as the unit points on l_2, l_2^*. For the coordinate systems $C = C(l_1, l_2, I_1, I_2)$ and $C^* = C(l_1^*, l_2^*, I_1^*, I_2^*)$, the corresponding orthogonal rates will be denoted by k and k^*. We want to find the relation between k and k^*.

For this purpose, let the number systems on the lines l_1, l_2 etc. be simply denoted as

$$N(l_1, O, I_1) = N_1, \quad N(l_2, O, I_2) = N_2,$$
$$N(l_1^*, O, I_1^*) = N_1^*, \quad N(l_2^*, O, I_2^*) = N_2^*.$$

Draw through I_1^* a parallel to l_2^*, meeting l_1 at A_1 and through I_2^* a parallel to l_1^*, meeting l_2 at A_2. Draw through A_2 a parallel to l_2^*, meeting l_1^* at B, through B and I_2^* two parallels to l_2, meeting l_1 at C and D respectively, and through A_1 a parallel to l_1^*, meeting l_2^* at E.

Obviously, the coordinates of I_1^* in C and in C^* are $(1, 1)$ and $(1, 0)^*$ respectively, denoted simply as

$$I_1^* = (1, 1) = (1, 0)^*.$$

From this, one knows that the slope of the line $l_1^* = OI_1^*$ in C is

$$m_1 = 1$$

and the slope of the line $l_2^* = OI_2^*$ which is perpendicular to l_1^* is

$$m_2 = -\frac{1}{k}.$$

As the second coordinate of I_2^* in C is clearly 1, the first coordinate of I_2^* in C should be $-k$, i.e.,

$$I_2^* = (-k, 1) = (0, 1)^*.$$

Now we determine the coordinates of A_1 in C^*. By construction, we know that $OI_2^*I_1^*A_1$ and $OI_1^*A_1E$ are parallelograms. Thus E is the symmetric point of I_2^* with respect to O, so E corresponds to the number -1 in N_2^*. It follows that the coordinates of A_1 in C^* are

$$A_1 = (1, -1)^*.$$

Hence the slope of the line l_1 in the coordinate system C^* is

$$m_1^* = -1.$$

Next, we determine the coordinates of A_2 in C^*. By construction, $OI_2^*A_2B$ is a parallelogram. Hence, according to Sect. 1.2 the diagonals OA_2 and I_2^*B of $\square OI_2^*A_2B$ bisect. As I_2^*D and BC are both parallel to l_2, O is the midpoint of the pair (CD). It is known from the above that the first coordinate of I_2^* in C is $-k$. Therefore, D corresponds to $-k$ in N_1, so C corresponds to k in N_1. As $BC, I_1I_1^* \perp l_1$, one has $BC \parallel I_1I_1^*$. Since I_1 and I_1^* correspond to 1 in N_1 and in N_1^* respectively, C and B correspond to k in both N_1 and N_1^*. From this, we know that

$$A_2 = (k, 1)^*,$$

so that the slope of the line $l_2 = OA_2$ in the coordinate system C^* is

$$m_2^* = \frac{1}{k}.$$

Now, the lines l_1 and l_2 are perpendicular, so their slopes $m_1^* = -1$ and $m_2^* = 1/k$ in C^* satisfy the relation

$$k^* \cdot (-1) \cdot \frac{1}{k} + 1 = 0.$$

This implies that

$$k^* = k.$$

If we take two points distinct from O on l_1^* and l_2^* as the unit points, then by our earlier proof, the corresponding orthogonal rates k^* and k differ by a non-zero square factor.

Clearly, any two orthogonal coordinate systems can be obtained by a repeated application of the above. Hence, in all cases the orthogonal rates are only different by a non-zero square factor and the theorem is proved. □

We have already explained the algebraic expression for the orthogonal relation between two lines in an orthogonal coordinate system. We have also introduced the orthogonal rate corresponding to the coordinate system, as well as a set of numbers – the concept of orthogonal square set – as a characteristic of orthogonal geometry. Next we shall discuss the algebraic expression for the symmetric relation.

Consider an orthogonal coordinate system $C(l_1, l_2, I_1, I_2)$ with orthogonal rate k. Let

$$L_u\colon \quad u_1x_1 + u_2x_2 + u_3 = 0$$

be a line in which u_1, u_2 are not simultaneously zero. When discussing the symmetric relation for L_u, we suppose that L_u is a non-isotropic line. So, in what follows,

$$ku_1^2 + u_2^2 \neq 0.$$

Now, for a point

$$A = (a_1, a_2),$$

let us determine the symmetric point

$$A^* = (a_1^*, a_2^*)$$

of A with respect to L_u as follows.

First of all, if A lies on L_u, i.e.,

$$a_1u_1 + a_2u_2 + u_3 = 0,$$

then $A^* = A$, i.e.,

$$a_1^* = a_1, \quad a_2^* = a_2.$$

Suppose otherwise A does not lie on L_u. Then A^* and A are distinct and

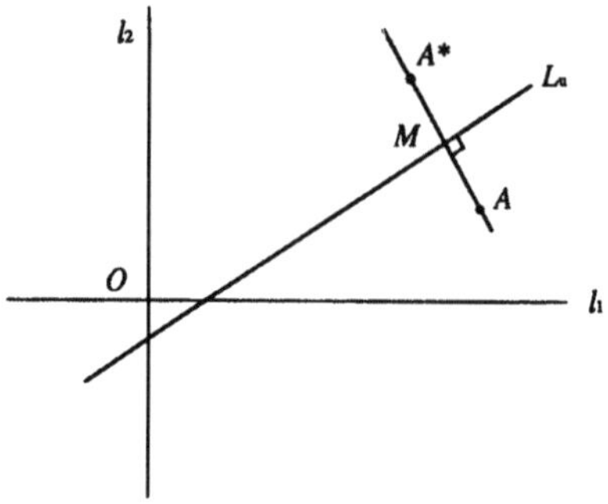

Fig. 2.22

determine a line AA^* having the equation

$$AA^*:\quad (a_2^*-a_2)(x_1-a_1)=(a_1^*-a_1)(x_2-a_2).$$

As AA^* is perpendicular to L_u, by Theorem 1′ we have

$$k(a_2^*-a_2)u_1-(a_1^*-a_1)u_2=0.$$

Since the midpoint of the pair (AA^*) is

$$M=\left(\frac{a_1^*+a_1}{2},\frac{a_2^*+a_2}{2}\right),$$

which lies on the line L_u, substitution of M into the equation of L_u leads to

$$u_1(a_1^*+a_1)+u_2(a_2^*+a_2)+2u_3=0.$$

Solving the last two equations, we get

$$a_1^*=\frac{-(ku_1^2-u_2^2)a_1-2ku_1u_2a_2-2ku_1u_3}{ku_1^2+u_2^2},$$

$$a_2^*=\frac{-2u_1u_2a_1+(ku_1^2-u_2^2)a_2-2u_2u_3}{ku_1^2+u_2^2}.$$

Clearly, when A lies on L_u, i.e., $u_1a_1+u_2a_2+u_3=0$, the above two equations become $a_1^*=a_1, a_2^*=a_2$. Hence, they are always satisfied no matter whether or not A lies on L_u. We may rewrite them in matrix form:

$$\begin{bmatrix} a_1^* \\ a_2^* \\ 1 \end{bmatrix} = \begin{bmatrix} -\dfrac{ku_1^2 - u_2^2}{ku_1^2 + u_2^2} & \dfrac{-2ku_1u_2}{ku_1^2 + u_2^2} & \dfrac{-2ku_1u_3}{ku_1^2 + u_2^2} \\ -\dfrac{2u_1u_2}{ku_1^2 + u_2^2} & \dfrac{ku_1^2 - u_2^2}{ku_1^2 + u_2^2} & \dfrac{-2u_2u_3}{ku_1^2 + u_2^2} \\ 0 & 0 & 1 \end{bmatrix} \begin{bmatrix} a_1 \\ a_2 \\ 1 \end{bmatrix},$$

in which

$$ku_1^2 + u_2^2 \neq 0.$$

We also call the correspondence relation

$$\begin{cases} A \to A^* \\ (a_1, a_2) \to (a_1^*, a_2^*) \end{cases}$$

a *reflection* with respect to line L_u, denoted by R_u. The above matrix equation may also be written as

$$A^* = R_u A,$$

where

$$R_u = \begin{bmatrix} -\dfrac{ku_1^2 - u_2^2}{ku_1^2 + u_2^2} & \dfrac{-2ku_1u_2}{ku_1^2 + u_2^2} & \dfrac{-2ku_1u_3}{ku_1^2 + u_2^2} \\ -\dfrac{2u_1u_2}{ku_1^2 + u_2^2} & \dfrac{ku_1^2 - u_2^2}{ku_1^2 + u_2^2} & \dfrac{-2u_2u_3}{ku_1^2 + u_2^2} \\ 0 & 0 & 1 \end{bmatrix}.$$

Now, let

$$L_v: \quad v_1(x_1 - a_1) + v_2(x_2 - a_2) = 0$$

be a line through $A = (a_1, a_2)$ (v_1, v_2 are not simultaneously 0). According to the last section, the reflection of L_v with respect to L_u, i.e., its symmetry with respect to L_u, is a line, denoted as

$$R_u(L_v) = L_v^*.$$

Let us determine the equation of L_v^* as follows. Take a point

$$B = (b_1, b_2) = (a_1 + v_2, a_2 - v_1)$$

on L_v, where

$$v_2 = b_1 - a_1, \quad -v_1 = b_2 - a_2.$$

Then the symmetric point $B^* = (b_1^*, b_2^*)$ of B with respect to L_u is determined

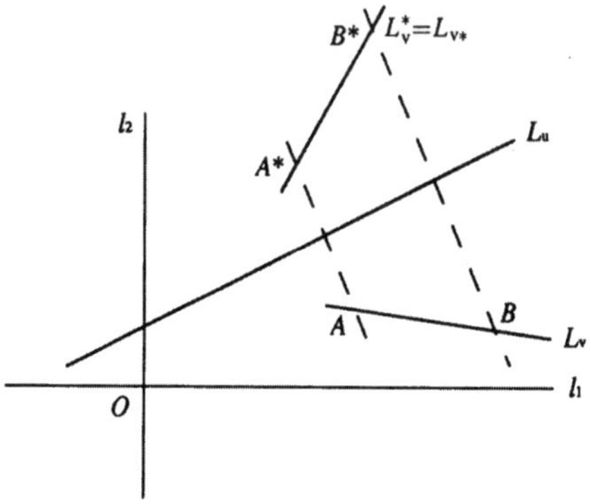

Fig. 2.23

by the following two equations:

$$(ku_1^2+u_2^2)b_1^* = -(ku_1^2-u_2^2)b_1 - 2ku_1u_2b_2 - 2ku_1u_3,$$
$$(ku_1^2+u_2^2)b_2^* = -2u_1u_2b_1 + (ku_1^2-u_2^2)b_2 - 2u_2u_3.$$

From b_1^*, b_2^* and the previous a_1^*, a_2^*, we obtain

$$(ku_1^2+u_2^2)(b_1^*-a_1^*) = 2ku_1u_2v_1 - (ku_1^2-u_2^2)v_2,$$
$$(ku_1^2+u_2^2)(b_2^*-a_2^*) = -(ku_1^2-u_2^2)v_1 - 2u_1u_2v_2.$$

Since L_v^* is determined by the two points A^* and B^*, its equation is

$$(b_2^*-a_2^*)(x_1-a_1^*) - (b_1^*-a_1^*)(x_2-a_2^*) = 0.$$

As $ku_1^2+u_2^2 \neq 0$, we may set

$$v_1^* = -(ku_1^2+u_2^2)(b_2^*-a_2^*),$$
$$v_2^* = (ku_1^2+u_2^2)(b_1^*-a_1^*).$$

In this case L_v^* may also be written as

$$L_v^* = L_{v*}:\quad v_1^*(x_1-a_1^*) + v_2^*(x_2-a_2^*) = 0,$$

where

$$v_1^* = (ku_1^2-u_2^2)v_1 + 2u_1u_2v_2,$$
$$v_2^* = 2ku_1u_2v_1 - (ku_1^2-u_2^2)v_2.$$

For any two lines L_u, L_v, not perpendicular to each other, we define a function

$$T(u,v) = \frac{u_1v_2-u_2v_1}{ku_1v_1+u_2v_2} = -T(v,u).$$

Now $ku_1v_1 + u_2v_2 \neq 0$, i.e., L_u, L_v are not perpendicular to each other, so $T(u, v)$ is well defined. Therefore, there is a relation

$$T(u, v) + T(u, v^*) = 0$$

between L_u, L_v and the reflection L_{v*} of L_v with respect to L_u, where L_u, L_{v*} are also not perpendicular to each other and thus $T(u, v^*)$ is well defined.

2.4 (Unordered) metric geometry

In (unordered) orthogonal geometry, two intersecting lines do not necessarily have a symmetric axis. For instance, if one of the two lines is isotropic but the other is not, then they cannot have a symmetric axis. In fact, even when two intersecting lines are non-isotropic, the existence of a symmetric axis cannot be derived from the original axioms of orthogonal geometry. It is proved in this section that the existence of symmetric axes is in essence equivalent to the inclusion of the concepts of congruence and metricity in the geometry. Even though there is no order relation in orthogonal geometry so that congruence and metricity concepts do not have all the properties associated with them in ordinary geometry, a rather important part of the properties (e.g., the Kou-Ku theorem) can be retained.

For this end, we introduce the following axiom.

Axiom of symmetric axes S. Any two non-isotropic intersecting lines have a symmetric axis.

By Property 5 in Sect. 2.2, two intersecting lines have exactly two symmetric axes. We call an (unordered) orthogonal geometry satisfying the axiom of symmetric axes an (unordered) *metric geometry*. We call it so because one can introduce some metric properties as long as the axiom of symmetric axes is assumed, see below for details.

First, we discuss the orthogonal rate and orthogonal square set in unordered metric geometry.

Take any two intersecting orthogonal lines l_1 and l_2 as the coordinate axes (see Fig. 2.24) and any point I_1 distinct from the intersection point O as the unit point on l_1. As l_1, l_2 are evidently non-isotropic, by the axiom of symmetric axes

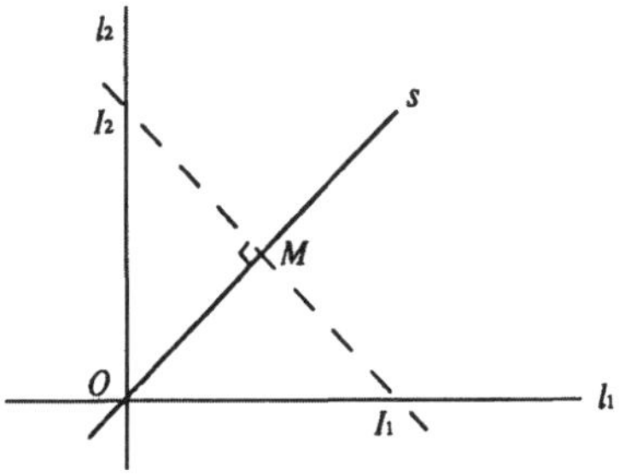

Fig. 2.24

we may take a symmetric axis of l_1, l_2, say s, and construct the symmetric point I_2 of I_1 with respect to s. Then I_2, lying on l_2, will be taken as the unit point on l_2. In this case the orthogonal rate k of the orthogonal coordinate system $C(l_1, l_2, I_1, I_2)$ is 1. We show this as follows.

The symmetric axis s is the perpendicular bisector of the pair $(I_1 I_2)$ and thus passes through the midpoint $M = (\frac{1}{2}, \frac{1}{2})$ of $(I_1 I_2)$. Therefore, the slope of s is $m = \frac{1}{2}/\frac{1}{2} = 1$. On the other hand, $I_1 = (1, 0)$, $I_2 = (0, 1)$, so the slope of $I_1 I_2$ is $m' = -1$. As s is perpendicular to $I_1 I_2$, their slopes satisfy the relation $kmm' + 1 = 0$. This yields $k = 1$.

Consider now two intersecting orthogonal lines as the coordinate axes and two points, which are on these two lines and are symmetric with respect to one symmetric axis of these two lines, as the unit points. The coordinate system formed in this way is called a *Desarguesian* coordinate system in unordered metric geometry and the corresponding coordinates of points are called *Desarguesian* coordinates. According to the last section, the above result can be stated as follows.

Theorem 1. The orthogonal square set of (unordered) metric geometry consists of non-zero square numbers. In particular, the orthogonal rate of a Desarguesian coordinate system is 1.

It follows from Sect. 2.3 and this theorem that as for a Desarguesian coordinate system, a necessary and sufficient condition for two lines

$$L_u:\quad u_1x_1 + u_2x_2 + u_3 = 0,$$
$$L_v:\quad v_1x_1 + v_2x_2 + v_3 = 0$$

(u_1, u_2 are not simultaneously 0 and neither are v_1, v_2) to be perpendicular to each other is

$$u_1v_1 + u_2v_2 = 0,$$

while a necessary and sufficient condition for L_u to be isotropic is

$$u_1^2 + v_1^2 = 0.$$

As for two lines L_u, L_v with slopes $m_u, m_v \neq 0$ or ∞, the condition can also be written as $m_u m_v + 1 = 0$. These are all the same as in ordinary geometry.

Next, we introduce the concept of *congruence*.

If A_1, A_2 are respectively the symmetric points of two points B_1, B_2 with respect to a non-isotropic line, then we say that the pair (A_1A_2) is *congruent* to the pair (B_1B_2). If (A_1A_2) is congruent to (B_1B_2) and (B_1B_2) is congruent to (C_1C_2), then the pair (A_1A_2) is also *congruent* to the pair (C_1C_2). When (A_1A_2) is congruent to (B_1B_2), we write

$$(A_1A_2) \equiv (B_1B_2).$$

The relation of congruence has some simple properties shown below.

Property 1. The relation of congruence among pairs of points is an equivalence relation.

Proof. The symmetry and transitivity of the relation of congruence can be obtained directly from the definition. We now prove the reflexivity as follows.

If A_1 is distinct from A_2 and the line l connecting A_1, A_2 is non-isotropic, then it is known from the definition that $(A_1A_2) \equiv (A_1A_2)$, for the symmetric points of A_1, A_2 with respect to l are themselves. In any case we can always take a non-isotropic line l in the plane. Letting the symmetric points of A_1, A_2 with respect to l be A_1', A_2' respectively, by definition we have

$$(A_1A_2) \equiv (A_1'A_2').$$

Since, on the other hand, the symmetric points of A_1', A_2' with respect to l are A_1, A_2, we also have

$$(A_1'A_2') \equiv (A_1A_2).$$

Hence, by symmetry in the definition, $(A_1A_2) \equiv (A_1A_2)$. □

Property 2. For any pair (A_1A_2) of points

$$(A_1A_2) \equiv (A_2A_1).$$

Proof. If A_1, A_2 are distinct and the connecting line A_1A_2 is non-isotropic, then we may take the perpendicular bisector l' of (A_1A_2) as a symmetric axis. As A_1, A_2 are respectively symmetric to A_2, A_1, we have $(A_1A_2) \equiv (A_2A_1)$. See Fig. 2.25.

Suppose now A_1A_2 is isotropic. Then we may take the midpoint M of (A_1A_2) and construct a non-isotropic line l through M using Axiom O 4. Construct through M a perpendicular l' to l using Axiom O 2 and construct the symmetric point A_1' of A_1 with respect to l. As $l' \parallel A_1A_1'$ and M is the midpoint of (A_1A_2), l' passes through the midpoint of $(A_1'A_2)$. Furthermore, l passes

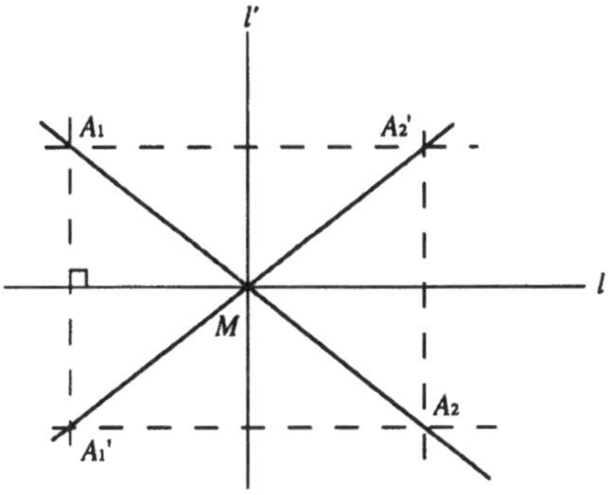

Fig. 2.25

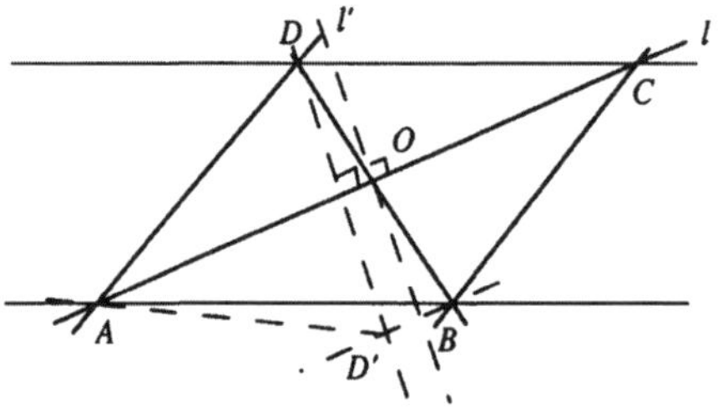

Fig. 2.26

through the common midpoint of (A_1A_1') and (A_1A_2), so $l \parallel A_1'A_2$. Therefore, $l' \perp A_1'A_2$ and thus the symmetric point of A_1' with respect to l' is A_2. Similarly, when the symmetric point of A_2 with respect to l is A_2', A_2' lies on the line $A_1'M$ and M is the midpoint of $(A_1'A_2')$. Also, the symmetric point of A_2' with respect to l' is A_1. Hence, with respect to the symmetric axes l, l', we have $(A_1A_2) \equiv (A_1'A_2')$, $(A_1'A_2') \equiv (A_2A_1)$. Therefore, $(A_1A_2) \equiv (A_2A_1)$. □

Property 3. Two opposite sides of a parallelogram $ABCD$ are congruent to each other, i.e.,

$$(AB) \equiv (CD), \quad (AD) \equiv (BC).$$

Proof. According to Sect. 1.2, the diagonals AC and BD of $\square ABCD$ bisect one another at their intersection point O. That is, O is the common midpoint of the pairs (AC) and (BD) (Fig. 2.26).

Suppose first that at least one of AC, BD (say AC) is non-isotropic. Let $AC = l$ and construct through O a perpendicular l' to l. Let the symmetric point of D with respect to l be designated by D'. Then, as in the proof of Property 2 the symmetric point of D' with respect to l' is identical to B. Now we have

$$(AD) \equiv (AD')$$

with respect to the symmetric axis l, and

$$(AD') \equiv (CB)$$

with respect to the symmetric axis l'. Hence $(AD) \equiv (CB)$ or, by Property 2, $(AD) \equiv (BC)$. Similarly, we also have $(AB) \equiv (CD)$.

Suppose now the diagonals AC and BD are all isotropic (Fig. 2.27). According to Sect. 2.2, the three sides of a triangle cannot all be isotropic, so the lines AB, BC, CD, AD cannot all be isotropic. Hence, we can construct a line l through O which, if perpendicular to AB and CD, will meet them. Let us denote the points of intersection by E and F. Construct through C and D respectively two perpendiculars to AB, meeting it at points A' and B'. Then E is the midpoint of both (AA') and (BB') because O is the midpoint of both (AC) and (BD). Since the symmetric points of A, B with respect to l are A', B' respectively, we have

$$(AB) \equiv (A'B').$$

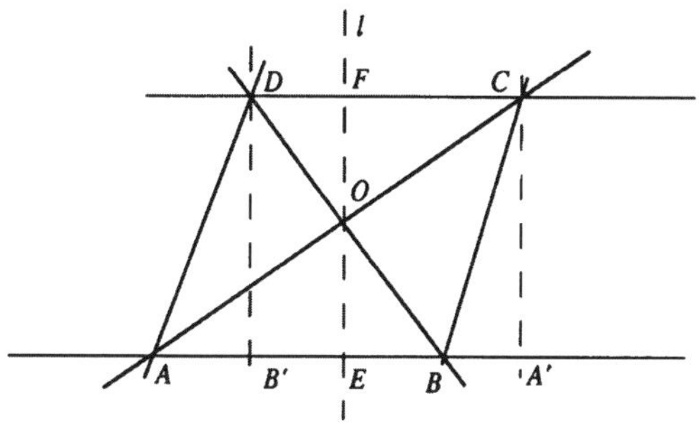

Fig. 2.27

Moreover, the two opposite sides $A'B'$ and CD of the parallelogram $A'B'DC$ are respectively perpendicular to $A'C$ and $B'D$. From this it is easy to show

$$(A'B') \equiv (CD).$$

In virtue of the transitivity of congruence, $(AB) \equiv (CD)$. Similarly, we have $(BC) \equiv (AD)$. □

The above definition and properties of congruence (except Theorem 1) do not actually use the axiom of symmetric axes and can thus still be regarded as the contents of orthogonal geometry. However, for some results below, the axiom of symmetric axes will be indispensable.

First of all, a pair of points on an isotropic line apparently cannot be congruent to a pair of points on a non-isotropic line. Thus the relation of congruence may be restricted to the pairs of points on non-isotropic lines. On the other hand, because of the axiom of symmetric axes any two non-isotropic lines always have symmetric axes, so we can consider the problem of congruence for pairs of points lying on any two non-isotropic lines.

For this purpose, we first take a fixed non-isotropic line l in the plane and two points I, J on l with midpoint O. For these fixed l, I, J, we say that (IJ) is a *standard pair of points*. Then we may restate the axiom of symmetric axes in the following form.

Axiom of transposition δ′. After fixing a point O' on any non-isotropic line l', there are exactly two points I' and J' with O' as their midpoint such that

$$(I'J') \equiv (J'I') \equiv (IJ).$$

If, for instance, a non-isotropic line l' passes through O but it is distinct from l and the chosen O' is the same as O, then, by setting s, s' to be the two symmetric axes of l, l', the two points I', J' are identical to the symmetric points of I, J with respect to s, s'. If l' does not pass through O and is not parallel to l, we can draw $l'' \parallel l'$ through O, and take $O'' = O$ on l'' and I'', J'' as the symmetric points of I, J with respect to s, s'. While drawing through

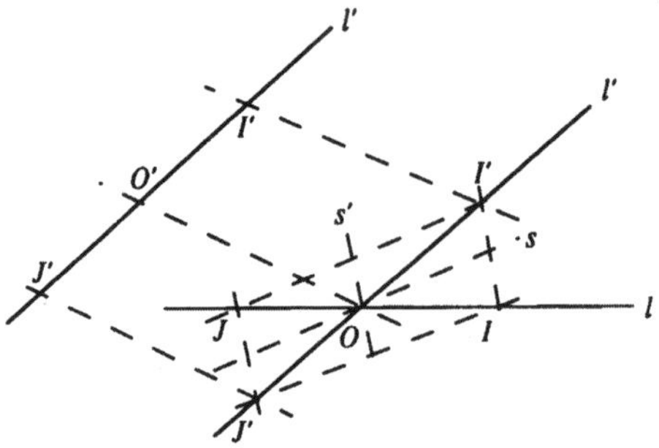

Fig. 2.28

I'', J'' parallels to OO', meeting l' at I', J', one sees that I', J' are the two points on l' with O as their midpoint in the axiom of transposition.

Now fix a line l and take thereon a standard pair (IJ) of points with midpoint O. Then we may determine Desarguesian number systems $N(l, O, I)$ and $N(l, O, J)$ on l. Similarly, for any line l', a point O' on it and a pair $(I'J')$ of points congruent to (IJ) with O' as their midpoint, we can determine the number systems $N(l', O', I')$ and $N(l', O', J')$, too. Such number systems are said to be *matchable*, with O and O' respectively as their *centers*.

Consider any two points A, B on l. Let A', B' be two points on another non-isotropic line l' such that $(A'B') \equiv (AB)$. If A, B correspond respectively to the numbers a, b in the number system $N(l, O, I)$, then they correspond to $-a, -b$ in the number system $N(l, O, J)$. For the matchable number systems $N(l', O', I')$ and $N(l', O', J')$ on l', it is easy to prove that there are c, d such that A', B' correspond to $a' = a + c$, $b' = b + c$ or $a' = -a + d$, $b' = -b + d$. Hence, in any case we have

$$(b - a)^2 = [(-b) - (-a)]^2 = (b' - a')^2.$$

In other words, the square of difference of the two corresponding numbers of a congruent pair of points in any matchable number system on the line where the two points lie is a constant. We call this constant the *square of distance* of the pair (AB), denoted as $\overline{\overline{AB}}$. This result can be summarized as the following important property.

Property 4. Non-isotropic congruent pairs of points have one and the same square of distance. The square of distance of a pair of points is 0 when the two points are the same, and is non-zero – more precisely, is the square of a non-zero number in the number field associated with the geometry – when the two points are distinct from each other. In the latter case the non-zero number is determined not uniquely but only up to sign.

Here, by a non-isotropic pair of points we mean that the two points, when distinct, do not lie on an isotropic line. The square of distance is with respect to a fixed standard pair of points.

We call a triangle with two perpendicular sides which together with the third one are all non-isotropic a *Kou-Ku triangle*. We call the third side of a Kou-Ku triangle its *hypotenuse* and the intersection point of the two perpendicular sides its *vertex*. The theorem about Kou-Ku triangles below is the main theorem of this section. It is also the main benefit of introducing the axiom of symmetric axes or of transposition.

Theorem 2 (Kou-Ku theorem). Let the vertex of a Kou-Ku triangle ABC be C. Then there is a relation

$$\overline{\overline{AB}} = \overline{\overline{AC}} + \overline{\overline{BC}}$$

among the squares of distances of the pairs (AB), (AC), (BC).

Before coming to the proof of this theorem, let us first prove the following lemma.

Lemma. Let the vertex of a Kou-Ku triangle ABC be C. Construct through C a perpendicular to AB, meeting AB at a point D. If we introduce two matchable number systems with A as their center on the lines AB and AC, and let C, B, D correspond to c, b, d respectively, then we have the relation

$$bd = c^2.$$

Proof. See Fig. 2.29. Let the two matchable number systems with center A on $AB = l$ and $AC = l'$ be $N = N(l, A, I)$ and $N' = N(l', A, I')$ respectively. Let s be a symmetric axis of l, l' and suppose that I is symmetric to I' with respect to s. Moreover, let the symmetric point of D with respect to s be D' on l' and the symmetric point of C with respect to s be C' on l. Then CC' and DD' are both perpendicular to s. Furthermore, the symmetric line of CD with respect to s is $C'D'$, so they intersect at a point E on s. By Property 3 of Sect. 2.2 or the axiom of the orthocenter, it is known from $CD \perp l$ that $C'D' \perp l'$, so that $C'D' \parallel BC$. Now through C draw a parallel to $I'C'$, meeting l at a point, say F. By applying Pascal's theorem to the intersecting lines l, l' and the two sets of points F, C', B and I', C, D' thereon, it follows that $FD' \parallel BI'$. By hypothesis,

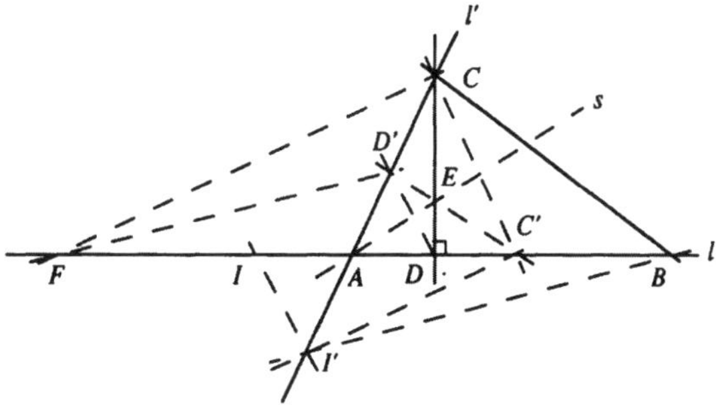

Fig. 2.29

B, C', D correspond respectively to b, c, d in the number system N. Similarly, C, D' correspond respectively to c, d in the number system N'. By the definition of multiplication in number systems, point F corresponds to both c^2 and bd in the number system N, i.e., $c^2 = bd$. □

Proof of the Kou-Ku theorem. Let $AB = l$, $AC = l'_1$, $BC = l'_2$ as in Fig. 2.30. Draw a perpendicular through C to l, meeting l at a point D, and construct the midpoint M of the pair (AB) and the symmetric point E of D with respect to M. Form the matchable number systems $N_1 = N(l, A, I_1)$ on l and $N'_1 = N(l'_1, A, I'_1)$ on l'_1. Let the symmetric point of I_1 with respect to M be I_2. Furthermore, form the matchable number systems $N_2 = N(l, B, I_2)$ on l and $N'_2 = N(l'_2, B, I'_2)$ on l'_2.

Suppose that

B, D	correspond to b_1, d_1 in N_1,
C	corresponds to c_1 in N'_1,
A, D	correspond to a_2, d_2 in N_2,
C	corresponds to c_2 in N'_2.

Then, by the above lemma we have

$$b_1 d_1 = c_1^2,$$
$$a_2 d_2 = c_2^2.$$

Obviously,

$$a_2 = b_1.$$

Moreover, the corresponding number d_2 of D in N_2 is the corresponding number of E in N_1, while M corresponds to $a_2/2 = b_1/2$, in both N_1 and N_2. Since M is the midpoint of D and E, we have

$$a_2 = b_1 = d_1 + d_2.$$

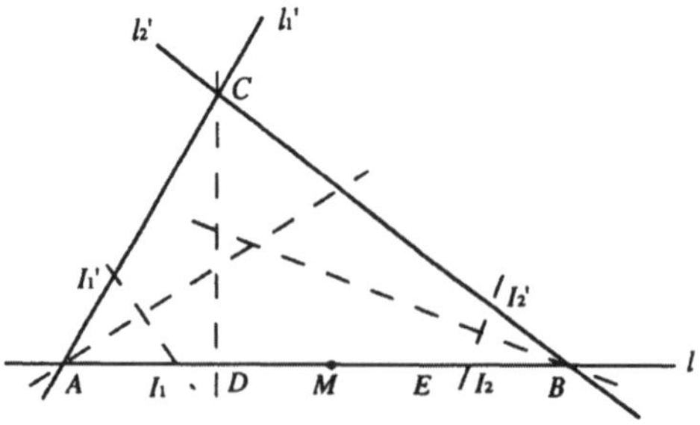

Fig. 2.30

From the definition of the square of distance, it follows that

$$c_1^2 = \overline{\overline{AC}}, \quad c_2^2 = \overline{\overline{BC}},$$
$$b_1^2 = a_2^2 = \overline{\overline{AB}}.$$

Combining the above equations, we have $\overline{\overline{AB}} = \overline{\overline{AC}} + \overline{\overline{BC}}$, which is what we wanted to prove. □

By the Kou-Ku theorem just proved we immediately obtain the following result.

Theorem 3. If the coordinates of two points P and Q in a Desarguesian coordinate system are

$$P = (x_1, x_2), \quad Q = (y_1, y_2)$$

respectively, then the square of distance of (PQ) is

$$\overline{\overline{PQ}} = (x_1 - y_1)^2 + (x_2 - y_2)^2,$$

where it is assumed that PQ is non-isotropic when P and Q are distinct from each other.

Until now we have only been concerned with the congruence of pairs of points but not the congruence of angles. Even though we have added the axiom of symmetric axes to the orthogonal axioms, there is still no concept of the length of a segment in unordered metric geometry. Hence, we have introduced only the square of distance of a pair of points but not the length of a segment nor the distance between two points. Also, rays and angles are not defined in this geometry. Therefore, we cannot speak about *four* intersectional angles of two intersecting lines. However, we might still say that two intersecting lines constitute a *total angle*. If these two lines are l and l' with intersection point O, we denote the total angle by $\measuredangle O(l, l')$ or by $\measuredangle O(l', l)$. If no confusion can arise, it is also simply denoted as $\measuredangle O$ or $\measuredangle(l, l')$. We shall extend this to the case of two coincident lines l and l' with a point O thereon. In that case, the total angle will be called a *straight angle*. When l and l' are perpendicular to each other and intersect at point O, the total angle formed is called a *right angle*. Whenever a total angle is referred to, it will always be restricted to the case in which l and l' are non-isotropic. Then, point O is called the *vertex* and l, l' the two *sides* of the total angle. According to the axiom of symmetric axes, the two sides of a total angle, when they do not coincide, i.e., the total angle is not a straight angle, have exactly two symmetric axes which will be called the *bisectors* of this total angle.

We say that two angles $\measuredangle O(l_1, l_2)$ and $\measuredangle O'(l_1', l_2')$ are *congruent* if there is a non-isotropic line s such that the symmetric point of O and the symmetric lines of l_1, l_2 with respect to s are O' and l_1', l_2' respectively. If $\measuredangle O(l_1, l_2)$ is congruent to $\measuredangle O'(l'_1, l'_2)$ and $\measuredangle O'(l'_1, l'_2)$ to $\measuredangle O''(l''_1, l''_2)$, we also define $\measuredangle O(l_1, l_2)$ to be *congruent* to $\measuredangle O''(l''_1, l''_2)$. If two angles $\measuredangle O(l_1, l_2)$ and $\measuredangle O'(l'_1, l'_2)$ are congruent to each other, we write

$$\measuredangle O(l_1, l_2) \equiv \measuredangle O'(l'_1, l'_2).$$

Suppose that for triangles ABC and $A'B'C'$ the following congruences hold:

$$(AB) \equiv (A'B'), \quad (AC) \equiv (A'C'), \quad (BC) \equiv (B'C'),$$
$$\measuredangle A(AB, AC) \equiv \measuredangle A'(A'B', A'C'),$$
$$\measuredangle B(BA, BC) \equiv \measuredangle B'(B'A', B'C'),$$
$$\measuredangle C(CA, CB) \equiv \measuredangle C'(C'A', C'B').$$

Then we say that $\triangle ABC$ is *congruent* to $\triangle A'B'C'$, and write simply

$$\triangle ABC \equiv \triangle A'B'C'.$$

Clearly we have the following property.

Property 5. $\measuredangle O(l_1, l_2) \equiv \measuredangle O(l_2, l_1)$, and the relation of congruence among angles is an equivalence relation. The relation of congruence among triangles is also an equivalence relation.

Theorem 4. Suppose that the sides of two triangles ABC and $A'B'C'$ are all non-isotropic and

$$(AB) \equiv (A'B'), \quad (AC) \equiv (A'C'), \quad (BC) \equiv (B'C').$$

Then these two triangles are congruent.

Proof. If A' and A are distinct and AA' is non-isotropic, we may construct the perpendicular bisector s of the pair (AA'). Let the symmetric points of B', C' with respect to s be B'', C''; then we have $\triangle A'B'C' \equiv \triangle AB''C''$. Suppose that B'' is distinct from B. If AB and AB'' do not coincide, we may take one of their angular bisectors as s'. If AB and AB'' coincide, construct their perpendicular s' through A. Let the symmetric point of C'' with respect to s' be C'''. Then

$$\triangle AB''C'' \equiv ABC'''.$$

Since the relation of congruence among triangles is an equivalence relation, to prove that $\measuredangle A \equiv \measuredangle A'$ it suffices to consider only the case in which A', B' of $\triangle A'B'C'$ coincide respectively with A, B. There is also no difficulty when AA' is an isotropic line.

Let us introduce a Desarguesian coordinate system with A as its origin and the line AB as l_1, one of the coordinate axes (see Fig. 2.31). Choose the unit

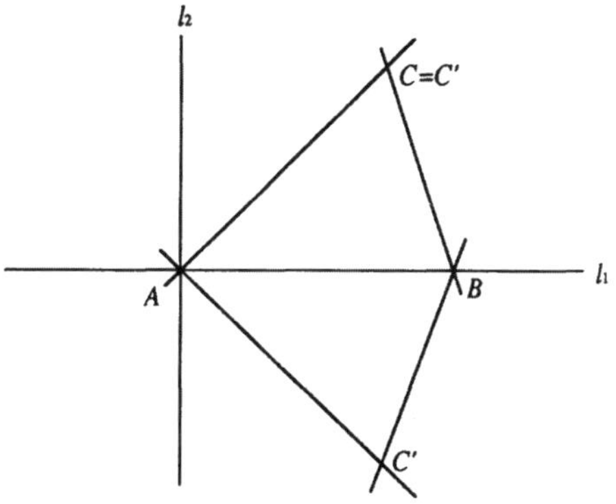

Fig. 2.31

points on the coordinate axes such that the number systems so determined are matchable with the standard number system. Let the coordinates of points be

$$B = (b, 0), \quad C = (c, d), \quad C' = (c', d').$$

Since the congruence of pairs of points corresponds to the equality of squares of distances, by Theorem 3 it follows from

$$(AC) \equiv (AC'), \quad (BC) \equiv (BC')$$

that

$$c^2 + d^2 = c'^2 + d'^2,$$
$$(c - b)^2 + d^2 = (c' - b)^2 + d'^2.$$

A subtraction of the above two equations yields

$$2cb = 2c'b.$$

As B does not coincide with A, we have $b \neq 0$. This implies that

$$c = c'.$$

It follows then from the first of the above two equations that $d^2 = d'^2$, i.e.,

$$d = d' \quad \text{or} \quad d = -d'.$$

In the former case C' coincides with C, and in the latter case C' is symmetric to C with respect to AB. Therefore, in any case we have

$$\triangle ABC \equiv \triangle ABC'.$$

The proof is thus complete. □

In contrast to Theorem 4, the other two theorems about the congruence of triangles in ordinary geometry do not hold here. On the other hand, though we cannot define the distance and length of a pair of points, we may define the square of distance as a certain measure of distance between two points which has some of the properties that the distance and length have in ordinary geometry. Similarly, we cannot define the size of a total angle, but we may still introduce a measure of the size of an angle having some of the properties it has in ordinary geometry. For instance, the square of the function T of two intersecting lines at the end of the last section may be taken as such a measure. Thus, unordered metric geometry retains some of the metric properties of ordinary geometry. Nevertheless, as there is no concept of order, these metric properties are not complete in comparison with those of ordinary geometry. The next section will discuss this problem further.

2.5 The axioms of order and ordered metric geometry

In the previous sections, we have established unordered orthogonal geometry and unordered metric geometry from the axioms of incidence H I, the axiom of parallels H IV, the axiom of infinity, Desargues' axioms D, and the orthogonal axioms O as well as the axiom of symmetric axes S. Although we can partially introduce the usual relation of congruence and metric properties in such geometries, as there is no concept of order, numbers in the geometry-associated field do not have positiveness and negativeness nor can their sizes be compared. Thus the corresponding relation of congruence and metric properties are not complete in comparison with those in ordinary geometry. For this reason we have to introduce an order or similar concept. These concepts and axioms can be introduced in different ways; while this book follows the original "Grundlagen der Geometrie" by Hilbert. There were several editions of Hilbert's book and the axioms of order were revised from time to time; the details were all slightly different but the final outcomes remain the same. Even though the axioms are not as completely independent of each other in the first edition as they are in later editions, in what follows we shall state them on the basis of the first edition since our goal is not the logical relation among axioms and their independence.

The axioms of order deal with three distinct points on a line, and include an undefined fundamental concept of one point lying *between* two other points. This concept satisfies the following axioms (in the plane).

Axioms of order H II

H II1. Let A, B, C be three distinct points on a line. If B lies between A and C, then B also lies between C and A.

H II2. For any two distinct points A and C on a line, there always exists another point B which lies between A and C, and another point D such that C lies between A and D.

H II3. Given any three distinct points A, B, C on a line, one and only one of the following three cases holds: B lies between A and C, A lies between B and C, and C lies between A and B.

Given two distinct points A and B, the totality of points on the line determined by A, B and between A and B is called a *segment*, denoted as $|AB|$. By Axiom H III1, it may also be denoted as $|BA|$. A and B are called the *end points* of $|AB|$ or $|BA|$. As usual, we may also define a *polygonal line* $|ABC \cdots KL|$ formed by the segments $AB, BC, \ldots, KL$ with A and L as end points and define a *polygon*, a *simple polygon* and so on. We do not explain them in further detail.

H II4 (Pasch's axiom). Let A, B, C be three points which are not collinear. If a line l passes through a point of the segment $|AB|$, then l passes through either point C, or a point on the segment $|AC|$, or a point on the segment $|BC|$.

By the axioms of incidence H I and axioms of order H II1–H II4, we can have some *separation* properties as listed below.

Separation property 1. Let a line l and a point O on l be given. Then all points on l distinct from O can be separated into two parts, called the *two sides* of O on l, such that O lies between A, B when A, B lie on *different sides*, and O does not lie between A, B when A, B lie on the *same side*. Each of the two parts is a *half-line* or a *ray* emanating from O on l.

Separation property 2. Given any line l in the plane, all points not on l can be separated into two parts, called the *two sides* of l in the plane such that any segment $|AB|$ and l have no common point when the two points A and B lie on the *same side*, and any polygonal line $|A \cdots B|$ connecting A, B and l must have common points when A and B lie on *different sides*.

Separation property 3. Any two half-lines l_1, l_2 emanating from a point O which together do not form a line are said to form an *angle* with O as its *vertex* and l_1, l_2 as its *sides*, denoted by $\measuredangle(l_1, l_2)$ or $\measuredangle(l_2, l_1)$. Then all points not *lying in the angle* (neither the vertex nor on the sides) can be separated by the angle into the *interior* and *exterior* parts. Any two points lying in the interior (exterior) of the angle can be connected by a segment (or polygonal segment) without meeting the angle. But if one of the two points lies in the interior and the other in the exterior of the angle, any polygonal segment connecting these two points meets the angle. Furthermore, the interior is the common part of the following two parts: one is that side of l_1 which contains points on l_2 and the other is that side of l_2 which contains points on l_1. As for the exterior of the angle, it consists of all points that lie neither in the interior of the angle nor in the angle.

Separation property 4. Given any simple polygon P, all points not on P can be separated into two parts, called the *interior* and *exterior* of P, such that there is a polygonal segment $|A \cdots B|$ connecting two points A, B that does not meet P when A, B lie in the interior or exterior, and any polygonal segment $|A \cdots B|$ connecting two points A, B should meet P when A, B lie respectively in the interior and the exterior. Furthermore, the difference between the interior and the exterior is that there are lines lying completely in the exterior but no line can lie completely in the interior of P.

Let P be a polygon. If for any two adjacent vertices of P, other vertices of P all lie on the same side of the line connecting these two adjacent vertices, we call this polygon a *convex polygon*. The segment connecting any two adjacent vertices is called a *diagonal segment* of this convex polygon. We then have the following.

Separation property 5. Any diagonal segment of a convex polygon lies in its interior.

The proofs of these separation properties are rather involved and some of them are not very easy. The reader may refer to the translators' notes in the Russian and Chinese editions of Hilbert's "Grundlagen der Geometrie," the book by Kerekjártó (1969), or the last chapter in vol. 2 of Veblen and Young (1918), etc.

Below we shall assume that besides all or part of the axioms in (unordered) metric geometry the plane also satisfies the axioms of order H II1–H II4. Such a geometry will be called an *ordered metric geometry*, or an ordered Pascalian geometry, or an ordered orthogonal geometry, or so forth. In these ordered geometries, one can easily obtain some corollaries from the separation properties.

Corollary 1. Let two sets of points A_1, B_1, C_1 and A_2, B_2, C_2, pairwise distinct and distinct from the possible intersection points of their connecting lines, lie on two distinct lines l_1, l_2 respectively. If $A_1A_2 \parallel B_1B_2 \parallel C_1C_2$, then whether or not B_1 lies between A_1 and C_1 (or C_1 lies between A_1 and B_1) depends on whether or not B_2 lies between A_2 and C_2 (or C_2 lies between A_2 and B_2). If l_1 meets l_2 at a point O while C_1, C_2 coincide with O, then in the case $A_1A_2 \parallel B_1B_2$, the same conclusion holds.

Proof. Let us consider the former case (Fig. 2.32). As $A_1A_2 \parallel B_1B_2$, the two points A_1, A_2 lie on the same side of the line B_1B_2. Similarly, C_1, C_2 lie also on the same side of B_1B_2. Hence, whether or not A_1, C_1 lie on different sides of B_1B_2 depends on whether or not A_2, C_2 lie on different sides of B_1B_2. That is, whether or not B_1 lies between A_1 and C_1 depends on whether or not B_2 lies between A_2 and C_2. The other cases are analogous. □

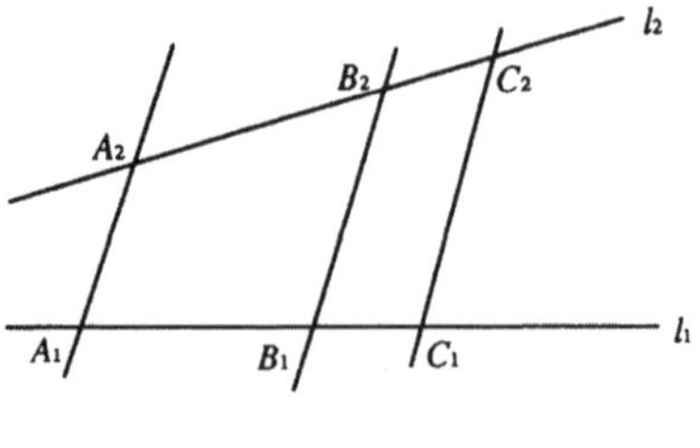

Fig. 2.32

Corollary 2. Triangles and parallelograms are all convex polygons.

Proof. This is immediate from the definition. □

Corollary 3. The midpoint of two distinct points lies between these two points.

Proof. Omitted. □

Now fix two points O and I on a line l and take them as 0 and 1 in order to determine a Desarguesian number system $N = N(l, O, I)$. We define those numbers in N whose corresponding points lie on the same side of O on l as I to be *positive numbers* and those whose corresponding points lie on the other side of O on l to be *negative numbers*. In particular, 1 is a positive number and by Corollary 3, -1 is a negative number. By Corollary 1, if we take points O' and I' on another line l' as 0 and 1 to form a Desarguesian number system $N' = N(l', O', I')$, then under the canonical isomorphism $N \approx N'$, the positive numbers in N correspond to the positive numbers in N' and the negative numbers in N correspond to the negative numbers in N'. On this basis we can define *positiveness* and *negativeness* for numbers other than 0 and the *absolute value* of any number in the number field associated with the geometry.

The following theorem can be considered the main theorem of this section.

Theorem 1. Under the above definition of positive and negative numbers, the number field N associated with ordered Pascalian geometry forms an *ordered field*. Namely, it has the following three properties:

1. If a is a positive number, then $-a$ is a negative number. If, on the contrary, a is a negative number, then $-a$ is a positive number.
2. If a and b are both positive numbers, then $a+b$ is also a positive number.
3. If a and b are both positive numbers or both negative numbers, then ab is a positive number. If one of a, b is positive and the other is negative, then ab is a negative number.

Proof. 1. Take O and I on the line l as 0 and 1 to define a Desarguesian number system $N = N(l, O, I)$. Let $a \neq 0$ and $-a \neq 0$ be considered as numbers in N and correspond respectively to two points A and B on l. Then O is the midpoint of A, B. By Corollary 3, A, B lie on different sides of O on l. So of the corresponding numbers a and $-a$, one is positive and the other is negative. Since 1 is positive and -1 is negative, this property may also be obtained by seeing $-a = (-1) \cdot a$ from the third property.

2. We prove the second property by assuming the third.

Take O, I on l and define the number system N as before. See Fig. 2.33. Let a, b on l correspond to points A, B and let C correspond to $a+b$. Construct a parallelogram $OADE$. By definition, $BCDE$ is also a parallelogram.

When $a = b$, A is the midpoint of O, C, so A, C lie on the same side of 0 and thus c, a are both positive numbers. Therefore, we may suppose that $a \neq b$ and, without loss of generality, A lies between O and B. Since O, E lie on the same side of the line AD while O, B lie on different sides of AD, points B, E

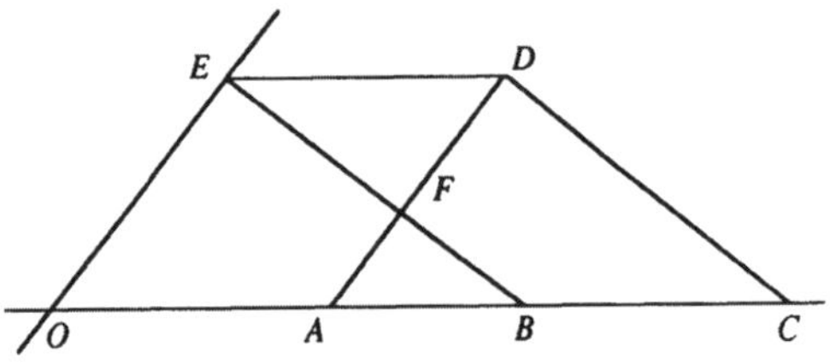

Fig. 2.33

lie on different sides of AD and the segment $|BE|$ meets line AD at a point, say F. As the segment $|EF|$ does not contain any point of l, i.e., E, F lie on the same side of l and so do E, D, the points D, F lie on the same side of l. Therefore, the segment $|DF|$ does not meet l, or A does not lie between D and F. Similarly, one may show that the segment $|BF|$ does not meet line DE and thus B, F lie on the same side of DE, and so do A, B. Since, moreover, A, F lie on the same side of DE, D does not lie between A and F but F lies between A and D. It follows from Corollary 1 that B lies between A and C. Hence on l, C and A, B all lie on the same side of O, which implies that $a+b$ is a positive number.

3. Let l, O, I, N be as before and A, B on l correspond respectively to the numbers a, b. Construct arbitrarily another line l' through O and take a point I' distinct from O on l'. Draw through B a parallel to II', meeting l' at a point B', and through B' a parallel to AI', meeting l at a point C. By the definition of multiplication, the corresponding number c of C is ab, i.e., $c = ab$.

By Corollary 1, whether or not A, C lie on the same side of O on l depends on whether or not I', B' lie on the same side of O; whereas, whether or not I', B' lie on the same side of O depends on whether or not I, B lie on the same side of O on l. Therefore, whether a, c are both positive or negative depends on whether b is positive or negative. The third property is proved (see Fig. 2.34). □

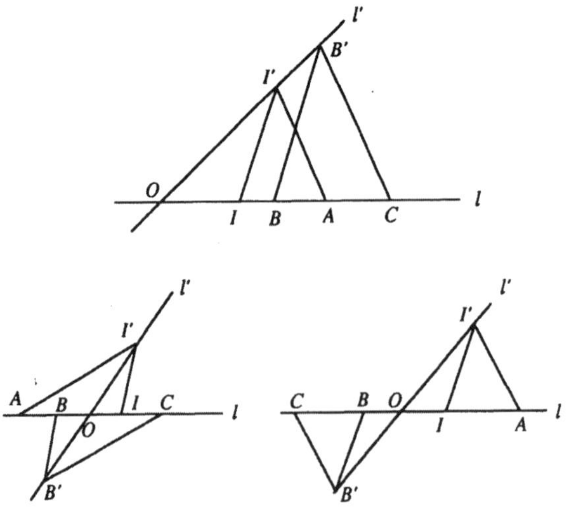

Fig. 2.34

From this theorem one immediately obtains the following result.

Theorem 2. Introduce the concepts of "greater than" > and "less than" < in the number field N associated with ordered Pascalian geometry as follows:

$$a > 0 \Longleftrightarrow a \text{ is a positive number;}$$

$$a < 0 \Longleftrightarrow a \text{ is a negative number;}$$

$$b > a \Longleftrightarrow b - a > 0 \Longleftrightarrow a < b \Longleftrightarrow a - b < 0.$$

Then the number field satisfies Axioms N 1–N 17 of Sect. 1.4.

Another important theorem is

Theorem 3. There are no isotropic lines in ordered metric geometry.

Proof. Let us introduce a Desarguesian coordinate system. If there exists an isotropic line L, we may suppose that its equation is

$$L: \quad u_1x_1 + u_2x_2 + u_3 = 0,$$

where u_1, u_2 are not simultaneously 0 (actually neither is 0). As the orthogonal rate of the Desarguesian coordinate system is 1, we have

$$u_1^2 + u_2^2 = 0,$$

so $-1 = (u_1/u_2)^2$. No matter whether u_1/u_2 is positive or negative, by the third property in Theorem 1, $(u_1/u_2)^2$ is always a positive number. But -1 is a negative number, which leads to a contradiction. Hence there are no isotropic lines. □

We have defined the congruence of pairs of points and of total angles. However, unlike the axiom system of Hilbert in ordinary geometry we introduced congruence without treating it as a fundamental concept nor mentioning the axioms of congruence. In fact, there is no necessity to do so because in this ordered metric geometry the concept of congruence of segments and of angles can be considered as derived from other axioms. To explain this, let us consider a line l which, by Theorem 3, is not isotropic. Now, let A, B, C be three points on an arbitrary line and B lie between A and C. We designate the symmetric points of A, B, C with respect to l as A', B', C'. Then, according to Sect. 2.2, A', B', C' are collinear. In addition, as $AA' \parallel BB' \parallel CC'$, it follows from Corollary 1 that B' lies between A' and C', and thus the reflection of the segment $|AC|$ with respect to l is also a segment $|A'C'|$. Similarly, the reflection of a half-line or a ray with respect to l is also a half-line or a ray, and the reflection of an angle with respect to l is also an angle. Therefore we can naturally define the *congruence of segments* and *congruence of angles*. When they can be obtained by means of some reflections, we shall still use the symbol $\equiv$ for congruence as before. As for $\triangle ABC$ and $\triangle A'B'C'$, we say that they are *congruent*, i.e.,

$$\triangle ABC \equiv \triangle A'B'C',$$

if

$$|AB| \equiv |A'B'|, \quad |AC| \equiv |A'C'|, \quad |BC| \equiv |B'C'|$$

and

$$\measuredangle BAC \equiv \measuredangle B'A'C', \quad \measuredangle ABC \equiv \measuredangle A'B'C',$$
$$\measuredangle ACB \equiv \measuredangle A'C'B',$$

where $\measuredangle BAC$ denotes the angle formed by the two half-lines emanating from A and passing through B and C, and similar for the others. Then Theorem 4 in Sect. 2.4 can be generalized to the following form.

Theorem 4. If for $\triangle ABC$ and $\triangle A'B'C'$

$$|AB| \equiv |A'B'|, \quad |AC| \equiv |A'C'|, \quad |BC| \equiv |B'C'|,$$

then these two triangles are congruent (according to the new definition above):

$$\triangle ABC \equiv \triangle A'B'C'.$$

In unordered metric geometry, even though there is a theorem of type $s.s.s$ about congruence of two triangles in correspondence with that in ordinary geometry, theorems corresponding to the types $a.s.a$ and $s.a.s$ do not hold. However, in ordered metric geometry it is easy to prove that the latter two theorems still hold. In fact, under the above definition of congruence of segments, of angles, and of triangles it is easy to prove the following.

Theorem 5. In ordered metric geometry, according to the above definition of congruence all axioms of congruence H III in Hilbert's axiom system are satisfied.

After fixing a standard pair of points in unordered metric geometry, any pair (AB) of points has a definite number in the associated number field $\boldsymbol{K}$ as its correspondence. This number is called the square of distance of (AB), denoted by $\overline{\overline{AB}}$. It is the square of a number in $\boldsymbol{K}$ and can be taken as $+c$ and $-c$, two distinct values (A, B are assumed to be distinct). Since numbers in $\boldsymbol{K}$ do not have positiveness and negativeness, we have no way to distinguish between the two values $+c$ and $-c$. However, in ordered metric geometry the non-zero numbers in the associated number field $\boldsymbol{K}$ have been separated into positive and negative ones. Thus we may speak about the *absolute value* of a number, denoted by using the usual symbol $|\cdot|$. Therefore, for the square of distance

$\overline{\overline{AB}} = (+c)^2 = (-c)^2$ we may take $|+c| = |-c| = |c|$ and define it to be the *distance* between the two points A and B or the *length* of the segment $|AB|$, denoted as $\overline{AB}$, i.e.,

$$\overline{\overline{AB}} = \overline{AB}^2.$$

Hence, in ordered metric geometry points and lines are taken as the fundamental objects and incidence, parallelism, orthogonality, and order or betweenness are taken as the fundamental relations. Starting from some relevant axioms, including the axioms of incidence H I, the axiom of parallels H IV, the axiom of infinity, Desargues' axioms D, the orthogonal axioms O, the axiom of symmetric axes S, and the axioms of order H II, we may introduce the concepts of congruence, distance, length, and so on which have the familiar properties of congruence and measure in ordinary geometry.

2.6 Ordinary geometry and its subordinate geometries

In the previous sections, we have established various kinds of geometries starting from the axioms. In view of this, it is necessary to give the concept *geometry* itself a strict mathematical definition. If one wishes to make clear the significance, goal, and essence of geometry, what geometry is, what geometry intends to study, and other such problems, one is thrown into problems of philosophy, history, and sociology and enters the realm of opinion and argument. For the aim of this book, we restrict ourselves to the descriptive manner of the axiomatization of geometry, i.e., we define geometry by adopting the formal manner of Hilbert. Of course, this is not the only way, and is not necessarily the best in all circumstances. It is not suitable for some of the most active branches of modern geometry, such as algebraic geometry, differential geometry, and topology. But for this book, in which the mechanization problem of elementary geometries is the theme, such an approach starting from axioms is proper.

For the aim of this book, we first give the following:

Definition 1. A *geometry* **G** is an aggregate consisting of three kinds of entities E, R, A. E is a set consisting of *sets of fundamental objects:* $E = \{E_1, \ldots, E_e\}$. R is a set of *fundamental relations* among the fundamental objects: $R = \{R_1, \ldots, R_r\}$. A is a set of *axioms:* $A = \{A_1, \ldots, A_a\}$, in which each axiom A_i is of the form:

(S) Let

$$e_1 \in E_{i_1}, \ldots, e_s \in E_{i_s}$$

be some fundamental objects and suppose there are some fundamental relations

$$R_v(e_{s_{1v}}, \ldots, e_{s_{kv}}), \quad v = v_1, \ldots, v_m,$$

among these fundamental objects. Then there are some other fundamental relations

$$R_u(e_{t_{1u}}, \ldots, e_{t_{lu}}), \quad u = u_1, \ldots, u_n$$

among these fundamental objects.

Due to different choices of fundamental objects E, fundamental relations R and axioms A, there are various kinds of geometries **G**. For a fixed geometry **G**, in terms of some fundamental objects that satisfy some fundamental relations, we may define new relations among these objects. We call such kinds of relations *derived relations* or *inferred relations*. Similarly, we may define *derived objects* or *inferred objects*, *derived concepts* or *inferred concepts*, etc.

Let us take Desarguesian (plane) geometry as an example. In this case, the set of fundamental objects E consists of two sets: E_1 and E_2. The elements of E_1 are called *points* and the elements of E_2 are called *lines*. The set of fundamental relations R consists of two elements R_1 and R_2. R_1 is called the *relation of incidence* and R_2 the *parallel relation*. The relation $R_1(e_1, e_2)$ means that e_1 is a point in E_1, e_2 is a line in E_2 and *point* e_1 *lies on line* e_2. The relation $R_2(e, e')$ means that e, e' are two distinct lines in E_2 and e, e' are *parallel* to each other. The axiom set A consists of Hilbert's axioms of incidence H I, the axiom of infinity D_∞, Desargues' axioms D_1, D_2 and the (sharper) axiom of parallels H IV, etc. For (unordered) orthogonal geometry, besides the above fundamental objects and relations, there is one more fundamental relation – the orthogonal relation – and some so-called orthogonal axioms are added to the axiom set. In this orthogonal geometry, for some fundamental objects such as two points $A, B \in E_1$ and a line l satisfying some relations, we may define some derived relations such as the *symmetry* of A and B with respect to l. In this geometry, moreover, we can define derived objects including triangles, parallelograms, and Kou-Ku triangles. The other geometries we have mentioned before can all be precisely defined in this manner.

In Sect. 1.1, we listed the fundamental objects, fundamental relations, and five groups of axioms that are satisfied by elementary (plane) geometry proposed in Hilbert's book "Grundlagen der Geometrie." The fundamental objects are points and lines. There are four kinds of fundamental relations: the relations of incidence, order, congruence, and parallelism. And the five groups of axioms: those of incidence, order, congruence, parallels, and continuity, denoted as H I–H V respectively. According to the strict definition of geometry above, these fundamental objects, relations, and axioms constitute a geometry. This special geometry is commonly called *Euclidean geometry* in the literature. The name originates from the Euclid's "Elements" written in ancient Greece, even though this geometry was prevalent in other ancient nations. There are big differences in the way in which the subject is presented, but in essence this geometry is such a geometry that all people over the world have consistently observed and used in daily life since ancient times. In comparison with other geometries such as non-Euclidean geometry which is widely known only among scientists or even only among some specialists of mathematics, we rename such a commonly called Euclidean geometry *ordinary geometry* in this book.

Ordinary geometry is not irrelevant to those geometries that we have previously mentioned. To explain some logical incidence among them, let us first give a strict mathematical definition for the concept *geometric theorem* as follows.

Definition 2. Let the fundamental objects and relations e_i, E_{i_k}, R_v, R_u, etc. mentioned in the sentence (S) (see above) belong to a given geometry **G**. Then (S) is called a *geometric sentence* or *assertion* in geometry **G** or we say that the sentence (S) is *meaningful* in geometry **G**. In this case, the relations R_v are called the *hypothesis part* and R_u the *conclusion part* of this sentence.

Definition 3. In a geometry **G**, if a meaningful geometric sentence can be deduced from axioms in **G** according to the general logical rules, then we say that (S) is a *theorem* in geometry **G** or that the *theorem (S) holds*. Otherwise we say that (S) is not a theorem in geometry **G** or that the *theorem* (S) does not hold.

Certainly, every axiom of **G** is a theorem in **G**. The word *theorem* in the above definition has two different usages. This is to follow convention and is unlikely to lead to confusion.

There is no difficulty in precisely formulating the above definition of *sentence* and *theorem* by using the language and symbols of mathematical logic. However, to avoid having the statement be excessively lengthy and tedious, we do not proceed in that way. It is necessary to remark that the definition does not involve any *existential* problems. Hence, in the precise logical formulation the existential quantifier $\exists$ should not occur, because the existence of various fundamental objects appearing in the conclusion of a sentence or a theorem has been already assumed in the hypothesis.

Now, we may make some comparison among different geometries as follows.

Definition 4. Let **G** and **G′** be two geometries. If every meaningful sentence in **G′** is also a meaningful sentence in **G** while each axiom in **G′** is a theorem in **G**, we say that the geometry **G′** is *subordinate* to the geometry **G** or **G′** is a *subordinate geometry* of **G**, symbolically

$$\mathbf{G}' \to \mathbf{G}.$$

According to Definition 4, the significance that the geometry **G′** is subordinate to the geometry **G** is only in logic, i.e., the fundamental relations among fundamental objects in **G′** remain in **G** and the theorems in **G′** also remain true in **G**. The concrete significance of the fundamental objects and relations, axioms, and theorems as well as such questions as of whether or not there are inclusion relations among sets is not under consideration. These formal definitions and language, stemming from Hilbert, exclude various additional factors irrelevant to logical proof and are appropriate and effective for the logical problem of how to prove geometric theorems.

According to the above definition of geometries and their reciprocal subordinate relations, it may be seen that all (plane) geometries previously introduced are the subordinate geometries of ordinary (plane) geometry. First, the fundamental objects all consist of the same two classes: points and lines, and the fundamental relations (or derived relations) contain the following kinds:

R_1 Relation of incidence;
R_2 Parallel relation;
R_3 Orthogonal relation;
R_4 Relation of congruence;
R_5 Relation of symmetry;
R_6 Relation of order.

Among those above, some relations such as the orthogonal relation and the relation of congruence are fundamental relations in some geometries but not in others, that is, they may be defined from other fundamental relations and axioms, or the so-called derived relations.

Among the (plane) geometries (the word plane will be omitted below) defined on the basis of these fundamental objects, relations and some axioms, we have mentioned the following:

Desarguesian geometry;
Pascalian geometry;
(Unordered) orthogonal geometry;
(Unordered) metric geometry;
Ordered Pascalian geometry;
Ordered metric geometry;
Ordinary geometry.

We may also add some fundamental relations or axioms to the geometries listed above or combine them to form other geometries. For example, we can add the relation and axioms of order to Desarguesian geometry and orthogonal geometry and call the geometries formed in this way ordered Desarguesian geometry and ordered orthogonal geometry.

There are simple subordinate relations among the above-mentioned geometries. For instance, each axiom in Pascalian geometry, if not an axiom in (unordered) orthogonal geometry, must be a theorem in this geometry (e.g., the Pascalian axiom, see Sect. 2.2). Thus Pascalian geometry is subordinate to (unordered) orthogonal geometry. In fact, all these geometries are subordinate to ordinary geometry (Fig. 2.35).

The various geometries discussed above are all restricted to the case of plane and thus the fundamental objects are only of two classes: points and lines. This restriction naturally does not have to be imposed. We may consider spatial and higher dimensional geometries, in which besides points and lines, there are other fundamental objects such as planes and hyperplanes. Furthermore, we may also consider other geometries that are subordinate to ordinary geometry and whose fundamental objects are not restricted to points and lines. For example, the line geometry which takes lines as its fundamental objects (Plücker, Klein, and others), various circle geometries whose fundamental objects are either points and circles or oriented vectors and oriented circles and whose fundamental

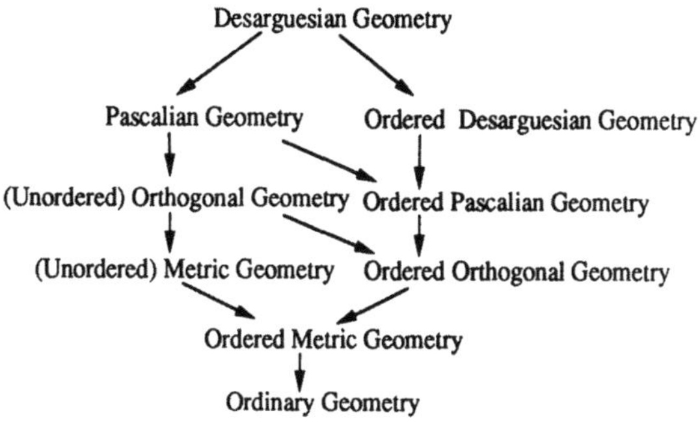

Fig. 2.35

relations are incidence, orthogonality, and tangency or directional tangency (Möbius, Laguerre, and others), the spherical geometry with points and spheres as its fundamental objects (S. Lie and others), and so on. For the choice of axiom systems, there is plenty of room. Besides so-called projective geometry, there are various non-Euclidean geometries (Lobachevsky, Bolyai, Riemann) resulting from the modification of the axiom of parallels. In virtue of the revision of Archimedes' and other axioms (Dehn and others) there are various semi-Euclidean geometries and non-Legendrean geometry, and so on. Avoiding the axiom of parallels, we have natural geometry and absolute geometry (Bachmann) and so forth. If no restriction is imposed, such geometries may be defined as one pleases.

3 Mechanization of theorem proving in geometry and Hilbert's mechanization theorem

3.1 Comments on Euclidean proof method

Generally speaking, Euclid's "Elements" is the origin of axiomatization of mathematics. However, from the viewpoint of rigor, it has many blemishes. This has been recognized for thousands of years. In the middle and late nineteenth century a critical movement attesting the very foundation of mathematics began. Mathematicians made a comprehensive analysis of the axioms besides the independence problem of the axiom of parallels in the "Elements." This movement was initiated in Germany and Italy by Pasch and Peano, followed by others and epitomized later on by Hilbert. He classified the axioms of Euclidean geometry into five groups (cf. Sect. 1.1) as the basis and starting point of all theorem proving so that Euclidean geometry has had a rigorous foundation ever since. However, even though there is a rigorous axiom system as the basis of all reasoning and proving, by using the Euclidean proof method for geometric theorems it is still impossible to reach an extent of rigor without flaws. This opposes all conventional understandings, as it seems that nobody has ever precisely pointed out or even recognized this fact. This section focuses on explaining this issue.

We are saying that by using the Euclidean method to prove theorems in geometry it is impossible to logically reach an extent of rigor. The crux in question lies in the following.

In Euclidean geometry, the statement of axioms and theorems usually involves an implicit assumption – the considered figures must be *normal*, *generic* cases but not *abnormal*, *degenerate* ones. For example, the statement that two straight lines are parallel implies that they are two distinct lines, not two coincident lines. Similarly, speaking about the intersection of two lines implies that they are neither parallel nor coincident lines. Also, the construction of a triangle implies that it is an ordinary, real triangle – its three vertices are distinct from each other, they are not collinear, and so forth. We can make various restrictions on the statements of definitions and theorems as we did for the definitions of parallel lines and triangles in Chap. 1, but the statements will become very verbose. It may be unclear what kind of non-degenerate restrictions are appropriate. Also, there is no precise definition for the word *degenerate*, so it is difficult to determine in advance whether or not an isosceles triangle or a right triangle should be included as a degenerate triangle.

It is particularly serious that the proof of a theorem is usually valid only for the normal, generic cases, but not for the abnormal, degenerate cases. In the degenerate cases, it may be necessary to change part or even all of the proof. Sometimes, the theorem itself may be meaningless or even invalid in the degenerate cases. The following examples may serve as illustrations.

Example 1 (Theorem). The diagonals AC and BD of any parallelogram $ABCD$ bisect one another (see Sect. 1.2).

This statement implies some assumptions about the configuration, i.e., $ABCD$ is a real quadrilateral, the points A, B, C, D are distinct, the parallel opposite sides AB and CD lie on two non-coincident parallel lines, and so do AD and BC (the definition of parallelogram in Sect. 1.2 has intentionally excluded the degenerate cases).

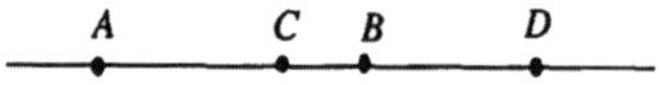

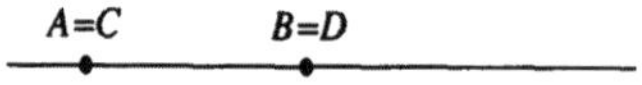

In some degenerate cases, such as where A, B, C, D coincide, the theorem may still be considered true by a proper interpretation of the terms used in its statement. In the case where A, B, C, D are distinct but collinear, though it is still possible to make the hypothesis of the theorem meaningful by a proper interpretation (for instance, the coincidence of AB and CD can be regarded as a degenerate case of $AB \parallel CD$), the conclusion will lose its meaning. If we interpret the bisection of AC and BD as the pairs of points (AC) and (BD) having a common midpoint, then the theorem falls into fallacies. If A and B are distinct while C degenerates to coincide with A and D to coincide with B, then the theorem is always false, regardless of the interpretation.

Example 2 (Desargues' axiom D_1). Suppose the three pairs of the corresponding sides of

$$\triangle ABC \quad \text{and} \quad \triangle A'B'C'$$

are all parallel to each other, i.e.,

$$AB \parallel A'B', \quad AC \parallel A'C', \quad BC \parallel B'C'.$$

Then either the three connecting lines AA', BB', CC' of the corresponding vertices are parallel to each other or they all intersect at the same point.

In the statement of this theorem, the triangles should be seen as normal and real triangles and the parallel lines should be understood as non-coincident, real

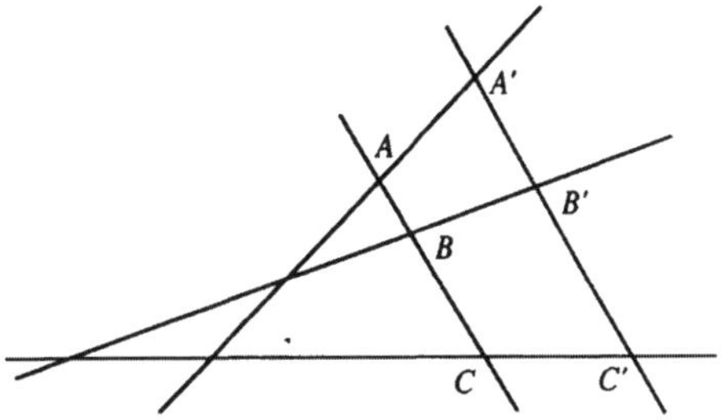

Fig. 3.1

parallel lines (see the definitions of parallel lines and triangles in Sects. 1.1 and 1.2).

As an example of a degenerate case, let us consider $\triangle ABC$ where the three points A, B, C are distinct but collinear, and $A'B'$, AB are parallel but not coincident. Then A', B', C' are collinear and the theorem is obviously false (see Fig. 3.1). For other degenerate cases, whether the theorem is meaningless or false can be ascertained only after a detailed, careful analysis.

Example 3 (Desargues' axiom D_2). If two pairs of the corresponding sides of $\triangle ABC$ and $\triangle A'B'C'$ are parallel to each other, e.g.,

$$AB \parallel A'B', \quad AC \parallel A'C',$$

and the three lines AA', BB', CC' connecting the corresponding vertices are either parallel to each other or intersect at the same point, then the third pair of the corresponding sides of these two triangles are also parallel to each other.

In the same degenerate case as discussed in Example 2, i.e., where A, B, C are collinear, the theorem may be considered (vacuously) true. However, the theorem is no longer true if the parallel lines $AB \parallel A'B'$ degenerate to coincide (see Fig. 3.2).

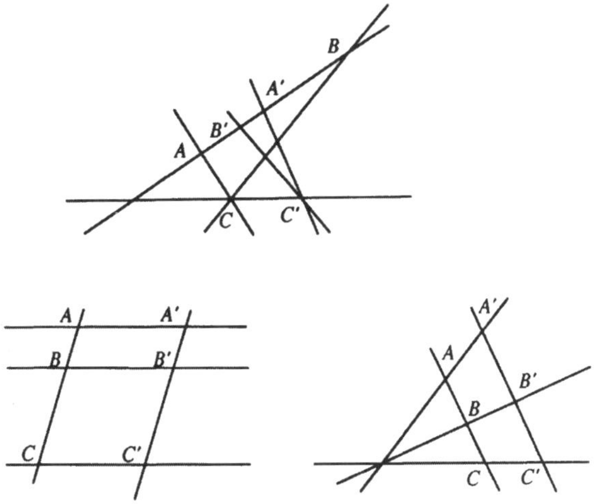

Fig. 3.2

The above examples indicate that an axiom or theorem often contains implicit, not clearly mentioned assumptions or restrictions on its genericity and applicability. The axiom or theorem may lose its meaning or even become invalid in those cases which go beyond these implicit assumptions. So in the proof of each theorem, whenever introducing auxiliary lines or applying given axioms and known theorems, we must always carefully verify whether the applied theorems are within the restrictions on their applicability and consider every possible degenerate case separately. This makes the proofs of theorems lengthy and tedious, but in order to meet the usual standard of rigor, the over-elaborate consideration is indispensable. For a concrete example, one may see the proof of Hessenberg's theorem in Sect. 2.2. We have repeatedly applied the theorem that the diagonals of any parallelogram bisect one another and Desargues' two axioms to many proofs in Chap. 1. These theorems and axioms may be meaningless or false in some degenerate cases as has been shown in Examples 1–3. Therefore we must be very careful each time when we use them so that the degenerate cases do not occur. This explains why we consumed so much space for some proofs in Chap. 1.

In the proof of a theorem, even though the configuration of the hypothesis at the outset is located in a generic, non-degenerate position, we are still unable to determine ahead of time whether or not the degenerate cases will occur when applying other theorems in the proof process. Not only is the verification of every applied theorem cumbersome and difficult, but it is actually also impossible to guarantee that the degenerate cases (in which the theorem is meaningless or false) do not happen in the proof process. On the other hand, we have no effective means to judge how much to restrict the statement of a theorem (to be proved) in order to ensure the truth of the theorem. These problems make it impossible for the Euclidean method of theorem proving to meet the requirements of necessary rigor. Therefore, even though there is a rigorous axiom system as the basis of theorem proving in geometry, *following the Euclidean method it is still impossible to reach rigor from the logical point of view*.

In order to make theorem proving in geometry rigorous, we must adopt a method that is different from the traditional Euclidean proof method.

In contrast to the Euclidean method, the method of *mechanical theorem proving in geometry* we are proposing not only reduces the amount of mental work but also really satisfies the requirements of rigor.

3.2 The standardization of coordinate representation of geometric concepts

According to the analysis in the preceding section, the various problems on the loss of rigor in geometry theorem proving are mainly caused by the degenerate cases. The traditional Euclidean proof method in geometry is unable to handle such situations. But, by imposing a number system and a coordinate system on the basis of the axiom system so that geometric statements are made algebraic as had been done from Descartes to Hilbert, one can effectively treat the degenerate cases and thus reach the true rigor required for theorem proving.

First, the word "degenerate" is in contrast with "generic," where the word "generic" has a strict definition, usually seen in algebraic geometry as *generic point* and *generic plane*. For our purpose, we do not adopt the strict definition from algebraic geometry but understand its meaning as follows.

A point is a *generic point* if some of its coordinates are parameters or indeterminates and a configuration is in a *generic position* if some of the coordinates of points in the configuration are parameters or indeterminates. The configuration *degenerates* if these parameters or indeterminates are specialized to values such that some algebraic expressions among some of the coordinates vanish.

Geometry theorems treated by the Euclidean proof method usually imply the assumption that the considered figures are in certain generic positions. To achieve a strict proof, one needs not only to show a theorem to be true in the generic case but also to explicitly point out the degenerate cases for which the theorem is false, and the region on which the theorem is true. The former can be expressed by means of the parameter coordinates, while the latter can be expressed by using some inequations involving the coordinates, i.e., by means of non-degeneracy conditions.

In Sect. 1.7 and in Chap. 2, in the process of expressing some fundamental geometric relations in terms of the quantitative relations among coordinates, we have added many restrictions which make the exposition very cumbersome. In order to identify degenerate cases, we first slightly modify the definitions of some fundamental geometric relations so that the quantitative relations among coordinates can both capture the generic and degenerate cases and make the latter easily distinguishable from the former. The fundamental geometric relations to be discussed are as follows.

Parallel relation

Let the coordinates of four points A_i in a certain coordinate system be (x_i, y_i), $i = 1, \ldots, 4$. Set

$$\alpha = (x_2 - x_1)(y_4 - y_3) - (y_2 - y_1)(x_4 - x_3).$$

Definition 1. We say that A_1A_2 is *parallel* to A_3A_4, which is denoted as

$$A_1A_2 \parallel A_3A_4,$$

if

$$\alpha = 0.$$

In case the points $A_1, \ldots, A_4$ are distinct from each other and A_1A_2, A_3A_4 are two distinct lines, the definition of parallelism here coincides with that given in Chap. 1, whereas the new definition also contains the degenerate cases, while the one in Chap. 1 does not. In order to distinguish the definitions, we call the parallelism previously defined *geometric parallelism*, and designate the term *algebraic parallelism* (or *parallelism* for short) for $\alpha = 0$, which contains the geometric parallelism and various degenerate cases such as A_1 coincides with

A_2; A_3 coincides with A_4; and A_1, A_2 are distinct, A_3, A_4 are distinct while the lines A_1A_2 and A_3A_4 coincide. The degenerate cases may be expressed by means of equations and the non-degenerate cases by means of inequations. For instance, $A_1 = A_2$ means that $x_1 = x_2$ and $y_1 = y_2$, and $A_1 \neq A_2$ means that $x_1 \neq x_2$ or $y_1 \neq y_2$.

Collinear relation

Let the coordinates of three points A_i in a coordinate system be (x_i, y_i), $i = 1, 2, 3$. Set

$$\beta = \begin{vmatrix} x_1 & y_1 & 1 \\ x_2 & y_2 & 1 \\ x_3 & y_3 & 1 \end{vmatrix}$$
$$= (x_2y_3 - x_3y_2) + (x_3y_1 - x_1y_3) + (x_1y_2 - x_2y_1).$$

Definition 2. We say that the three points A_1, A_2, A_3 lie *on the same line* or are *collinear*, or A_3 lies *on* the line A_1A_2, or A_1 lies *on* A_2A_3, or A_2 lies *on* A_1A_3, if $\beta = 0$.

In case the three points A_1, A_2, A_3 are distinct from each other, the definition here coincides with the definition previously given, but the new definition also contains the degenerate cases that some of the three points coincide. In order to distinguish this definition from the previous one where the three points are distinct, we call the previously defined collinearity *geometric collinearity*. While speaking about collinearity of three points A_1, A_2, A_3 hereafter, we shall mean that either the three points are distinct and collinear, or two of them coincide, or all three coincide.

Orthogonal relation

Let the coordinates of four points A_i in an orthogonal coordinate system be (x_i, y_i), $i = 1, \ldots, 4$.

Definition 3. Let the orthogonal rate of the orthogonal coordinate system be k. The two lines A_1A_2 and A_3A_4 are said to be *perpendicular*, which is denoted as

$$A_1A_2 \perp A_3A_4,$$

if

$$k(y_2 - y_1)(y_4 - y_3) + (x_2 - x_1)(x_4 - x_3) = 0.$$

In particular, in a Descartes coordinate system, $k = 1$.

Evidently, in case A_1, A_2 are distinct and A_3, A_4 are distinct, the new definition coincides with the previous definition of perpendicularity of two lines, but the new one also contains the degenerate case that A_1 coincides with A_2 or A_3 coincides with A_4. To distinguish them from each other, we call the previously defined perpendicularity *geometric perpendicularity.*

Order relation

In any ordered geometry (see Sect. 2.6), we may take an arbitrary coordinate system and define an order relation as follows.

Definition 4. Let three points $A_i = (x_i, y_i)$, $i = 1, 2, 3$, be on the same line. We say that A_2 lies *between* A_1 and A_3 if at least one of the following cases holds:

(1) $x_1 < x_2 < x_3$;
(2) $x_1 > x_2 > x_3$;
(3) $y_1 < y_2 < y_3$;
(4) $y_1 > y_2 > y_3$.

This definition has excluded the degenerate cases that some of the points A_i coincide.

Definition 5. Let

$$L: \quad u_1x_1 + u_2x_2 + u_3 = 0$$

be a line, where (x_1, x_2) are the coordinates of the moving point and u_1, u_2 are not simultaneously 0. If two points $A = (a_1, a_2)$ and $B = (b_1, b_2)$ do not lie on the line L, we say that A and B lie *on the same side* or *different sides* of L according to whether $u_1a_1 + u_2a_2 + u_3$ and $u_1b_1 + u_2b_2 + u_3$ have the same sign or different signs.

Metric property

We have introduced the concept of square of the distance between a pair of points in an unordered metric geometry. Moreover, the concepts of distance and length have been defined in an ordered metric geometry. The definition after standardization is stated as follows.

Definition 6. Take a Descartes coordinate system in metric geometry. For two arbitrary points $A = (a_1, a_2)$ and $B = (b_1, b_2)$, the scalar

$$\overline{\overline{AB}} = (b_1 - a_1)^2 + (b_2 - a_2)^2$$

is called the *square of the distance* between a pair (AB) or (BA) of points. If the metric geometry has an order, then there is one and only one real number $c \geq 0$ such that

$$c^2 = (b_1 - a_1)^2 + (b_2 - a_2)^2.$$

This number, denoted as $\overline{AB}$, is called the *distance* between the two points A

and B or the *length* of the *segment* $|AB|$:

$$\overline{AB}^2 = \overline{\overline{AB}}, \quad \overline{AB} \geq 0.$$

The concept of angles and their measure are much more complicated than those of point pairs and segments and are often ambiguous (without the help of the concept of continuity and related axioms). By appealing to the theorem that the three pairs of the corresponding angles of two triangles respectively are equal (i.e., congruent) if the three pairs of their corresponding sides respectively are equal, the degree of angles can be indirectly treated by using the length of segments. For theorem proving in geometry, we can avoid direct use of the measure of angles, so we do not give it a definition.

Symmetry and reflection

In orthogonal geometry, we have already defined the relations of symmetry and reflection with respect to a non-isotropic line (see Sect. 2.3). Now, we restate the definition as follows.

Definition 7. Take an orthogonal coordinate system in orthogonal geometry with orthogonal rate k. Let the line

$$L_u: \quad u_1x_1 + u_2x_2 + u_3 = 0$$

be non-isotropic, i.e.,

$$ku_1^2 + u_2^2 \neq 0.$$

The *symmetric point* of a point $A = (a_1, a_2)$ with respect to L_u is defined to be $A^* = (a_1^*, a_2^*)$, where

$$(ku_1^2 + u_2^2)a_1^* = -(ku_1^2 - u_2^2)a_1 - 2ku_1u_2a_2 - 2ku_1u_3,$$
$$(ku_1^2 + u_2^2)a_2^* = -2u_1u_2a_1 + (ku_1^2 - u_2^2)a_2 - 2u_2u_3.$$

The *symmetric line* of a line

$$L_v: \quad v_1(x_1 - a_1) + v_2(x_2 - a_2) = 0$$

passing through point $A = (a_1, a_2)$ with respect to L_u is defined to be

$$L_v^*: \quad v_1^*(x_1 - a_1^*) + v_2^*(x_2 - a_2^*) = 0$$

which passes through the point A^*, where

$$v_1^* = (ku_1^2 - u_2^2)v_1 + 2u_1u_2v_2,$$
$$v_2^* = 2ku_1u_2v_1 - (ku_1^2 - u_2^2)v_2.$$

The above A^* and L_v^* are also called the *reflections* of A and L_v respectively, with respect to line L_u.

Angular bisectors

In unordered orthogonal geometry or unordered metric geometry, there is only the concept of total angles but not the concept of angles. For a total angle, in unordered orthogonal geometry either there are exactly two bisectors or there is none. It is ensured that in unordered metric geometry there are exactly two bisectors for any total angle, but they cannot be distinguished from each other. Now we give a standardized definition of angular bisectors as follows.

Definition 8. Let the orthogonal rate of an orthogonal coordinate system be k. For three arbitrary points $A_i = (a_i, b_i)$, $i = 0, 1, 2$, we say that the following equation defines the *bisectors* of the angle $\measuredangle A_0(A_0A_1, A_0A_2)$ or $\measuredangle A_1A_0A_2$:

$$\begin{aligned}
&[(b_1 - b_0)(x - a_0) - (a_1 - a_0)(y - b_0)] \cdot [(a_2 - a_0)(x - a_0) \\
&\qquad + k(b_2 - b_0)(y - b_0)] \\
&= [(a_2 - a_0)(y - b_0) - (b_2 - b_0)(x - a_0)] \cdot [(a_1 - a_0)(x - a_0) \\
&\qquad + k(b_1 - b_0)(y - b_0)].
\end{aligned}$$

In the *non-degenerate* case that A_1 and A_2 are both distinct from A_0, one sees that this equation determines two lines which are in accordance with the usual concept of angular bisectors. If the geometry is ordered, we may further distinguish these two angular bisectors into an *internal* angular bisector and an *external* angular bisector according to the order relation above.

Derived concepts and derived relations

Among the above definitions, there are both fundamental relations, such as parallel relation and orthogonal relation, and derived relations, such as symmetry and congruence. We may also consider other derived concepts and relations, for instance, slope, area, midpoint, circle, radius, (internal, external) tangency of circles, orthogonality of circles as well as tangent lines and singular points of curves. These derived concepts and relations may be added if necessary and will not be listed here. The following two points should be noted.

(1) When giving a standardized definition via the coordinate system (of a subordinate geometry), the non-degenerate cases should agree with their original

geometric meaning. In the degenerate cases, the definition may be arbitrary, but once fixed, it should not be altered.

(2) The definition should not vary with different choices of coordinate systems (of the subordinate geometry), i.e., the algebraic expressions (or relations) in the definition should be invariant (relations) or semi-invariant (relations) under the coordinate transformation of the subordinate geometry.

A detailed discussion of (2) goes far beyond the scope of this book, while from this perspective the corresponding invariant theory is far from being developed to a mature extent. However, the invariance of the algebraic expressions (relations) listed above is evident.

3.3 The mechanization of theorem proving and Hilbert's mechanization theorem about pure point of intersection theorems in Pascalian geometry

Descartes pointed out that every proof of a theorem in Euclidean geometry requires some new and often ingenious ideas. In contrast, he proposed the scheme of mechanizing the reasoning procedure to reduce the amount of work for problem solving (cf. Kline 1972: chap. 15). According to this scheme, for a certain class of geometric theorems, no matter how they are stated concretely, we would have a common proof method which is applicable to all of them. This method proceeds for the proof of every theorem according to a mechanical reasoning procedure and can ensure, after a finite number of steps, whether or not the theorem is true. Furthermore, when the theorem is proved to be true in the *generic* case, the method can mechanically find all *degeneracy* conditions that may make the theorem meaningless or false during the mechanical proof process. Such a method that can be used to prove a certain class of theorems is called a *mechanical method*. The statement that there is a mechanical method for proving a certain class of theorems is itself a theorem, called a *mechanization theorem*.

In Chap. 1 we introduced, starting from some axioms, the number system and coordinate system based roughly on the model exhibited in Hilbert's "Grundlagen der Geometrie," which follows a path from axiomatization to algebraization to coordinatization. On the way, we have to prove many theorems. The proof of these theorems all follows the Euclidean method: individual theorem by individual method, yielding an individual proof. As we have pointed out in Sect. 3.1, this method actually cannot reach the extent of necessary rigor. Moreover, when using the Euclidean method one is in need of some new and often ingenious ideas for every proof as pointed out by Descartes. Although the ingenious ideas are fascinating and make the method aesthetically attractive, from the viewpoint of effectiveness this method has many limitations. Maybe this is one of the major reasons why Descartes proposed the scheme of mechanical proving; it is the reason why we investigate mechanical proving in this book. Owing to the appearance of electronic computers, the scheme of mechanical proving has been promoted from the status of idle dream to the reality of accomplishment in practice. This chapter provides a theoretical basis for the scheme of mechanical proving, while the later chapters will present concrete examples with hand-calculation and computer experiments. The author will supply more examples in other relevant books in order to illustrate the efficiency of this method.

As we have indicated in Sect. 3.1, geometry theorems are true only under some conditions, i.e., the so-called *non-degeneracy conditions*. These conditions are not usually explicitly expressed in the hypothesis of a theorem. In fact, it is also difficult to determine at the beginning under which conditions the theorem will be true. For this reason, we give some explicit definitions for *theorem*, *proof*, *mechanized proof* and *mechanization theorem*.

Definition 1. If a meaningful geometric sentence (S) in a geometry **G** will be true after the addition of some conditions meaningful in the geometry as *subsidiary hypotheses*, then we call the sentence (S) a *conditional theorem* or say that the *theorem* (S) *is generically true*. The subsidiary conditions are called the *non-degeneracy conditions* for the theorem to be true.

Definition 2. For a *generically* true theorem (S), if there is a process starting from the hypothesis of (S), proceeding step by step according to axioms in geometry **G** and logical inference rules, meanwhile adding some *subsidiary conditions* to the hypothesis, and finally arriving at the conclusion of (S), then this process is called a *proof* of the conditional theorem.

Definition 3. If for a certain class of meaningful geometric sentences **S** in geometry **G**, there is a mechanical procedure such that for any sentence in **S**, starting from its hypothesis and proceeding with logical inference, in every step one can determine mechanically the next step, give the subsidiary conditions which must be added according to the procedure, and, after a finite number of steps, the conclusion of the sentence or its opposite is finally reached, then we say that *there is* a *mechanical method* for proving this class of theorems *S* or the proof of this class of theorems is *mechanizable*. The mechanical procedure is called a *mechanical method* for proving this class of theorems.

Definition 4. If **S** in the above definition contains all meaningful geometric sentences in geometry **G**, we say that theorem proving in geometry **G** is *mechanizable*, or simply that **G** is *mechanizable*.

The assertion "theorem proving in geometry **G** is mechanizable" or "the proof of a certain class of theorems in **G** is mechanizable" is itself a theorem that has to be proved, and of which the proof requires the exhibition of the mechanical procedure mentioned in the definition. This type of theorem will be commonly called a *mechanization theorem*.

Of course, the *subsidiary hypotheses* or *subsidiary conditions* in the above definitions cannot be arbitrary. For example, the negation of one of the original hypotheses obviously cannot be added as a subsidiary condition. What we call subsidiary conditions are such conditions that supply only a negligible (polynomial) equality relation among the coordinates, after the original geometric hypotheses have been expressed in a chosen coordinate system as (polynomial) equality and/or inequality relations among the coordinates according to Sect. 3.2.

In other words, if we consider the coordinate variables in a fixed order as a point in a vector space, all hypothesis (polynomial) equality relations (and/or inequality relations) will define a set V of points in this vector space, which is commonly called an *algebraic variety* (or *algebraic set* if there are inequality relations). Then the supplied (polynomial) equality relation corresponding to the subsidiary condition should define a *true* subset of V. Using the terminology of algebraic geometry, we have no difficulty in giving a precise definition, for which the reader may refer to Chap. 4.

In mathematical logic, a terminology analogous to mechanizability is *decidability*. As in mathematical logic where many mathematical problems and theories have proved undecidable, we have the following:

Conjecture. Desarguesian geometry is non-mechanizable, that is, it is impossible to find a mechanical method for proving all theorems in this geometry.

In geometric reasoning and proving, we expect to discover and prove true theorems rather than obtain the negative conclusion that a theorem is not true or cannot be proved. In fact, our desire is not extravagant but fair and reasonable. It can be completely satisfied. We will show that as long as the Pascalian axiom for intersecting lines is added to Desarguesian geometry so that the geometry-associated number sfield becomes commutative, the geometries usually encountered in mathematical research such as those we have mentioned in Chap. 2 and will mention in Chap. 6 are all mechanizable. The geometries which are conjectured to be non-mechanizable are probably so only due to some artificial, impractical restrictions. In this chapter, we present a simple example for which theorem proving is mechanizable. This example originates from Hilbert and will be called *Hilbert's mechanization theorem.*

Although the main contents of Hilbert's book are in laying down the rigorous foundation of geometry and in elaborating the logical relations among some important geometric axioms and facts, the book spends most of its space on expounding how to advance towards algebraization from axiomatization, and thus paves a way for the mechanization of theorem proving in geometry. Actually, one of the earliest mechanization theorems appeared in Hilbert's "Grundlagen der Geometrie," though he did not describe the theorem with ideas of mechanization, and it is even difficult to say whether Hilbert himself was aware of this aspect.

What is here called Hilbert's mechanization theorem appeared at the end of chap. 6 of "Grundlagen der Geometrie." In the first edition of his book, this result was written only in italics without being listed as a theorem. It was formally numbered as theorem 62 in later editions. The original statement of this theorem runs as follows.

Theorem 62. Every pure point of intersection theorem that holds in a plane geometry in which Axioms I, 1–3, II, IV* and Pascal's Theorem are valid takes, through the construction of suitable auxiliary points and lines, the form of a combination of finite number of Pascalian configurations.

Let us first explain some phrases occurring in the theorem one by one according to the original interpretation in Hilbert's book.

The so-called "Axioms I, 1–3" stand for the first group of axioms of incidence, "II" for the second group of axioms of order and "IV*" for the fourth group of axiom of parallels in sharper form in a plane, i.e., Axioms H I, H II and the (sharper) axiom of parallels H IV in Sect. 1.1 of this book.

The so-called "Pascal's Theorem" is Pascal's theorem for intersecting lines in Sect. 2.1. The so-called "Pascalian configuration" is a configuration corresponding to the contents of Pascal's theorem, including two intersecting lines l, l', three points A, B, C on l, three points A', B', C' on l' and $AB' \parallel A'B$, $BC' \parallel B'C$.

The so-called "pure point of intersection theorem" is a "theorem that contains an assertion about the common locus of points and lines and the parallelism of lines without the use of other relations such as congruence and perpendicularity."

Following this sentence in Hilbert's book a more detailed explanation for this class of theorems was given as follows:

"Every such pure point of intersection theorem in a plane geometry can be put in the following form:

Choose an arbitrary set of a finite number of points and lines. Then draw in a prescribed manner any parallels to some of these lines. Choose any points on some of the lines and draw any lines through some of these points. Then, after having constructed connecting lines and points of intersection as well as parallels through the points already existing in the prescribed manner, a definite set of finitely many lines is eventually reached, about which the theorem asserts that they either pass through the same point or are parallel."

In Sect. 3.4, we shall illustrate by examples that the two explanations about pure point of intersection theorems (*PIP theorems* for short) are in fact not equivalent. To distinguish between the two types of theorems, we shall call the latter, the special pure point of intersection theorems, *PIP theorems of Hilbert's type* or *of constructive type* and simply use *PIP theorems* for the general pure point of intersection theorems.

Moreover, following Theorem 62 in his book Hilbert made a supplement:

"In proving the point of intersection theorem with the aid of Pascal's Theorem it is then no longer necessary to revert to the congruence and continuity axioms."

This indicates that Hilbert had the same spirit for Theorem 62 as for the whole book, with logical relations of dependence among axioms and theorems in mind. But from the explanation in the paragraph preceding Theorem 62, which in essence is equivalent to a proof of the theorem, we see that Hilbert already gave a mechanical method, at least for proving PIP theorems of Hilbert's type.

In the previous chapter, we indicated how to partially replace the axioms of order and continuity by an axiom of infinity and Desargues' axioms in Hilbert's axiom system, and called a geometry, which satisfies the (plane) axioms of incidence H I, the (sharper) axiom of parallels H IV, the axiom of infinity D_∞ and the Pascalian axiom for intersecting lines, a (plane) Pascalian *geometry*. Now, based on the idea furnished by Hilbert in his proof, we may restate Theorem 62 in the following alternative form.

Hilbert's mechanization theorem. There is a mechanical method for proving pure point of intersection theorems of Hilbert's type in (plane) Pascalian geometry. This method can mechanically find all the degeneracy conditions that may make a theorem false during the proof process.

In the next section, we shall give some mechanized proofs of several PIP theorems on the basis of Hilbert's method. The proof of Hilbert's mechanization theorem itself will be given in Sect. 3.5.

3.4 Examples for Hilbert's mechanical method

Before giving the proof of Hilbert's mechanization theorem, in this section we illustrate by examples the mechanical method used by Hilbert and explain the difference between PIP theorems of Hilbert's type and the general PIP theorems.

Example 1 (Theorem). The diagonals of any generic parallelogram bisect one another.

Let the generic parallelogram be $\square ABCD$ (see Fig. 3.3). Denote the intersection point of the diagonals AC and BD by E. Choose a coordinate system with A as its origin O, B on the first axis l_1 and the second axis arbitrary. Such a choice is not material and does not affect explaining the method. It is only intended to simplify the calculation and to avoid unnecessary complication.

To reflect the genericity of the parallelogram $ABCD$, the coordinates of the points B and D may be chosen as $(u_1, 0)$ and (u_2, u_3) respectively, where u_1, u_2, u_3 are all considered as parameters. Then the points C and E are no longer generic but are constrained by the geometric conditions. Let their coordinates be $C = (x_1, x_2)$, $E = (x_3, x_4)$. Then these x_i, according to the geometric hypotheses, should satisfy the following algebraic relations:

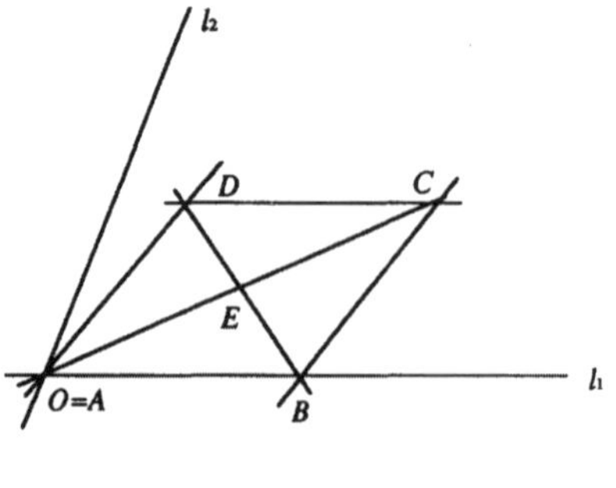

Fig. 3.3

$$BC \parallel AD \Longleftrightarrow (x_1 - u_1)(u_3 - 0) - (x_2 - 0)(u_2 - 0) = 0$$

$$\Longleftrightarrow u_3(x_1 - u_1) - u_2x_2 = 0, \tag{1}$$

$$CD \parallel AB \Longleftrightarrow (x_1 - u_2)(0 - 0) - (x_2 - u_3)(u_1 - 0) = 0$$

$$\Longleftrightarrow u_1(x_2 - u_3) = 0, \tag{2}$$

$$E \text{ lies on } AC \Longleftrightarrow \begin{vmatrix} x_3 & x_4 & 1 \\ x_1 & x_2 & 1 \\ 0 & 0 & 1 \end{vmatrix} = 0 \Longleftrightarrow x_2x_3 - x_1x_4 = 0, \tag{3}$$

$$E \text{ lies on } BD \Longleftrightarrow \begin{vmatrix} x_3 & x_4 & 1 \\ u_2 & u_3 & 1 \\ u_1 & 0 & 1 \end{vmatrix} = 0$$

$$\Longleftrightarrow u_3x_3 + (u_1 - u_2)x_4 - u_1u_3 = 0. \tag{4}$$

The expressions (1)–(4) above constitute the hypothesis of this theorem. The conclusion to be proved consists of

$$E \text{ is the midpoint of } AC \Longleftrightarrow \begin{cases} 2x_3 - x_1 = 0, & (5) \\ 2x_4 - x_2 = 0, & (6) \end{cases}$$

$$E \text{ is the midpoint of } BD \Longleftrightarrow \begin{cases} 2x_3 - (u_1 + u_2) = 0, & (7) \\ 2x_4 - u_3 = 0. & (8) \end{cases}$$

The problem of proving this theorem amounts to a deduction of the conclusion (5)–(8) from the hypothesis (1)–(4). The mechanized proof consists in designing a mechanical method which decides whether (5)–(8) can be deduced from (1)–(4) in a finite number of steps, and finds all possible degeneracy conditions that are sufficient to make the theorem meaningless or false during the deduction process.

Such a mechanical method proceeds by first fixing an order of the variables, such as x_1, x_2, x_3, x_4, then making some necessary elimination for (1)–(4) in order to find each x_i successively (according to the above order) and determining meanwhile the non-degeneracy conditions which appear in the process. Finally, substitute the x's into (5)–(8) in order to verify whether the equations are satisfied.

In detail, we first find x_1, x_2 from (1)–(2) and obtain

$$x_1 = u_1 + u_2, \tag{9}$$

$$x_2 = u_3. \tag{10}$$

The non-degeneracy condition to be observed is

$$u_1u_3 \neq 0. \tag{11}$$

Substituting the obtained x_1 and x_2 into (3), we have

$$u_3x_3 - (u_1 + u_2)x_4 = 0. \tag{12}$$

From (12) and (4) we find x_3 and x_4:

$$x_3 = \frac{1}{2}(u_1 + u_2), \tag{13}$$

$$x_4 = \frac{1}{2}u_3. \tag{14}$$

The corresponding non-degeneracy condition is still (11), or written as

$$u_1 \neq 0, \tag{15}$$

$$u_3 \neq 0. \tag{16}$$

The geometric meanings of the non-degeneracy conditions can be easily explained:

$$u_1 \neq 0 \Longleftrightarrow B \text{ and } A \text{ do not coincide},$$

$$u_3 \neq 0 \Longleftrightarrow D \text{ does not lie on } AB.$$

Denote the expressions on the left-hand side of the conclusion equations (5)–(8) by $g_1, \ldots, g_4$ respectively. Substituting all the x's found from (9), (10), (13) and (14) into each g_i and calculating its value directly, we can verify that $g_i = 0$. In other words, under the non-degeneracy conditions (15) and (16), the theorem holds true. The calculational procedure above yields a proof of the theorem.

Thus, we arrive at the following conclusion:

So long as the vertices B and A of $\square ABCD$ do not coincide and the vertex D does not lie on line AB, the diagonals AC and BD bisect one another.

If necessary, we may check either of the two degeneracy cases individually, or consider each case as a new theorem and prove its truth by the same procedure as above.[1] We do not explain this in more detail.

Though this example is rather simple, it is very typical. The mechanical procedure used is applicable to theorems that are much more complex than this one, but the number of calculations will be greatly increased. Therefore, we have to use modern equipment such as an electronic computer. In the following we shall give a few more complex examples to explain this point.

---

1 For example, Pascal's theorem about the inscribed hexagon of a conic in projective geometry, in the cases that some vertices of the hexagon *degenerate* to coincide, may be proved as other theorems.

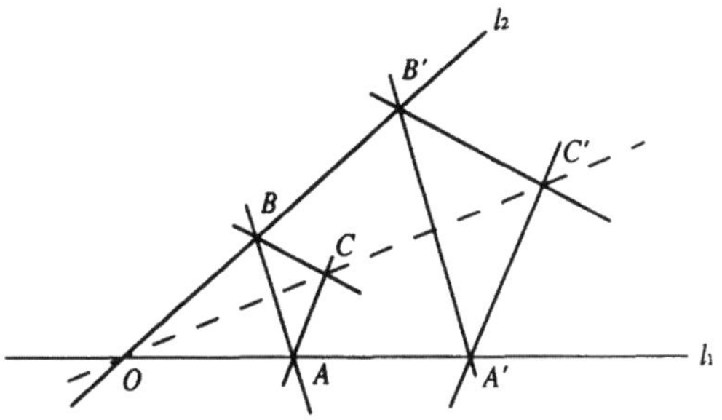

Fig. 3.4

Example 2 (Desargues' theorem). Suppose the three pairs of the corresponding sides of two generic triangles ABC and $A'B'C'$ respectively are parallel to each other and both AA' and BB' pass through a point O. Then CC' also passes through point O (cf. Fig. 3.4).

We want to prove this theorem and determine the subsidiary conditions about genericity. The theorem has already been introduced as an axiom to set up the Desarguesian number system and coordinate system. Here we prove it. This is only to illustrate the mechanical method and is independent of the problem of cyclical proof.

Let us assume that the lines AA' and BB' do not coincide. To simplify the calculation, we take these two lines as the axes l_1 and l_2 and O as the origin. Let the coordinates of points be

$$\begin{aligned} A &= (u_1, 0), & A' &= (u_2, 0), \\ B &= (0, u_3), & C &= (u_4, u_5), \\ B' &= (0, x_1), & C' &= (x_2, x_3), \end{aligned}$$

where $u_1, \ldots, u_5$ are considered as parameters and x_1, x_2, x_3 are constrained by the geometric conditions.

The hypothesis of this theorem is expressed by the following algebraic relations with the corresponding geometric conditions indicated:

$$A'B' \parallel AB \Longleftrightarrow u_1x_1 - u_2u_3 = 0, \tag{17}$$

$$A'C' \parallel AC \Longleftrightarrow (u_4 - u_1)x_3 - u_5(x_2 - u_2) = 0, \tag{18}$$

$$B'C' \parallel BC \Longleftrightarrow u_4(x_3 - x_1) - (u_5 - u_3)x_2 = 0. \tag{19}$$

The conclusion of the theorem is given by

$$O \text{ is on } CC' \Longleftrightarrow u_4x_3 - u_5x_2 = 0. \tag{20}$$

From (17), we obtain

$$x_1 = \frac{u_2u_3}{u_1}, \tag{21}$$

with the non-degeneracy condition

$$u_1 \neq 0 \tag{22}$$

introduced. Substituting x_1 in (21) into (19) and solving (18) and (19) for x_2 and x_3, we have

$$x_2 = \frac{u_2 u_4}{u_1}, \tag{23}$$

$$x_3 = \frac{u_2 u_5}{u_1}. \tag{24}$$

The non-degeneracy condition is that the coefficient determinant of (18) and (19) does not vanish, i.e.,

$$u_1 u_3 - u_1 u_5 - u_3 u_4 \neq 0. \tag{25}$$

Denote the left-hand side of the conclusion equation (20) by g. Substituting all the x's in (21), (23) and (24) into g and simplifying it, we have $g = 0$. Therefore, the theorem is true when the non-degeneracy conditions (22) and (25) are satisfied.

The geometric meanings of the non-degeneracy conditions are

$$u_1 \neq 0 \Longleftrightarrow A \text{ and } O \text{ do not coincide},$$

$$u_1 u_3 - u_1 u_5 - u_3 u_4 \neq 0 \Longleftrightarrow A, B, C \text{ do not lie on the same line}.$$

Hence we arrive at the following conclusion:

Under the generic or non-degeneracy conditions that AA' and BB' do not coincide and intersect at a point O, A does not coincide with O and A, B, C are not collinear, the theorem holds true.

Whether or not the theorem holds in the two degeneracy cases may be further verified if necessary. For instance, in the degeneracy case that A, B, C are collinear, i.e.,

$$u_1 u_3 - u_1 u_5 - u_3 u_4 = 0,$$

u_4 and u_5 can no longer be parameters at the same time. In this case, let us change, without loss of generality, u_5 to x_0. Then the geometric hypotheses are

$$\begin{aligned}
&- u_1 x_0 - u_3 u_4 + u_1 u_3 = 0,\\
&u_1 x_1 - u_2 u_3 = 0,\\
&(u_4 - u_1)x_3 - x_0(x_2 - u_2) = 0,\\
&u_4(x_3 - x_1) - (x_0 - u_3)x_2 = 0.
\end{aligned}$$

From them, one obtains

$$x_0 = \frac{u_3(u_1 - u_4)}{u_1},$$

$$x_1 = \frac{u_2 u_3}{u_1},$$

$$x_2 = \text{arbitrary},$$

$$x_3 = -\frac{u_3}{u_1}(x_2 - u_2),$$

with the non-degeneracy conditions

$$u_1 \neq 0 \Longleftrightarrow A \text{ does not coincide with } O,$$

$$u_1 - u_4 \neq 0 \Longleftrightarrow AC \nparallel BB'$$

introduced.

Substituting all the x's into the expression to be proved, we verify that

$$\begin{aligned} g &= u_4 x_3 - x_0 x_2 \\ &= -u_4 \cdot \frac{u_3}{u_1}(x_2 - u_2) - \frac{u_3(u_1 - u_4)}{u_1} \cdot x_2 \\ &= -\frac{u_3}{u_1} \cdot (u_1 x_2 - u_2 u_4) \not\equiv 0. \end{aligned}$$

Therefore, in this case the theorem is no longer true unless other degeneracy conditions, e.g., $u_1 = 0$ or $u_1 - u_4 = 0$, are added. This may be in contrast with Example 3 in Sect. 3.1.

Example 3 (Desargues' theorem). Suppose the three lines AA', BB', CC' connecting the corresponding vertices of two generic triangles ABC and $A'B'C'$ are concurrent at O and two pairs of the corresponding sides respectively are parallel to each other, e.g.,

$$AB \parallel A'B', \quad AC \parallel A'C'.$$

Then the third pair of the corresponding sides are also parallel to each other: $BC \parallel B'C'$.

As in Example 2, we assume that the two lines AA' and BB' do not coincide, and are taken as the coordinate axes with O as the origin. Let the coordinates of points be also as in Example 2. Then the hypothesis consists of

$$A'B' \parallel AB \Longleftrightarrow u_1 x_1 - u_2 u_3 = 0,$$

$$A'C' \parallel AC \Longleftrightarrow (u_4 - u_1)x_3 - u_5(x_2 - u_2) = 0,$$

$$CC' \text{ passes through } O \Longleftrightarrow u_4 x_3 - u_5 x_2 = 0,$$

and the conclusion is

$$B'C' \parallel BC \Longleftrightarrow g \equiv u_4(x_3 - x_1) - (u_5 - u_3)x_2 = 0.$$

Solving the hypothesis equations for the x's, we get

$$x_1 = \frac{u_2u_3}{u_1}, \quad x_2 = \frac{u_2u_4}{u_1}, \quad x_3 = \frac{u_2u_5}{u_1},$$

where the non-degeneracy conditions are

$$u_1 \neq 0 \Longleftrightarrow A \text{ does not coincide with } O,$$

$$u_5 \neq 0 \Longleftrightarrow C \text{ does not lie on } AA'.$$

Substituting all the x_i into g, we have

$$g = u_4 \cdot \left(\frac{u_2u_5}{u_1} - \frac{u_2u_3}{u_1}\right) - (u_5 - u_3) \cdot \frac{u_2u_4}{u_1} = 0.$$

Hence the theorem in question is true under the above non-degeneracy conditions.

Upon verifying further, one finds that in the case $u_5 = 0$, i.e., C lies on AA', the theorem is no longer true if no other degeneracy conditions are added. This may be in contrast with Example 2 in Sect. 3.1.

From the above three examples, we see that the mechanical method used can cause the various degenerate cases in which the truth of the theorem may be affected to naturally appear one by one, and the degenerate cases can be treated individually in the same way as necessary. From the three examples in Sect. 3.1, we see on the contrary that the discovery and treatment of degenerate cases using the usual Euclidean method are practically blindfold and also almost powerless.

To explain the mechanical method of this section further, we shall give another example which is less trivial.

The linear Pascalian axiom, which is very important for the algebraization of geometry, is a degenerate case of Pascal's theorem in projective geometry. Pascal's original *theorem* says that a necessary and sufficient condition for the six vertices of a hexagon to lie on a conic is that the intersection points of the three pairs of opposite sides of the hexagon are collinear. In this case, the hexagon (with a fixed order of vertices) is called a Pascalian *hexagon* and the line on which the intersection points of the three pairs of opposite sides lie is called a Pascalian *line* of the hexagon. As usual, the statement of this theorem implies some assumptions about genericity. The degenerate cases are rather complicated. In addition, the statement of the theorem is related to conic and is beyond the scope of Hilbert's mechanical method for PIP theorems. However, we can easily transform the theorem into a PIP theorem in the following way.

Suppose $A_1A_2A_3A_4A_5A_6$ is a generic Pascalian hexagon with the intersection points of the three pairs of opposite sides

$$A_1A_2 \wedge A_4A_5, \quad A_2A_3 \wedge A_5A_6, \quad A_3A_4 \wedge A_6A_1$$

collinear (where $\wedge$ stands for intersection). According to Pascal's original theorem, the six points $A_1, \ldots, A_6$ should lie on a common conic. Therefore, if we make an arbitrary permutation for these six points, by Pascal's original theorem the obtained hexagon $A_{i_1}A_{i_2}\ldots A_{i_6}$ is also a Pascalian hexagon, i.e., the points of intersection

$$A_{i_1}A_{i_2} \wedge A_{i_4}A_{i_5}, \quad A_{i_2}A_{i_3} \wedge A_{i_5}A_{i_6}, \quad A_{i_3}A_{i_4} \wedge A_{i_6}A_{i_1}$$

are collinear. In this way, we can avoid involving conics and get different PIP theorems in terms of different permutations. We pick out one of them for illustration.

Example 4 (Theorem). If $A_1A_2A_3A_4A_5A_6$ is a generic Pascalian hexagon, then $A_1A_4A_3A_2A_5A_6$ is also a Pascalian hexagon. In other words, for six points $A_1, A_2, \ldots, A_6$, if

$$P = A_1A_2 \wedge A_4A_5, \quad Q = A_2A_3 \wedge A_5A_6, \quad R = A_3A_4 \wedge A_6A_1$$

are collinear in the generic case, then the points of intersection

$$P' = A_1A_4 \wedge A_2A_5, \quad Q' = A_3A_4 \wedge A_5A_6, \quad R' = A_2A_3 \wedge A_6A_1$$

are also collinear.

For the sake of simplicity, we take A_6 as the origin O and the lines A_6A_1 and A_6A_5, which are supposed to be distinct, as the two coordinate axes l_1 and l_2. The coordinates of points are chosen as

$$\begin{aligned}
&A_1 = (u_1, 0), \quad A_2 = (u_2, u_3),\\
&A_3 = (u_4, u_5), \quad A_4 = (u_6, u_7),\\
&R = A_3A_4 \wedge A_6A_1 = (x_1, 0),\\
&Q = A_2A_3 \wedge A_5A_6 = (0, x_2),\\
&Q' = A_3A_4 \wedge A_5A_6 = (0, x_3),\\
&R' = A_2A_3 \wedge A_6A_1 = (x_4, 0),\\
&P = A_1A_2 \wedge A_4A_5 = (x_5, x_6),\\
&A_5 = (0, x_7),\\
&P' = A_1A_4 \wedge A_2A_5 = (x_8, x_9).
\end{aligned}$$

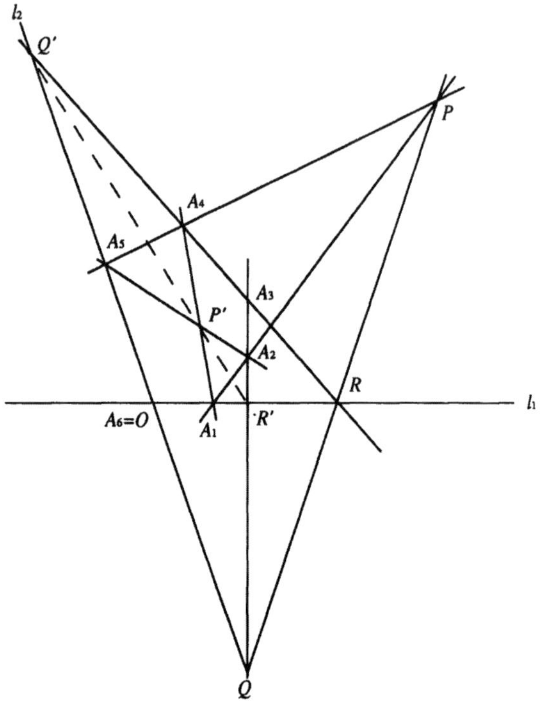

Fig. 3.5

The geometric conditions in the hypothesis are

$$R \text{ lies on } A_3A_4 \Longleftrightarrow (u_5-u_7)x_1+u_4u_7-u_5u_6=0,$$
$$Q \text{ lies on } A_2A_3 \Longleftrightarrow (u_2-u_4)x_2-u_2u_5+u_3u_4=0,$$
$$Q' \text{ lies on } A_3A_4 \Longleftrightarrow (u_4-u_6)x_3-u_4u_7+u_5u_6=0,$$
$$R' \text{ lies on } A_2A_3 \Longleftrightarrow (u_3-u_5)x_4+u_2u_5-u_3u_4=0,$$
$$P \text{ lies on } A_1A_2 \Longleftrightarrow u_3u_5+(u_1-u_2)x_6-u_1u_3=0,$$
$$P, Q, R \text{ are collinear} \Longleftrightarrow x_2x_5+x_1x_6-x_1x_2=0,$$
$$P \text{ lies on } A_4A_5 \Longleftrightarrow (u_6-x_5)x_7+u_7x_5-u_6x_6=0,$$
$$P' \text{ lies on } A_1A_4 \Longleftrightarrow u_7x_8+(u_1-u_6)x_9-u_1u_7=0,$$
$$P' \text{ lies on } A_2A_5 \Longleftrightarrow (u_3-x_7)x_8-u_2x_9+u_2x_7=0.$$

The geometric relation to be proved is

$$P', Q', R' \text{ are collinear} \Longleftrightarrow g \equiv x_3x_8+x_4x_9-x_3x_4=0.$$

Solving the hypothesis equations for all the x's, we obtain

$$x_1 = \frac{u_5u_6 - u_4u_7}{u_5 - u_7} = R_1(u),$$

$$x_2 = \frac{u_2u_5 - u_3u_4}{u_2 - u_4} = R_2(u),$$

$$x_3 = \frac{u_4u_7 - u_5u_6}{u_4 - u_6} = R_3(u),$$

$$x_4 = \frac{u_3u_4 - u_2u_5}{u_3 - u_5} = R_4(u),$$

$$x_5 = \frac{u_1u_3 - (u_1 - u_2)x_2}{u_3x_1 - (u_1 - u_2)x_2} \cdot x_1 = R_5(u),$$

$$x_6 = \frac{u_3x_2(x_1 - u_1)}{u_3x_1 - (u_1 - u_2)x_2} = R_6(u),$$

$$x_7 = \frac{u_7x_5 - u_6x_6}{x_5 - u_6} = R_7(u),$$

$$x_8 = \frac{u_2[u_1u_7 + (u_6 - u_1)x_7]}{(u_1 - u_6)(u_3 - x_7) + u_2u_7} = R_8(u),$$

$$x_9 = \frac{u_1u_3u_7 + (u_2 - u_1)u_7x_7}{(u_1 - u_6)(u_3 - x_7) + u_2u_7} = R_9(u).$$

The non-degeneracy conditions introduced are

$$u_5 - u_7 \neq 0 \Longleftrightarrow A_3A_4 \text{ is not parallel to } l_1,$$

$$u_2 - u_4 \neq 0 \Longleftrightarrow A_2A_3 \text{ is not parallel to } l_2,$$

$$u_4 - u_6 \neq 0 \Longleftrightarrow A_3A_4 \text{ is not parallel to } l_2,$$

$$u_3 - u_5 \neq 0 \Longleftrightarrow A_2A_3 \text{ is not parallel to } l_1,$$

$$\alpha \equiv u_3x_1 - (u_1 - u_2)x_2 \neq 0 \Longleftrightarrow QR \text{ is not parallel to } A_1A_2,$$

$$x_5 - u_6 \neq 0 \Longleftrightarrow PA_4 \text{ is not parallel to } l_2,$$

$$\beta \equiv (u_1 - u_6)(u_3 - x_7) + u_2u_7 \neq 0 \Longleftrightarrow A_1A_4 \text{ is not parallel to } A_2A_5.$$

There are simple geometric interpretations for some of the non-degeneracy conditions. Consider, for instance, the third non-degeneracy condition. Since A_3A_4 and l_2 contain a common point Q', the condition that A_3A_4 is not parallel to l_2 means that they do not coincide, or A_3, A_4, A_5, A_6 are not collinear. Clearly, degeneracy cases of this kind are all short of geometric interest and not worth being further considered.

In case the non-degeneracy conditions are satisfied, we substitute all the $x_i = R_i(u)$ into the conclusion expression g. By tedious yet direct calculation, we may verify that

$$g = x_3x_8 + x_4x_9 - x_3x_4 = R_3R_8 + R_4R_9 - R_3R_4 = 0,$$

and thus the theorem holds true.

Example 5. The theorem is the same as the preceding one.

If we take A_6 as the origin O, A_6A_1 and A_5A_6 as the two coordinates axes l_1 and l_2 as in Example 4, but the coordinates of points as

$$\begin{aligned}
&A_1 = (u_1, 0), && A_2 = (u_2, u_3),\\
&A_3 = (u_4, u_5), && A_5 = (0, u_6),\\
&A_4 = (u_7, x_1), && Q = (0, x_2),\\
&R = (x_3, 0), && R' = (x_4, 0),\\
&P = (x_5, x_6), && Q' = (0, x_7),\\
&P' = (x_8, x_9), &&
\end{aligned}$$

then the hypothesis relations become

$$\begin{aligned}
A_4, A_5, P \text{ are collinear} &\Longleftrightarrow (x_1 - u_6)x_5 - u_7(x_6 - u_6) = 0,\\
P, Q, R \text{ are collinear} &\Longleftrightarrow x_2x_5 + x_3x_6 - x_2x_3 = 0,\\
A_3, A_4, R \text{ are collinear} &\Longleftrightarrow (x_1 - u_5)x_3 - u_4x_1 + u_5u_7 = 0,\\
A_1, A_2, P \text{ are collinear} &\Longleftrightarrow u_3x_5 + (u_1 - u_2)x_6 - u_1u_3 = 0,\\
A_2, A_3, Q \text{ are collinear} &\Longleftrightarrow (u_2 - u_4)x_2 - u_2u_5 + u_3u_4 = 0,
\end{aligned}$$

and some others. If we proceed with elimination so that the x's are successively introduced, we shall get an expression of the form

$$Ax_1^2 + Bx_1 + C = 0$$

and some others (under certain non-degeneracy conditions), where A, B, C are polynomials in $u_1, \ldots, u_7$. Under the non-degeneracy condition $A \neq 0$, one obtains an equation of degree 2 in x_1. Therefore, we are unable to prove this same theorem by using the previous method.

The difference between this example and the last example is that in the previous example the theorem was transformed to a PIP theorem of constructive type, where in this example it was not. In other words, the points of the geometric configuration in Example 4 can be introduced constructively step by step in the following order:

Take first an arbitrary point A_6 as the origin O.
Take then two arbitrary lines l_1, l_2 through O as the coordinate axes.
Take an arbitrary point $A_1 = (u_1, 0)$ on l_1.
Construct through A_1 an arbitrary line and take thereon an arbitrary point $A_2 = (u_2, u_3)$.
Construct through A_2 an arbitrary line and take thereon an arbitrary point $A_3 = (u_4, u_5)$.
Construct through A_3 an arbitrary line and take thereon an arbitrary point $A_4 = (u_6, u_7)$.
Let the line A_3A_4 meet l_1 at the point $R = (x_1, 0)$.
Let the line A_2A_3 meet l_2 at the point $Q = (0, x_2)$.
Let the line A_3A_4 meet l_2 at the point $Q' = (0, x_3)$.
Let the line A_2A_3 meet l_1 at the point $R' = (x_4, 0)$.
Extend the line QR to meet A_1A_2 at the point $P = (x_5, x_6)$.
Extend the line PA_4 to meet l_2 at the point $A_5 = (0, x_7)$.
Extend the lines A_1A_4 and A_2A_5 to meet at the point $P' = (x_8, x_9)$.
The conclusion is: P', Q', R' are collinear.

Obviously, such a statement is equivalent to the original theorem, but has become one of a constructive type.

3.5 Proof of Hilbert's mechanization theorem

In this section we give a proof of Hilbert's mechanization theorem stated in Sect. 3.3. The contents of this mechanization theorem are directed to a certain class of PIP theorems in Pascalian geometry. The number field associated with Pascalian geometry will be denoted by N. The proof of the mechanization theorem consists in designing a mechanical method for proving this class of PIP theorems. As indicated in Example 5 of Sect. 3.4, the PIP theorems will be restricted to those which have been transformed into a *constructive type*. The feature of this class of PIP theorems is that the points and lines involved in the statement of each theorem are introduced or constructed one by one according to some definite steps. Moreover, at each step the construction proceeds by taking an arbitrary point, constructing an arbitrary line, letting two lines meet at a point or constructing a parallel.

Below we shall present the mechanical procedure for proving the class of PIP theorems of constructive type.

First, take an arbitrary coordinate system. Then the points and lines involved in a theorem may all be represented by pairs of numbers and linear equations in terms of the coordinates of points. The numbers in the pairs and the coefficients of equations are all taken from the number field N associated with the geometry and will not be further explained.

We shall avoid using equations of lines and represent lines by pairs of points. Moreover, among the points and lines, some are arbitrarily chosen and the others are constructed in a certain manner. Therefore, in the coordinate representation of points, some coordinates are arbitrary and are denoted by u_i, whereby u_i are considered as parameters, and the others are determined in a certain manner according to the prescribed geometric conditions and denoted by x_j. When no distinction between these two kinds of coordinates is necessary, they will be denoted by α_k.

In the considered PIP theorems, the following constructions are involved.

1. A point is arbitrarily given or chosen.

This point will be represented as (u_i, u_j).

2. A line is arbitrarily given or chosen.

We may arbitrarily choose two points (u_i, u_j) and (u_k, u_l) with the line determined by these two points.

3. Draw an arbitrary line through a point (α_i, α_j) already constructed.

We may take instead an arbitrary point (u_k, u_l) and represent the line by the points (α_i, α_j) and (u_k, u_l).

4. Draw the connecting line of two points already constructed.

As the line has been determined by two points thereon, this construction needs no further consideration.

5. Choose an arbitrary point on a line already drawn.

If the line is determined by two points (α_i, α_j) and (α_k, α_l) already constructed, then the arbitrary point chosen thereon may be represented either as (u_r, x_s) or as (x_s, u_r) satisfying the equation

$$\begin{vmatrix} u_r & x_s & 1 \\ \alpha_i & \alpha_j & 1 \\ \alpha_k & \alpha_l & 1 \end{vmatrix} = 0 \quad \text{or} \quad \begin{vmatrix} x_s & u_r & 1 \\ \alpha_i & \alpha_j & 1 \\ \alpha_k & \alpha_l & 1 \end{vmatrix} = 0.$$

6. Construct an arbitrary parallel to a line already drawn.

As the line is determined as before by two points thereon already constructed, say (α_i, α_j) and (α_k, α_l), the parallel to be constructed will be determined instead as follows. Take first an arbitrary point (u_m, u_n), and then a point (u_r, x_s) or (x_s, u_r) such that the connecting line of this point to the point (u_m, u_n) is parallel to the line determined by (α_i, α_j) and (α_k, α_l), so that x_s satisfies the equation

$$(\alpha_k - \alpha_i)(x_s - u_n) - (\alpha_l - \alpha_j)(u_r - u_m) = 0$$

or

$$(\alpha_l - \alpha_j)(x_s - u_m) - (\alpha_k - \alpha_i)(u_r - u_n) = 0.$$

7. Draw a line through a point (α_m, α_n) already constructed parallel to a line already constructed.

Let the line already constructed be determined by two points (α_i, α_j) and (α_k, α_l), then the line to be drawn will be taken to be one determined instead by (α_m, α_n) and another point (u_r, x_s) or (x_s, u_r) satisfying the equation

$$(\alpha_k - \alpha_i)(x_s - \alpha_n) - (\alpha_l - \alpha_j)(u_r - \alpha_m) = 0$$

or

$$(\alpha_l - \alpha_j)(x_s - \alpha_m) - (\alpha_k - \alpha_i)(u_r - \alpha_n) = 0.$$

8. Construct the intersection point of two intersecting lines already constructed.

Let the two lines be determined respectively by two pairs of points (α_i, α_j), (α_k, α_l) and (α_p, α_q), (α_r, α_s) already constructed. Then the point of intersection may be taken as (x_g, x_h), satisfying the system of equations

$$(\alpha_j - \alpha_l)x_g' - (\alpha_i - \alpha_k)x_h + \alpha_i\alpha_l - \alpha_j\alpha_k = 0,$$

$$(\alpha_q - \alpha_s)x_g - (\alpha_p - \alpha_r)x_h + \alpha_p\alpha_s - \alpha_q\alpha_r = 0.$$

Denote the coefficient determinant of this system of equations by

$$D = \begin{vmatrix} \alpha_j - \alpha_l & \alpha_k - \alpha_i \\ \alpha_q - \alpha_s & \alpha_r - \alpha_p \end{vmatrix}.$$

Now introduce the non-degeneracy condition

$$D \neq 0.$$

Then under this condition, the original system is equivalent to the following system of two equations

$$Dx_g + (\alpha_r - \alpha_p)(\alpha_i\alpha_l - \alpha_j\alpha_k) - (\alpha_k - \alpha_i)(\alpha_p\alpha_s - \alpha_q\alpha_r) = 0,$$

$$-Dx_h + (\alpha_q - \alpha_s)(\alpha_i\alpha_l - \alpha_j\alpha_k) - (\alpha_j - \alpha_l)(\alpha_p\alpha_s - \alpha_q\alpha_r) = 0.$$

In case $D = 0$, either the original system of equations is contradictory and thus the theorem is meaningless, or one of x_g, x_h (or both of them) may take arbitrary value and may thus be considered as a parameter u, while the other x is determined by one of the original two equations.

9. Construct the intersection point of a line already constructed and a line through a point (α_m, α_n) already constructed and parallel to a second line already constructed.

Let the two lines already constructed be determined respectively by two pairs of points (α_i, α_j), (α_k, α_l) and (α_p, α_q), (α_r, α_s). Represent the point to be constructed by (x_g, x_h). Then x_g, x_h satisfy the system of equations

$$(\alpha_k - \alpha_i)(x_h - \alpha_n) - (\alpha_l - \alpha_j)(x_g - \alpha_m) = 0,$$

$$(\alpha_q - \alpha_s)x_g - (\alpha_p - \alpha_r)x_h + \alpha_p\alpha_s - \alpha_q\alpha_r = 0.$$

As before, introduce the non-degeneracy condition

$$D \neq 0.$$

Then the system of equations is equivalent to a system of two equations of the form

$$Dx_g + \cdots = 0,$$

$$Dx_h + \cdots = 0.$$

In case $D = 0$, either the original equations are contradictory and thus the theorem is meaningless, or one of x_g, x_h (or both of them) may be taken as a parameter while the other x is determined by one of the original two equations.

10. Construct the intersection point of two lines passing through each of the two points already constructed and parallel respectively to each of the two lines already constructed.

This is similar to 8 and 9.

Since we only consider PIP theorems of constructive type, the points and lines in a theorem will occur one after another in a definite order of succession. It follows that the coordinates of points involved (with lines replaced by two points thereon, see above) can be arranged in a definite order in accordance with their ordering of appearance in the construction as follows:

$$u_1 \prec u_2 \prec \cdots \prec u_m,$$

$$x_1 \prec x_2 \prec \cdots \prec x_n.$$

In particular, whenever a new point is introduced by the construction 8, 9, or 10, its coordinates attributed will be two x's in succession, say (x_g, x_{g+1}).

According to 1–10, we may successively introduce the equations for all x_i and the associated non-degeneracy conditions as follows.

Consider the first step of construction in the statement of a theorem and assume that some parameters u_i have already been introduced. We may have three cases according as the construction is one of 1–4, or of 5–7, or of 8–10.

If the construction is one of 1–4, some new u_i should be introduced.

If the construction is one of 5–7, then an equation of the following form will be obtained:

$$A_1 x_1 + B_1 = 0,$$

where A_1 and B_1 are both polynomials in the variables $u_1, \ldots, u_m$, the parameters already introduced, with coefficients in N. By the commonly used denotation, that is

$$A_1, B_1 \in N[u_1, \ldots, u_m].$$

When A_1 is identically equal to 0 but B_1 is not, the non-degeneracy condition

$$B_1 \neq 0$$

is introduced. In this case, the procedure terminates and reports that the hypothesis of the theorem is self-contradictory under the above non-degeneracy condition. In case both A_1 and B_1 are identically equal to 0, we introduce a new parameter u_{m+1} and set

$$x_1 = u_{m+1}.$$

In case A_1 is not identically equal to 0, we introduce a non-degeneracy condition

$$A_1 \neq 0,$$

and solve the equation for x_1:

$$x_1 = -\frac{B_1}{A_1}.$$

If the construction is one of 8–10, then the following two equations will be obtained:

$$A_{11}x_1 + A_{12}x_2 + B_1 = 0,$$
$$A_{21}x_1 + A_{22}x_2 + B_2 = 0,$$

where all $A_{11}, A_{12}, A_{21}, A_{22}, B_1, B_2 \in N[u_1, \ldots, u_m]$.

Consider the coefficient determinants

$$P = \begin{vmatrix} A_{11} & A_{12} \\ A_{21} & A_{22} \end{vmatrix}, \quad Q_1 = \begin{vmatrix} A_{12} & B_1 \\ A_{22} & B_2 \end{vmatrix}, \quad Q_2 = \begin{vmatrix} B_1 & A_{11} \\ B_2 & A_{21} \end{vmatrix}.$$

In case P is not identically equal to 0, i.e.,

$$\text{rank of } \begin{bmatrix} A_{11} & A_{12} \\ A_{21} & A_{22} \end{bmatrix} = 2,$$

we introduce a non-degeneracy condition

$$P \neq 0,$$

and solve the equations for x_1, x_2:

$$x_1 = -\frac{Q_1}{P}, \quad x_2 = -\frac{Q_2}{P}.$$

When P is identically equal to 0 but not all of $A_{11}, \ldots, A_{22}$ are identically 0, i.e.,

$$\text{rank of } \begin{bmatrix} A_{11} & A_{12} \\ A_{21} & A_{22} \end{bmatrix} = 1,$$

there are two polynomials $a_1, a_2 \in N[u_1, \ldots, u_m]$, not all 0, such that

$$a_1 A_{11} + a_2 A_{21} \equiv 0,$$
$$a_1 A_{12} + a_2 A_{22} \equiv 0.$$

In this case, from the original system of equations we obtain

$$a_1 B_1 + a_2 B_2 = 0.$$

Now, if $a_1B_1 + a_2B_2$ is not identically equal to 0, then we introduce a non-degeneracy condition

$$a_1B_1 + a_2B_2 \neq 0,$$

and report that the hypothesis of the theorem is self-contradictory under this non-degeneracy condition and the procedure terminates.

When P and $a_1B_1 + a_2B_2$ are both identically equal to 0 but not all of $A_{11}, \ldots, A_{22}$ are identically 0, for instance, A_{11} is not identically equal to 0, we introduce a new parameter u_{m+1} and a non-degeneracy condition

$$A_{11} \neq 0$$

and solve the equations for x_1, x_2:

$$x_1 = -\frac{A_{12}u_{m+1} + B_1}{A_{11}}, \quad x_2 = u_{m+1}.$$

When $A_{11}, \ldots, A_{22}$ are all identically equal to 0 but not both of B_1 and B_2 are identically 0, we introduce the non-degeneracy condition(s)

$$B_1 \neq 0 \quad \text{and/or} \quad B_2 \neq 0,$$

and report that the hypothesis of the theorem is contradictory under the above non-degeneracy condition(s) and the procedure terminates.

Finally, if $A_{11}, \ldots, A_{22}, B_1, B_2$ are all identically equal to 0, we introduce two new parameters u_{m+1}, u_{m+2} and set

$$x_1 = u_{m+1}, \quad x_2 = u_{m+2}.$$

If no contradiction for the hypothesis has occurred (i.e., the procedure has not already terminated), proceed to the second step of the construction according to the statement of the theorem.

Now suppose that the constructions according to the statement of the theorem have proceeded up to a certain step, but not the last, meanwhile some new parameters

$$u_{m+1}, \ldots, u_{m+s}$$

have been introduced, some non-degeneracy conditions

$$D_1 \neq 0, \ldots, D_j \neq 0$$

have been obtained and $x_1, \ldots, x_i$ have been found:

$$x_1 = Q_1/P_1, \ldots, x_i = Q_i/P_i,$$

in which

$$D_1, \ldots, D_j, P_1, \ldots, P_i, Q_1, \ldots, Q_i \in N[u_1, \ldots, u_m, u_{m+1}, \ldots, u_{m+s}]$$

and $P_1, \ldots, P_i$ are the *power products* (i.e., products of non-negative powers) of $D_1, \ldots, D_j$.

Suppose no contradiction for the hypothesis did occur (and thus the procedure did not terminate yet). Then proceed to the next step of construction. According as the construction is one of 1–4, or of 5–7, or of 8–10, either some parameters are introduced, or an equation of the form

$$A_{i+1}x_{i+1} + B_{i+1} = 0,$$

is obtained, or two equations of the form

$$A_{i+1,1}x_{i+1} + A_{i+1,2}x_{i+2} + B_{i+1} = 0,$$

$$A_{i+2,1}x_{i+1} + A_{i+2,2}x_{i+2} + B_{i+2} = 0,$$

are obtained, where the A, B's are all polynomials in the variables $u_1, \ldots, u_{m+s}$, $x_1, \ldots, x_i$ with coefficients in N.

Now substituting all the rational expressions of $x_1, \ldots, x_i$ previously found into these equations (where $P_1 \neq 0, \ldots, P_i \neq 0$ in the non-degeneracy cases $D_1 \neq 0, \ldots, D_j \neq 0$), we obtain, after the reduction of fractions to a common denominator, an equation

$$A^*_{i+1}x_{i+1} + B^*_{i+1} = 0$$

or a system of two equations

$$A^*_{i+1,1}x_{i+1} + A^*_{i+1,2}x_{i+2} + B^*_{i+1} = 0,$$

$$A^*_{i+2,1}x_{i+1} + A^*_{i+2,2}x_{i+2} + B^*_{i+2} = 0,$$

in which the A^*, B^*'s are all polynomials in $N[u_1, \ldots, u_{m+s}]$.

In the same way as before, it is known that either under some newly introduced non-degeneracy conditions

$$D_{j+1} \neq 0 \quad (\text{or } D_{j+1} \neq 0,\ D_{j+2} \neq 0),$$

the hypothesis of the theorem is self-contradictory and the procedure terminates, or one can introduce some new parameter u_{m+s+1} (or parameters u_{m+s+1}, u_{m+s+2}) and some non-degeneracy conditions as above and solve the equation (or equations) for x_{i+1} (or x_{i+1}, x_{i+2}) to get

$$x_{i+1} = Q_{i+1}/P_{i+1} \quad (\text{or } x_{i+1} = Q_{i+1}/P_{i+1},\ x_{i+2} = Q_{i+2}/P_{i+2}),$$

where all D_{j+1} (or D_{j+1}, D_{j+2}) and P_{i+1}, Q_{i+1} (or $P_{i+1}, P_{i+2}, Q_{i+1}, Q_{i+2}$) are polynomials in $u_1, \ldots, u_{m+s}$ and the introduced new parameter u_{m+s+1} (or parameters u_{m+s+1}, u_{m+s+2}), and P_{i+1} (or P_{i+1}, P_{i+2}) is (are) some power product(s) of $D_1, \ldots, D_j$ and the possibly introduced new polynomial D_{j+1} (or polynomials D_{j+1}, D_{j+2}).

If we still have not already arrived at the last, continue to proceed to the next step of construction. In this way, one of the following two cases finally must occur.

Case 1. In addition to the parameters $u_1, \ldots, u_m$ which have been specified in the statement of the theorem, some new parameters $u_{m+1}, \ldots, u_{m+t}$ (each of which is a coordinate) are introduced:

$$x_{i_{m+1}} = u_{m+1}, \ldots, x_{i_{m+t}} = u_{m+t}.$$

Moreover, some non-degeneracy conditions

$$D_1 \neq 0, \ldots, D_r \neq 0$$

are obtained, where all the D's are polynomials in the new and old parameters $u_1, \ldots, u_m$ and $x_{i_{m+1}}, \ldots, x_{i_{m+t}}$. Under these non-degeneracy conditions, the hypothesis of the theorem is self-contradictory and the procedure terminates.

Case 2. Some new parameters $u_1, \ldots, u_{m+t}$ and some non-degeneracy conditions are introduced as above, and under these non-degeneracy conditions all the x's are found as

$$x_1 = \frac{Q_1}{P_1}, \ldots, x_n = \frac{Q_n}{P_n},$$

where the P, Q's are all polynomials in the new and old parameters $u_1, \ldots, u_{m+t}$ and all the P's are some power products of $D_1, \ldots, D_r$.

The conclusion of the PIP theorem can be expressed as some equations

$$S_k(u_1, \ldots, u_m, x_1, \ldots, x_n) = 0,$$

where

$$S_k \in N[u_1, \ldots, u_m, x_1, \ldots, x_n].$$

Suppose now we are in Case 2. Then under the non-degeneracy conditions

$$D_1 \neq 0, \ldots, D_r \neq 0,$$

we substitute the found

$$x_1 = \frac{Q_1}{P_1}, \ldots, x_n = \frac{Q_n}{P_n}$$

into S_k and obtain

$$S_k = S_k(u_1, \ldots, u_m, \frac{Q_1}{P_1}, \ldots, \frac{Q_n}{P_n}) = \frac{R'_k}{R''_k},$$

where both R'_k and R''_k are polynomials in the new and old parameters $u_1, \ldots, u_{m+t}$ and R''_k is a power product of the P's and thus of the D's and, therefore, is not equal to 0 under the non-degeneracy conditions. Evidently, we can determine whether or not R'_k is identically equal to 0 by very complicated yet rather easy mechanical computation and thus can determine whether or not the theorem is true under the assumption that the non-degeneracy conditions are satisfied.

From the above, we draw the following conclusion:

For any PIP theorem of constructive type, one can always choose coordinates $u_1, \ldots, u_m, x_1, \ldots, x_n$, pick up some coordinates out of x's as new parameters according to a certain mechanical procedure and get, after a finite number of steps, a set of polynomials $D_1, \ldots, D_r$ in the new and old coordinates and determine that one and only one of the following cases holds under the assumption that the non-degeneracy conditions

$$D_1 \neq 0, \ldots, D_r \neq 0$$

are satisfied:

(1) The hypothesis of the theorem is self-contradictory;
(2) the theorem is true;
(3) the theorem is false.

In case the theorem is true, the above mechanical steps constitute a mechanized proof of the considered PIP theorem with the corresponding non-degeneracy conditions provided.

The proof of Hilbert's mechanization theorem is now complete. □

4 The mechanization theorem of (ordinary) unordered geometry

4.1 Introduction

This is the principal chapter of the book. It aims to prove that if the associated number system of a certain geometry is a number field, i.e., multiplication is commutative, then the proving of theorems whose hypotheses and conclusions can be expressed as polynomial *equality relations* is mechanizable. This class of theorems will be called *theorems of equality type*. In fact, this class contains most of the important theorems in elementary geometries though it excludes theorems involving order relations and thus polynomial *inequalities* in the algebraic relations. However, the latter does not often occur except in high school Euclidean geometry. In modern geometries such as algebraic geometry, the associated number fields are usually the complex field and arbitrary fields of characteristic 0, so no order relation is involved. Therefore it is compatible with the current state of geometry to restrict mechanical proving to theorems that can be expressed by means of only equalities.

According to Sects. 3.2 and 3.3, the mechanical proving of theorems of equality type may be reduced to the following mechanization problem stated in algebraic form.

Mechanization problem (algebraic form). Given a number field $\boldsymbol{K}$, a set of variables

$$x_1, \ldots, x_n,$$

a set of polynomials

$$f_i(x_1, \ldots, x_n) \in \boldsymbol{K}[x_1, \ldots, x_n], \quad i \in I$$

and any other polynomial

$$g(x_1, \ldots, x_n) \in \boldsymbol{K}[x_1, \ldots, x_n],$$

find a mechanical method which determines, after a finite number of steps, a finite set of *non-degeneracy conditions*

$$D_j(x_1, \ldots, x_n) \in \boldsymbol{K}[x_1, \ldots, x_n], \quad j \in J$$

and decides whether the following conclusion holds true for some extension field $\bar{K}$ of K:

(1) For arbitrary $a_1, \ldots, a_n \in \bar{K}$, if

$$f_i(a_1, \ldots, a_n) = 0, \quad i \in I,$$
$$D_j(a_1, \ldots, a_n) \neq 0, \quad j \in J, \quad \text{then } g(a_1, \ldots, a_n) = 0.$$

When (1) holds, we say that the equation

$$g(x_1, \ldots, x_n) = 0$$

is a *formal consequence* of the equations

$$f_i(x_1, \ldots, x_n) = 0, \quad i \in I$$

under the non-degeneracy conditions

$$D_j(x_1, \ldots, x_n) \neq 0, \quad j \in J.$$

The above non-degeneracy conditions D_j cannot be arbitrary. They must satisfy the requirement that each

$$D_j(x_1, \ldots, x_n) = 0$$

is not a formal consequence of

$$f_i(x_1, \ldots, x_n) = 0, \quad i \in I.$$

In other words, for each $j \in J$ there is at least one n-tuple of numbers

$$a_1, \ldots, a_n \in \bar{K}$$

such that

$$f_i(a_1, \ldots, a_n) = 0, \quad i \in I,$$

while

$$D_j(a_1, \ldots, a_n) \neq 0.$$

See the remarks on non-degeneracy conditions in Sect. 3.3.

In statement (1), if $\bar{K}$ is the geometry-associated number field K itself, then any n-tuple of numbers $a_1, \ldots, a_n \in \bar{K}$ corresponds to a geometric configuration satisfying the hypotheses of the considered theorem, and $g(a_1, \ldots, a_n) = 0$ is equivalent to saying that the conclusion of the theorem is true for this geometric configuration. Hence, the resolution of the mechanization problem of algebraic form in the case $\bar{K} = K$ shows that the corresponding theorem proving in geometry is mechanizable. That is, the mechanization theorem of the corresponding geometry holds true.

The following sections are devoted to solving the algebraic form of the mechanization problem. In our solution, however, the field $\bar{K}$ in (1) is not the original geometry-associated number field K, but an algebraically closed extension field of K. For the goal of mechanical proving in geometry, this is sufficient. If (1) holds for the extension field $\bar{K}$, then it holds for the original number field K as well, and thus the corresponding geometric theorem is true. If (1) does not hold for the extension field $\bar{K}$, even though we cannot conclude that it does not hold for K, and thus cannot conclude that the corresponding geometric theorem is false either, we can at least point out that the probability that this theorem is true is small, or at least that the theorem is not true after slightly enlarging the geometric configuration (for example, enlarging the real geometric configuration to the complex configuration). If we use the language of algebraic geometry and apply the theorem that the dimension of an algebraic variety does not depend on the base field, then, at least for ordinary geometry, the above case does not occur. However, we wish to obtain positive proofs of geometric theorems rather than negative disproofs. Therefore, it is not so important whether $\bar{K}$ and K are the same. In view of this point, we shall extend the definitions of mechanical methods and the mechanizability of theorem proving in geometry to include the above case. In detail, when $\bar{K}$ is taken as an extension field of the corresponding field associated with the geometry and the mechanization problem of the algebraic form has a solution, we shall still say that the corresponding theorem proving in geometry is *mechanizable*. When the considered geometric theorem is true, the algebraic mechanical procedure used gives a *mechanized proof* of this theorem. Under this wider notion, this chapter will give a solution to the mechanization problem in algebraic form which leads to the following.

Mechanization theorem. There is a mechanical method for proving theorems in various unordered geometries such as unordered Pascalian geometry, unordered orthogonal geometry, and unordered metric geometry (in which the Pascalian axiom for intersecting lines holds true) subordinate to ordinary geometry as described in Chap. 2.

The above-mentioned algebraic problem is actually a problem of algebraic geometry. Its solution relies upon algebraic geometry, a rather modern subject of mathematics. But much of modern algebraic geometry has been developed to a purely *existential* theory. Since the solution we require must be a mechanical method, we need a *constructive* theory – the purely existential one is not sufficient. Fortunately, J. F. Ritt (1932, 1950) has already developed such a *constructive* algebraic geometry, which fulfills our requirements. As Ritt's work is little known and his arguments are often analytic, in this chapter we shall rewrite Ritt's theory in a form that better fits our needs. Ritt's constructive theory of algebraic geometry is itself worthy of our attention. For example, the idea of dimension of an algebraic variety is one of the most fundamental and intuitive notions; and the dimension theorem on the sections of an algebraic variety by a hyperplane is one of the most fundamental and intuitive theorems in algebraic geometry. Nevertheless, in popular books of algebraic geometry, both the introduction of the notion and proof of the theorem are not simple and need to use some deep methods. Moreover, they are all existential. On the contrary, Ritt's theory gives not only an elementary and direct proof of the theorem, but also a constructive method for the determination of dimensions (cf. Sect. 4.6).

4.2 Factorization of polynomials

Let $\boldsymbol{K}$ be a fixed number field. We shall restrict ourselves, without further indication from now on, to the case in which $\boldsymbol{K}$ is of characteristic 0. As usual, the ring formed by all polynomials in the variables $x_1, x_2, \ldots, x_n$ with coefficients in $\boldsymbol{K}$ is denoted by $\boldsymbol{K}[x_1, x_2, \ldots, x_n]$. Similarly, if the coefficients are in some ring $\boldsymbol{A}$, the corresponding polynomial ring is denoted by $\boldsymbol{A}[x_1, x_2, \ldots, x_n]$.

Let $\bar{\boldsymbol{K}}$ be any number field containing $\boldsymbol{K}$, called an *extension field* of $\boldsymbol{K}$. For a θ in $\bar{\boldsymbol{K}}$ but not in $\boldsymbol{K}$, let us consider the sequence

$$1, \theta, \theta^2, \ldots, \theta^m, \ldots.$$

Denote by $\boldsymbol{K}(\theta)$ the totality of elements $f(\theta)/g(\theta)$ in $\bar{\boldsymbol{K}}$, in which f, g are both polynomials in θ with coefficients in $\boldsymbol{K}$ and $g(\theta)$ is not 0 in $\bar{\boldsymbol{K}}$. Under the operations of $\bar{\boldsymbol{K}}$, $\boldsymbol{K}(\theta)$ constitutes a number field containing $\boldsymbol{K}$, called a *simple extension field* obtained from $\boldsymbol{K}$ by adjoining θ. There are two cases as follows.

Case 1. In the sequence formed by the powers of θ, any finite number of terms are linearly independent over $\boldsymbol{K}$. In this case, θ is a *transcendental* number with respect to $\boldsymbol{K}$ and $\boldsymbol{K}(\theta)$ is a *transcendental extension field* obtained from $\boldsymbol{K}$ by adjoining θ.

Case 2. There is a finite number of terms in the sequence such that they are linearly dependent over $\boldsymbol{K}$. Let m be the minimal integer such that $1, \theta, \ldots, \theta^m$ are linearly dependent over $\boldsymbol{K}$. Then there are numbers $a_0, a_1, \ldots, a_m$, not all 0, in $\boldsymbol{K}$ such that

$$a_0 + a_1\theta + \cdots + a_m\theta^m = 0.$$

But, for any $n + 1$ ($n < m$) numbers $b_0, b_1, \ldots, b_n$, not all 0, we have

$$b_0 + b_1\theta + \cdots + b_n\theta^n \neq 0,$$

and thus $a_m \neq 0$ in particular. Now, θ is an *algebraic* number over $\boldsymbol{K}$, $\boldsymbol{K}(\theta)$ is an *algebraic extension field* obtained from $\boldsymbol{K}$ by adjoining θ and m is the degree of θ or $\boldsymbol{K}(\theta)$ over $\boldsymbol{K}$.

In the second case, we denote the polynomial $a_0 + a_1x + \cdots + a_mx^m$ by

$$P_\theta(x) = a_0 + a_1x + \cdots + a_mx^m \in \boldsymbol{K}[x].$$

Then $P_\theta(x)$ is obviously *irreducible* in $\boldsymbol{K}[x]$, i.e., it cannot be factorized as a product of two polynomials of degree at least 1 in $\boldsymbol{K}[x]$. In particular, we have $a_0 \neq 0$. Apparently, P_θ can be uniquely determined up to a non-zero factor in $\boldsymbol{K}$. We shall call any such polynomial $P_\theta(x)$ an *adjoining polynomial* of θ. We see that $P_\theta(x)$ is irreducible in $\boldsymbol{K}[x]$ and

$$P_\theta(\theta) = 0,$$

i.e., θ is a *solution* of $P_\theta(x) = 0$ in a certain extension field of $\boldsymbol{K}$.

Let $f(\theta)/g(\theta)$ be an arbitrary number in $\boldsymbol{K}(\theta)$, whence $f(x), g(x) \in \boldsymbol{K}[x]$ and $g(\theta) \neq 0$. Since $P_\theta(x)$ is irreducible in $\boldsymbol{K}[x]$, $g(x)$ and $P_\theta(x)$ cannot have a common divisor. By the division algorithm of polynomials, there are $\varphi(x), \psi(x), Q(x), R(x) \in \boldsymbol{K}[x]$ such that

$$\varphi(x)g(x) + \psi(x)P_\theta(x) = 1,$$

$$f(x)\varphi(x) = Q(x)P_\theta(x) + R(x),$$

where the degrees of $\varphi(x)$ and $R(x)$ in x are less than the degree of $P_\theta(x)$ in x (i.e. m), and the degree of $\psi(x)$ is less than the degree of $g(x)$ in x. As $P_\theta(\theta) = 0$,

$$\frac{f(\theta)}{g(\theta)} = f(\theta)\varphi(\theta) = R(\theta).$$

Therefore, an arbitrary number in $\boldsymbol{K}(\theta)$ can be represented as a polynomial of θ whose degree is less than or equal to $m - 1$. This representation is unique and can be constructively determined via algebraic operations.

Suppose now that the field $\boldsymbol{K}$ is the quotient field of a domain $\boldsymbol{A}$ which has a unit element. Then no matter whether $\boldsymbol{K}(\theta)$ is a transcendental extension or an algebraic extension, the set $\boldsymbol{A}[\theta]$ of polynomials in θ with coefficients in $\boldsymbol{A}$ is a domain which has a unit element and $\boldsymbol{K}(\theta)$ is the quotient field of $\boldsymbol{A}(\theta)$. According to Gauss' theorem, if a polynomial in $\boldsymbol{A}[x]$ can be factorized in $\boldsymbol{K}[x]$, it can also be factorized in $\boldsymbol{A}[x]$. Moreover, if $\boldsymbol{A}$ is a unique factorization domain (UFD), then $\boldsymbol{A}[x]$ is a unique factorization domain as well. Without further indication, domains considered hereafter are assumed to be UFDs. A simple concrete example is the ring of integers, i.e., $\boldsymbol{A} = \boldsymbol{Z}$.

The main questions discussed in this section are the following – suppose we know how to factorize elements of $\boldsymbol{A}$ into irreducible factors.

1. How can we factorize an arbitrary polynomial in $\boldsymbol{A}[z]$ into irreducible factors, where z is a variable?
2. How can we factorize an arbitrary polynomial in $\boldsymbol{A}[\theta][z] = \boldsymbol{A}[\theta, z]$ into irreducible factors, where $\boldsymbol{K}(\theta)$ is a simple extension of the quotient field of $\boldsymbol{A}$ by adjoining θ?

In most books on algebra, the answers to these two questions are often restricted to existence and possibility, but not concerned with constructive methods. This is not sufficient for the requirements of mechanization. In the first edition of van der Waerden's "Modern Algebra" (1930a), there were two paragraphs (§23 and §37) devoted to a mechanical method of solving the above problems in a finite number of steps. But in later editions, though the contents corresponding to §23 remained, those corresponding to §37 were completely left out. Because these problems are rather important for the mechanical method of theorem proving in geometry, and they cannot be easily found in common books, we shall give a detailed introduction below.

As the characteristic of $\boldsymbol{K}$ is 0, the domain $\boldsymbol{A}$ which has $\boldsymbol{K}$ as its quotient field must consist of infinitely many elements. Take, for example, a unit element e of $\boldsymbol{A}$. Then all ne (n any integer) constitute a set of infinitely many elements. In fact, if $\boldsymbol{A}$ contains only a finite number of elements, then for an arbitrary polynomial $f(z)$ in $\boldsymbol{A}[z]$ or $\boldsymbol{A}[\theta][z]$ there are only finitely many polynomials which are the factors of $f(z)$. Dividing $f(z)$ by each of these polynomials, we can determine whether or not $f(z)$ is irreducible in a finite number of steps and factorize it into irreducible factors in the reducible case. Therefore, when $\boldsymbol{A}$ is finite, it is easy to answer the above questions.

In what follows, we only consider the case when $\boldsymbol{A}$ consists of an infinite number of elements and the characteristic of the quotient field $\boldsymbol{K}$ is 0. Under this condition, question 2 can be separated into two cases according to whether θ is transcendental or algebraic. For the former, we may rewrite θ as another variable, say y. Then question 2 becomes that of factorizing polynomials in $\boldsymbol{A}[y,z]$. In combination with question 1, we can ask a more general question, i.e., the question of how to factorize polynomials in $\boldsymbol{A}[x_1,\ldots,x_n]$. We formulate the answers to these questions as the following two theorems.

Theorem 1. Let $\boldsymbol{A}$ be a domain with unit element and let a mechanical method which can uniquely factorize an arbitrary element in $\boldsymbol{A}$ into irreducible factors in a finite number of steps (determined up to an invertible factor in $\boldsymbol{A}$) be given. Then there is a mechanical method which can factorize an arbitrary polynomial in $\boldsymbol{A}[x_1,x_2,\ldots,x_n]$ into irreducible factors in a finite number of steps (determined up to an invertible factor in $\boldsymbol{A}$).

Theorem 2. Let $\boldsymbol{A}$ be the same as in Theorem 1, $\boldsymbol{K}$ be the quotient field of $\boldsymbol{A}$ and $\boldsymbol{K}(\theta)$ be an algebraic extension field formed by adjoining θ to $\boldsymbol{K}$, with

$$P_\theta(y)=a_0y^m+a_1y^{m-1}+\cdots+a_m\in\boldsymbol{A}[y]$$

the adjoining polynomial of θ, where $a_i\in\boldsymbol{A}$, $a_0a_m\neq 0$ and $a_0,a_1,\ldots,a_m$ do not have non-invertible common divisors. Then there is a mechanical method which can factorize an arbitrary polynomial in $\boldsymbol{A}[\theta][z]$ into irreducible factors in a finite number of steps (determined up to an invertible factor in $\boldsymbol{A}$).

Proof of Theorem 1. Let the given polynomial be

$$f(x_1,\ldots,x_n)\in\boldsymbol{A}[x_1,\ldots,x_n].$$

Take an integer m bigger than the degree of f in any x_i, introduce a new variable t and set

$$x_1 = t, x_2 = t^m, \ldots, x_n = t^{m^{n-1}}.$$

Then we have

$$x_1^{k_1} \cdots x_n^{k_n} = t^{k_1+k_2m+\cdots+k_nm^{n-1}}.$$

When $0 \leq k_i \leq m-1$, the above expression induces a one-to-one correspondence between the power products of x's whose degrees are less than m in each x_i on the left-hand side and the powers of t whose degrees are less than m^n on the right-hand side, so a one-to-one correspondence between polynomials in $\boldsymbol{A}[x_1, \ldots, x_n]$ of degree less than m in each x_i and polynomials in $\boldsymbol{A}[t]$ of degree less than m^n. Let f correspond to F:

$$f(x_1, \ldots, x_n) = F(t).$$

Suppose we know how to factorize $F(t)$ into irreducible factors

$$F(t) = F_1(t) \cdots F_s(t).$$

Then the irreducible factors of f are among the inverses $f_{i_1 \cdots i_m}(x_1, \ldots, x_n)$ of

$$F_{i_1}(t) \cdots F_{i_m}(t) \quad (m \leq s, 1 \leq i_1 < \cdots < i_m \leq s)$$

under the above one-to-one correspondence. Hence the factorization of polynomials in n variables $x_1, \ldots, x_n$ may be reduced to the factorization of polynomials in one variable with some simple division tests and substitutions.

Consider now a unary polynomial $F(t) \in \boldsymbol{A}[t]$ of degree r. By Gauss' theorem, the problem of factorizing $F(t)$ in $\boldsymbol{A}[t]$ is the same as that in $\boldsymbol{K}[t]$. If F can be factorized into two irreducible factors of degree greater than or equal to one in $\boldsymbol{K}[t]$ and thus in $\boldsymbol{A}[t]$: $F(t) = \varphi(t) \cdot \psi(t)$, where the degree of $\varphi(t)$ is less than or equal to the degree of $\psi(t)$, then the degree of $\varphi(t)$, say q, is less than or equal to $[r/2]$, the integer part of $r/2$. We need to provide a mechanical method which either determines that no such factor of F exists, i.e., F cannot be factorized and thus is irreducible, or finds a factor $\varphi(t)$ in a finite number of steps. In the latter case, we can get another factor $\psi(t)$ of $F(t)$ by using the division algorithm. Applying the same mechanical method again to $\varphi(t)$ and $\psi(t)$, we should arrive finally at an irreducible factorization of $F(t)$.

To this end, set $q = 1, 2, \ldots, [r/2]$ successively and determine if $F(t)$ has a polynomial factor of degree q. Since $\boldsymbol{A}$ is infinite, we can choose $q+1$ pairwise distinct elements $a_0, a_1, \ldots, a_q$ of $\boldsymbol{A}$, for instance, $a_k = ke$, where e is a unit element of $\boldsymbol{A}$. For arbitrary $c_i \in \boldsymbol{A}$, construct

$$\varphi(t) = \sum_{i=0}^{q} c_i \cdot \frac{(t-a_0)\cdots(t-a_{i-1})(t-a_{i+1})\cdots(t-a_q)}{(a_i-a_0)\cdots(a_i-a_{i-1})(a_i-a_{i+1})\cdots(a_i-a_q)} \in \boldsymbol{K}[t].$$

Then $\varphi(t)$ is a polynomial of degree q such that

$$\varphi(a_i) = c_i, \quad i = 0, \ldots, q.$$

If $\varphi(t)$ is a factor of $F(t)$ in $\boldsymbol{A}[t]$, then $\varphi(a_i) = c_i$ is a factor of $F(a_i) \in \boldsymbol{A}$. By assumption one knows that all possible factors of $F(a_i)$ can be determined by using the factorization method for $\boldsymbol{A}$. So there is only a finite number of such c_i's and they can be found by a mechanical method. Hence, those polynomials $\varphi(t)$ of degree q which are possible factors of $F(t)$ are finite in number and can be mechanically found. Dividing $F(t)$ by this finite number of polynomials respectively, we can mechanically determine whether or not $F(t)$ can be factorized and, if so, find a factor of it. In this way, the irreducible factorization of $F(t)$ can be accomplished in a finite number of steps. This proves the theorem. □

The proof of Theorem 2 is much more difficult than that of Theorem 1. We base our exposition on Trager's improved form (1976) of the proof given in §37 of van der Waerden (1930a). For this purpose, we first make some preparations as follows.

Let the adjoining polynomial of the algebraic extension field $\boldsymbol{K}(\theta)$ be

$$P_\theta(y) = y^m + a_1 y^{m-1} + \cdots + a_m \in \boldsymbol{K}[y].$$

By the extension theory of fields, $\boldsymbol{K}$ has an extension field $\tilde{\boldsymbol{K}}$ such that $P_\theta(y)$, as a polynomial in $\tilde{\boldsymbol{K}}[y]$, can be completely factorized into linear factors:

$$P_\theta(y) = (y - \theta_1)(y - \theta_2) \cdots (y - \theta_m),$$

where $\theta_1 = \theta$ and all $\theta_2, \ldots, \theta_m$ are called the *conjugate elements* of θ. Here it suffices to know the existence of θ_i (instead of considering their construction). What we actually need to know is that $(-1)^j a_j$ are elementary symmetric functions of θ_i.

Let $f(\theta, z)$ be a polynomial in $\boldsymbol{K}(\theta)[z] = \boldsymbol{K}[\theta, z]$. The product

$$\operatorname{Norm} f(\theta, z) = f(\theta_1, z) \cdot f(\theta_2, z) \cdots f(\theta_m, z)$$

is called the *norm* of f. In its expansion, each coefficient of the powers of z is a symmetric function of $\theta_1, \ldots, \theta_m$. Therefore, by the fundamental theorem of symmetric functions, all of them are polynomials in $a_1, \ldots, a_m$ and thus can be computed directly from f and P_θ, without needing to consider the exact form of θ_i.

Lemma 1. Let $f(\theta, z)$ be irreducible in $\boldsymbol{K}(\theta)[z]$. Then $\operatorname{Norm} f$ is a power of an irreducible polynomial in $\boldsymbol{K}[z]$.

Proof. Suppose the lemma does not hold. Then we have

$$\text{Norm}\, f = g(z) \cdot h(z),$$

where both g and h are polynomials of degree greater than or equal to 1 in $\boldsymbol{K}[z]$ and they have no common divisor in $\boldsymbol{K}[z]$. Since $f(\theta, z)$ is irreducible, it must be a factor of g or h in $\boldsymbol{K}(\theta)[z]$. Let $f(\theta, z)$ be a factor of g; then Norm $f(\theta, z)$ will be a factor of Norm $g = g^m$. Hence gh is a factor of g^m. This contradicts the assumption that g and h have no common divisor and the lemma is proved. □

Lemma 2. There is a mechanical method which determines, for an arbitrary square-free polynomial $f(\theta, z)$ in $\boldsymbol{K}(\theta)[z]$, a square-free polynomial $g(z)$ in $\boldsymbol{K}[z]$ in a finite number of steps such that $g(z)$ has $f(\theta, z)$ as a factor in $\boldsymbol{K}(\theta)[z]$.

Proof. First construct

$$\text{Norm}\, f = G(z) \in \boldsymbol{K}[z]$$

mechanically. Form the derivative $G'(z)$ of $G(z)$ and the greatest common divisor of $G(z)$ and $G'(z)$ by using the division algorithm. Then $g(z) = G(z)/D(z)$ is what we want to determine. Suppose $f(\theta, z)$ in $\boldsymbol{K}(\theta)[z]$ can be factorized into the product $f_1(\theta, z) \cdots f_r(\theta, z)$ of r distinct irreducible factors. Then by Lemma 1 we have

$$G(z) = \text{Norm}\, f = \text{Norm}\, f_1 \cdots \text{Norm}\, f_r = g_1^{s_1} \cdots g_r^{s_r},$$

in which $g_1, \ldots, g_r$ are irreducible polynomials in $\boldsymbol{K}[z]$ and $s_i > 0$. Hence

$$g(z) = g_{i_1} \cdots g_{i_k} \in \boldsymbol{K}[z],$$

where $g_{i_1}, \ldots, g_{i_k}$ are those g_i of $g_1, \ldots, g_r$ which are distinct from each other. As every f_i is a factor of $g_i^{s_i}$, it is also a factor of g_i and $g(z)$. On the other hand, all f_i are pairwise distinct, so the product f of the f_i is also a factor of $g(z)$. Clearly $g(z)$ is square-free. This completes the proof. □

Lemma 3. There is a mechanical method which determines, in a finite number of steps for an arbitrary square-free polynomial $g(z) \in \boldsymbol{K}[z]$, an integer s such that Norm $g(z - s\theta)$ is square-free in $\boldsymbol{K}[z]$.

Proof. Set $G_s = \text{Norm}\, g(z - s\theta)$ for $s = 1, 2, \ldots$. Form the derivative G_s' of G_s and find the greatest common divisor D_s of G_s and G_s' by using the division algorithm. Then there must be an s such that the degree of D_s in z is 0, i.e., G_s is square-free. For in some extension field $\tilde{\boldsymbol{K}}$ of $\boldsymbol{K}$, $g(z)$ can be factorized into linear factors $z - \beta_j$, where all β_j are pairwise distinct. Hence G_s will be factorized into linear factors $z - (\beta_j + s\theta_i)$, and thus there are only finitely many s such that G_s is not square-free. □

Lemma 4. Let $f(\theta, z) \in \boldsymbol{K}(\theta)[z]$ be square-free. Then there is a mechanical method which finds an integer s such that $\operatorname{Norm} f(\theta, z - s\theta)$ is square-free in $\boldsymbol{K}[z]$.

Proof. By Lemma 2, we can find a square-free polynomial $g(z) \in \boldsymbol{K}[z]$ having $f(\theta, z)$ as a factor in $\boldsymbol{K}(\theta)[z]$ by using a mechanical method. By Lemma 3, we can find an integer s such that $\operatorname{Norm} g(z - s\theta)$ is square-free in $\boldsymbol{K}[z]$ by using a mechanical method, too. Now $g(z - s\theta)$ has $f(\theta, z - s\theta)$ as a factor in $\boldsymbol{K}(\theta)[z]$. Since $\operatorname{Norm} g(z - s\theta)$ is square-free, $\operatorname{Norm} f(\theta, z - s\theta)$ is also square-free. □

Lemma 5. Let the norm $\operatorname{Norm} h(\theta, z)$ of $h(\theta, z) \in \boldsymbol{K}(\theta)[z]$ be square-free and its irreducible factorization in $\boldsymbol{K}[z]$ be

$$\operatorname{Norm} h = H_1(z) \cdots H_r(z).$$

Then the irreducible factorization of $h(\theta, z)$ in $\boldsymbol{K}(\theta)[z]$ is

$$h = \prod_{i=1}^{r} \gcd(h(\theta, z), H_i(z)),$$

where gcd denotes the greatest common divisor in $\boldsymbol{K}(\theta)[z]$ which can be determined by using the division algorithm.

Proof. Let the irreducible factorization of h in $\boldsymbol{K}(\theta)[z]$ be

$$h = h_1 \cdots h_s.$$

Then

$$\operatorname{Norm} h = \operatorname{Norm} h_1 \cdots \operatorname{Norm} h_s = H_1 \cdots H_r.$$

Since each h_i is irreducible in $\boldsymbol{K}(\theta)[z]$, by Lemma 1, $\operatorname{Norm} h_i$ is a power of an irreducible polynomial in $\boldsymbol{K}[z]$. Furthermore, $\operatorname{Norm} h$ is square-free, so $\operatorname{Norm} h_i$ itself is an irreducible polynomial and actually is one of $H_1, \ldots, H_r$, up to a non-zero factor in $\boldsymbol{K}$. For the same reason these irreducible polynomials $\operatorname{Norm} h_i$ are distinct from each other, so we have $s = r$ and we may assume, without loss of generality, that $\operatorname{Norm} h_1 = H_1, \ldots, \operatorname{Norm} h_r = H_r$. Evidently, h and $H_i = \operatorname{Norm} h_i$ have a common divisor h_i in $\boldsymbol{K}(\theta)[z]$, but they cannot have other h_j $(j \neq i)$ as their common divisors. Hence $\gcd(h, H_i) = h_i$ and the lemma is proved. □

Proof of Theorem 2. (Recall that $\boldsymbol{A}$ is infinite and $\boldsymbol{K}$ is the quotient field of $\boldsymbol{A}$.)

Let $f(\theta, z) \in \boldsymbol{K}(\theta)[z]$ be the polynomial to be factorized. First, form the greatest common divisor $d(\theta, z)$ of $f(\theta, z)$ and its derivative $f'(\theta, z)$ in $\boldsymbol{K}(\theta)[z]$. Then $g(\theta, z) = f(\theta, z)/d(\theta, z)$ is square-free in $\boldsymbol{K}(\theta)[z]$. By Lemma 4, we can find an integer s such that $\operatorname{Norm} h$ of $h(\theta, z) = g(\theta, z - s\theta)$ in $\boldsymbol{K}[z]$ is square-free. From the hypothesis one already knows a factorization method for $\boldsymbol{A}$, so by Theorem 1 we may factorize Norm h in $\boldsymbol{K}[z]$ into irreducible factors:

$$\operatorname{Norm} h = H_1(z) \cdots H_r(z).$$

Form the greatest common divisor of $H_i(z)$ and h in $\boldsymbol{K}(\theta)[z]$ by using the division algorithm

$$h_i(\theta, z) = \gcd(h(\theta, z), H_i(z)).$$

Then by Lemma 5, h has an irreducible factorization in $\boldsymbol{K}(\theta)[z]$:

$$h(\theta, z) = h_1(\theta, z) \cdots h_r(\theta, z).$$

Hence $g(\theta, z)$ has an irreducible factorization in $\boldsymbol{K}(\theta)[z]$:

$$g(\theta, z) = h(\theta, z + s\theta) = h_1(\theta, z + s\theta) \cdots h_r(\theta, z + s\theta).$$

Therefore, the irreducible factorization of the given polynomial $f(\theta, z) = g(\theta, z)d(\theta, z)$ can be obtained by some divisions. □

Theorems 1 and 2 above are both restricted to the case of finite extension fields of $\boldsymbol{K}$. If there is no such restriction, probably no mechanical method exists, see van der Waerden (1930b). The mechanical methods mentioned in Theorems 1 and 2 all originate from Kronecker (cf. Hermann 1926). A little thought shows that the efficiency of these methods is not especially high. In recent years, developments of computer science have led to the discovery of new, more efficient methods. These new methods employ the Chinese remainder theorem, p-adic number theory and other mathematical tools. The reader may refer to works by Berlekamp, Zassenhaus, P. S. Wang, and the relevant part of "The Art of Computer Programming" by Knuth (1969).

4.3 Well-ordering of polynomial sets

Example 5 in Sect. 3.4 indicates that even for a pure point of intersection (PIP) theorem in Pascalian geometry, if it is not expressed in the form of constructive type, during the processing of its hypothesis, polynomials of higher degree may occur and thus Hilbert's mechanical method cannot be used. For non-PIP theorems or theorems in other geometries, in general there will be polynomials of higher degree occurring in the hypotheses. We shall prove later that in this case there is still a mechanical method so long as the hypotheses and conclusions of the considered theorems can be expressed as polynomial equalities, i.e., theorems of equality type (see Sect. 4.1). Of course, this mechanical method is totally different from Hilbert's. While applying it to PIP theorems of constructive type in Pascalian geometry, we get two different mechanical methods. Before presenting this widely applicable mechanical method, this and the next section will make some preparations. The main basis for our exposition is the work by Ritt (1932, 1950). Many concepts, methods, and results originate in these books.

In what follows, we consider a basic field $\boldsymbol{K}$ of characteristic 0, and two sets of variables

$$u_1, \ldots, u_e \quad \text{and} \quad x_1, \ldots, x_N$$

with a fixed order

$$u_1 \prec u_2 \prec \cdots \prec u_e \prec x_1 \prec x_2 \prec \cdots \prec x_N.$$

Consider $u_1, \ldots, u_e, x_1, \ldots, x_N$ as the base of the $(e+N)$-dimensional linear space $\boldsymbol{K}^{e+N}$ over $\boldsymbol{K}$.

By a *polynomial* we shall mean one in the variables $u_1, \ldots, u_e, x_1, \ldots, x_N$ with coefficients in $\boldsymbol{K}$, i.e., an element in the polynomial ring $\boldsymbol{K}[u_1, \ldots, u_e, x_1, \ldots, x_N]$.

A *monomial*

$$\mu = au_1^{i_1} \cdots u_e^{i_e} x_1^{m_1} \cdots x_N^{m_N} \quad (a \in \boldsymbol{K})$$

will sometimes also be written as

$$\mu = aU^I X^M, \quad I = (i_1, \ldots, i_e), \quad M = (m_1, \ldots, m_N)$$

or

$$\mu = aZ^\alpha, \quad \alpha = (I, M) = (i_1, \ldots, i_e, m_1, \ldots, m_N),$$

where $i_1, \ldots, i_e, m_1, \ldots, m_N$ are non-negative integers. If $a \neq 0$, and the last non-zero component of the N-tuple $(m_1, \ldots, m_N)$ is m_p, we say that the monomial μ is of *class* p; otherwise we say that the monomial $\mu \neq 0$ is of *class* 0. In the latter case, there may occur u's, but no x's in μ.

For two s-tuples

$$\alpha = (a_1, \ldots, a_s), \quad \beta = (b_1, \ldots, b_s)$$

of non-negative integers, we say that α *precedes* β or β *follows* α, which we denote as

$$\alpha \prec \beta \quad \text{or} \quad \beta \succ \alpha,$$

if there is an index k such that

$$a_{k+1} = b_{k+1}, \ldots, a_s = b_s \quad \text{while} \quad a_k < b_k.$$

For two non-zero monomials

$$\lambda = au_1^{i_1} \cdots u_e^{i_e} x_1^{l_1} \cdots x_N^{l_N}, \quad a \neq 0,$$

$$\mu = bu_1^{j_1} \cdots u_e^{j_e} s_1^{m_1} \cdots x_N^{m_N}, \quad b \neq 0,$$

we say that λ *precedes* μ or μ *follows* λ, which we denote as

$$\lambda \prec \mu \quad \text{or} \quad \mu \succ \lambda,$$

if

$$(i_1, \ldots, i_e, l_1, \ldots, l_N) \prec (j_1, \ldots, j_l, m_1, \ldots, m_N).$$

Any non-zero polynomial F can be written in the form

$$F = a_1 Z^{\alpha_1} + a_2 Z^{\alpha_2} + \cdots + a_s Z^{\alpha_s},$$

where

$$a_i \in \boldsymbol{K},$$
$$a_1 \neq 0, \ldots, a_s \neq 0,$$
$$\alpha_1 \succ \alpha_2 \succ \cdots \succ \alpha_s.$$

In this case, $a_1 Z^{\alpha_1}$ is called the *leading term* of F and the class of Z^{α_1} is called the *class* of F.

If a non-zero polynomial F has its class $= p > 0$ and the leading term $a_1 Z^{\alpha_1}$ of F has degree m in x_p, then F can be written in the form

$$F = C_0 x_p^m + C_1 x_p^{m-1} + \cdots + C_m,$$

in which the C's are all polynomials in the u's and $x_1, \ldots, x_{p-1}$, containing none of $x_p, x_{p+1}, \ldots, x_N$, and $C_0 \neq 0$. The polynomial C_0 will then be called the *initial* of F. If the leading term of C_0 is c_0, then the leading term of F is clearly $c_0 x_p^m$.

Consider two non-zero polynomials F and G and any variable x_p. If the highest degree of x_p appearing in F is less than that in G, we say that F has a *lower rank* than G or G has a *higher rank* than F *with respect to* x_p. We say that F and G have the *same rank with respect to* x_p when neither of them is of higher rank than the other.

For two non-zero polynomials F and G we say that F has a *lower rank* than G or G has a *higher rank* than F, which is denoted as

$$F \prec G \quad \text{or} \quad G \succ F,$$

if either

(1) class $F <$ class G; or
(2) class $F =$ class $G = p > 0$, while the degree of x_p in F is less than that of x_p in G, or in other words, F has a lower rank than G with respect to x_p.

In case neither of F and G is of higher rank than the other, F and G will be said to be of the *same rank*, denoted as

$$F \sim G.$$

For instance, two non-zero polynomials whose classes are 0 have the same rank.

Let F be a polynomial of class $p > 0$. Any polynomial whose rank is lower than that of F with respect to x_p will be said to be *reduced* with respect to F. Clearly the initial of F is of class $< p$ and is thus reduced with respect to F.

Let F be of class $p > 0$ and written in the form

$$F = f_0 x_p^m + f_1 x_p^{m-1} + \cdots + f_m,$$

in which

$$f_i \in \boldsymbol{K}[u_1, \ldots, u_e, x_1, \ldots, x_{p-1}] \quad \text{and} \quad f_0 \neq 0.$$

Any non-zero polynomial G which is not reduced with respect to F can always be put in the form

$$G = g_0 x_p^M + g_1 x_p^{M-1} + \cdots + g_M,$$

in which

$$g_i \in \boldsymbol{K}[u_1, \ldots, u_e, x_1, \ldots, x_{p-1}, x_{p+1}, \ldots, x_N]$$

and

$$g_0 \neq 0, \quad M \geq m.$$

By the division algorithm of polynomials, we may get, in dividing G by F, an expression of the form

$$f_0^s G = QF + R,$$

where Q and R are both polynomials. In case $R \neq 0$, the degree of R in x_p is less than m so that R is reduced with respect to F. The integer s will be determined as the smallest which makes such an expression possible, so $s \leq M - m$. If G is already reduced with respect to F, then we can simply take $s = 0$, $Q = 0$, $R = G$ so that the above expression holds true. In any case, the (uniquely determined) polynomial R will be called the *remainder* of F with respect to G. The procedure for getting the remainder R from G will be called the *reduction* of G with respect to F.

It is particularly important for us that when R and G are considered as polynomials in x_p, all coefficients of the powers of x_p in R are linear sums of those in G, i.e., of the form

$$g_0 h_0 + \cdots + g_M h_M,$$

where all the coefficients $h_0, \ldots, h_M$ of the sum are polynomials in the coefficients $f_0, \ldots, f_m$ of the powers of x_p in F and thus polynomials in $u_1, \ldots, u_e$, $x_1, \ldots, x_{p-1}$. In later relevant books by this author in which we shall make a complexity analysis for the process of mechanical theorem proving, more explicit description about this representation will be given.

Consider now a sequence consisting of finite number of polynomials, say

$$\mathcal{A}: \quad A_1, A_2, \ldots, A_r.$$

This sequence is called an *ascending set* if either

$$r = 1 \quad \text{and} \quad A_1 \neq 0,$$

or

$$r > 1,$$

$$0 < \text{class}\, A_1 < \text{class}\, A_2 < \cdots < \text{class}\, A_r,$$

and A_j is, moreover, reduced with respect to A_i for each pair $j > i$.

Obviously, for any ascending set we always have $r \leq N$.

An ascending set is said to be *contradictory* if $r = 1$, $A_1 \neq 0$ and class $A_1 = 0$.

Consider two ascending sets $\mathcal{A}$ and

$$\mathcal{B}: \quad B_1, B_2, \ldots, B_s.$$

We say that $\mathcal{A}$ has a *higher rank* than $\mathcal{B}$ or $\mathcal{B}$ has a *lower rank* than $\mathcal{A}$, denoted as

$$\mathcal{A} \succ \mathcal{B} \quad \text{or} \quad \mathcal{B} \prec \mathcal{A},$$

if there is some $j \leq \min(r, s)$ such that

$$A_1 \sim B_1, \ldots, A_{j-1} \sim B_{j-1}, \quad A_j \succ B_j,$$

or

$$s > r \quad \text{and} \quad A_1 \sim B_1, \ldots, A_r \sim B_r.$$

If neither of the ascending sets $\mathcal{A}$ and $\mathcal{B}$ is of higher rank than the other, we say that $\mathcal{A}$ and $\mathcal{B}$ are of the *same rank*, denoted as

$$\mathcal{A} \sim \mathcal{B}.$$

In this case we have

$$r = s \quad \text{and} \quad A_1 \sim B_1, \ldots, A_r \sim B_r.$$

It is clear that ascending sets are partially ordered by rank. Hence for any collection of ascending sets we can introduce the notion of *minimal ascending set*, i.e., one whose rank is not higher than any other's in the collection if it exists. The following lemma is fundamental and will play an important role in this book.

Lemma 1. Let

$$\Phi_1, \Phi_2, \ldots, \Phi_q, \ldots$$

be a sequence of ascending sets Φ_q for which the ranks never increase, i.e., for any q we have either

$$\Phi_{q+1} \prec \Phi_q \quad \text{or} \quad \Phi_{q+1} \sim \Phi_q.$$

Then there is an index q' such that for any $q > q'$ we have

$$\Phi_q \sim \Phi_{q'}.$$

In other words, there is a q' such that any Φ_q for $q \geq q'$ is a minimal ascending set of the above sequence.

Proof. For each ascending set Φ_q let us denote the first polynomial and the number of polynomials in Φ_q by A_q and r_q respectively. Then

$$A_1, A_2, \ldots, A_q, \ldots$$

is a sequence of polynomials for which the ranks never increase, i.e., for any q we have either

$$A_{q+1} \prec A_q \quad \text{or} \quad A_{q+1} \sim A_q.$$

Consequently, for any q we have $\text{class}\, A_{q+1} \leq \text{class}\, A_q$ and in the case $\text{class}\, A_{q+1} = \text{class}\, A_q = p > 0$, the degree of A_{q+1} in x_p is less than or equal to the degree of A_q in x_p. As both class and degree are non-negative integers, there should be some index q_1 such that all A_q are of the same rank for $q \geq q_1$.

If there is some $q_1' \geq q_1$ such that all $r_q = 1$ for any $q \geq q_1'$, then the lemma is clearly true. Otherwise, there should be some $q_1' \geq q_1$ such that all $r_q \geq 2$ for any $q \geq q_1'$. Denote the second polynomial in such Φ_q by $A_q^{(1)}$. Then

$$A_{q_1'}^{(1)}, A_{q_1'+1}^{(1)}, \ldots, A_q^{(1)}, \ldots$$

will be a sequence of polynomials with non-increasing ranks. As before there will be some $q_2 \geq q_1'$ such that all $A_q^{(1)}$ are of the same rank for any $q \geq q_2 \geq q_1' \geq q_1$.

If all $r_q \leq 2$, then the lemma is already proved. Otherwise, there will be some $q_2' \geq q_2$ such that all $r_q \geq 3$ for any $q \geq q_2'$ and we may take the third polynomials in such Φ_q's, say $A_q^{(2)}$, to form a sequence of polynomials with non-increasing ranks. As for all q we have $r_q \leq N$, so proceeding in this way we should stop at some r and some q' such that for all $q \geq q'$ we have $r_q = r$ and the r-th polynomials taken from such Φ_q will all have the same rank. It follows that all such Φ_q's will have the same rank and the lemma is proved. □

From this lemma we obtain the following

Lemma 1′. If in a sequence of ascending sets the ranks are steadily decreasing, then such a sequence is composed of a finite number of ascending sets.

Suppose now we have a non-empty set $\Sigma = \{F_\alpha\}$ of non-zero polynomials. An ascending set $\mathcal{A}$ is said to *belong to* Σ if all polynomials in $\mathcal{A}$ belong to Σ. Since each non-zero polynomial F_α forms by itself an ascending set, there exist ascending sets which belong to Σ. In the collection of all ascending sets which belong to Σ, any minimal ascending set will be called a *basic set* of Σ.

The following lemma points out not only the existence of such basic sets but also an algorithmic method for determining the basic sets.

Lemma 2. Let Σ be a finite set of non-zero polynomials. Then Σ necessarily has basic sets and there is a mechanical method for determining such a basic set in a finite number of steps.

Proof. As Σ is finite, the existence of basic sets is quite evident. So the problem is to provide a mechanical method for the determination of such a basic set.

To show this let us find at the outset a polynomial, say A_1, of lowest rank from $\Sigma = \Sigma_1$. This can clearly be done in a mechanical manner. If class $A_1 = 0$, then A_1 alone already forms a basic set. Suppose therefore class $A_1 > 0$. Check whether there are polynomials other than A_1 in Σ_1 which are reduced with respect to A_1. If no such polynomial exists in Σ_1, then A_1 by itself forms a basic set of Σ_1. Otherwise, let Σ_2 be the subset of Σ_1 formed by all polynomials except A_1, which are reduced with respect to A_1. From the choice of A_1 all polynomials in Σ_2 will have ranks higher than that of A_1. Now, let A_2 be a polynomial in Σ_2 of lowest rank. If there is not any polynomial in Σ_2 which is different from A_2 and is already reduced with respect to A_2, then A_1, A_2 will form a basic set of Σ. Otherwise, let Σ_3 be the subset of Σ_2 consisting of all polynomials except A_2 which are already reduced with respect to A_2. Choose from Σ_3 a polynomial A_3 of lowest rank and proceed as before. As the classes of the polynomials $A_1, A_2, A_3, \ldots$ are steadily increasing and cannot become greater than N, we have to stop in a finite number of steps and get, finally, a basic set of Σ in a mechanical manner. The lemma is proved. □

Lemma 3. Let Σ be a finite set of non-zero polynomials with a basic set

$$\mathcal{A}: \quad A_1, A_2, \ldots, A_r,$$

of which class $A_1 > 0$. Let B be a non-zero polynomial reduced with respect to all the A's. Then the set Σ' obtained from Σ by adjoining B will have a basic set of rank lower than that of $\mathcal{A}$.

Proof. If class $B = 0$, then B alone will form a basic set of Σ' of rank lower than that of $\mathcal{A}$. Suppose therefore class $B = p > 0$. As B is already reduced with respect to all the A's, there should be some i such that $p >$ class A_{i-1} and $p \leq$ class A_i. Moreover, in the case $p =$ class A_i, the degree of B in x_p is less than the degree of A_i in x_p. Hence,

$$A_1, A_2, \ldots, A_{i-1}, B$$

will be an ascending set contained in Σ' with its rank lower than that of $\mathcal{A}$. The basic set of Σ' will have therefore *a fortiori* its rank lower than that of $\mathcal{A}$. □

Remark. If we assume the axiom of choice, then all the above lemmas can be extended to the case of arbitrary infinite sets of polynomials. But this is not in accordance with the idea of mechanization in this book. Hence, the description of above lemmas has been restricted to finite sets or countable sequences.

Suppose now

$$\mathcal{A}: \quad A_1, A_2, \ldots, A_r$$

is an ascending set as before with class $A_1 > 0$. Let class $A_i = p_i$ and the initial of A_i be I_i. Then

$$0 < p_1 < p_2 < \cdots < p_r,$$

and for each i we have

$$\text{class}\, I_i < p_i, \quad \text{and}$$

$$I_i \text{ is reduced with respect to } A_1, \ldots, A_{i-1}.$$

Let B be an arbitrary polynomial. Set $B = R_r$. We can successively form the remainders $R_{r-1}, \ldots, R_0$ of R_r with respect to polynomials in $\mathcal{A}$ starting from A_r to A_1 by using the division algorithm, so that we get:

$$I_r^{s_r} R_r = Q_r A_r + R_{r-1},$$

$$I_{r-1}^{s_{r-1}} R_{r-1} = Q_{r-1} A_{r-1} + R_{r-2},$$

$$\cdots$$

$$I_1^{s_1} R_1 = Q_1 A_1 + R_0.$$

Set $R_0 = R$. Then we have an expression of the form

$$I_1^{s_1} \cdots I_r^{s_r} B = Q_1' A_1 + \cdots + Q_r' A_r + R,$$

in which Q_i' are polynomials. The polynomial R is uniquely determined from B and the ascending set $\mathcal{A}$. We shall call R the *remainder* of B with respect to $\mathcal{A}$. We shall also call the above formula the *remainder formula.*

From the remark on remainder formula given before, one knows that the degree of an arbitrary term of R in x_{p_i} is less than that of A_i in x_{p_i}, i.e., R is reduced with respect to each polynomial A_i in $\mathcal{A}$. We shall say briefly that R is *reduced with respect to* $\mathcal{A}$ and call the above procedure of getting R from B and $\mathcal{A}$ the *reduction* of B with respect to $\mathcal{A}$. As the determination of the remainder of one polynomial with respect to the other is done mechanically by using the division algorithm, we have the following.

Lemma 4. Let B be a non-zero polynomial and $\mathcal{A}$ an ascending set of which the class of the first polynomial is greater than 0. Then there is an algorithmic method which determines the remainder R of B with respect to $\mathcal{A}$ in a finite number of steps. Denote the i-th term of $\mathcal{A}$ by A_i and its class by p_i. Then the degree of any term of the remainder R in x_{p_i} is less than the degree of A_i in x_{p_i} for each i.

After having introduced a number of notions such as class, rank, initial, leading term, reduction, remainder, ascending set, basic set, etc. and proved several simple lemmas, we now come to the theme of this section: the notion of *zeros* and the method of *well-ordering* polynomial sets.

For this purpose let us consider an arbitrary polynomial F. Let

$$(u_1^0, \ldots, u_e^0, x_1^0, \ldots, x_N^0)$$

be an $(e + N)$-tuple of numbers in the field $\boldsymbol{K}$ which, when substituted for $u_1, \ldots, u_e, x_1, \ldots, x_N$ in F, are such that $F = 0$. This $(e+N)$-tuple of numbers, considered as the coordinates of a point in the linear space $\boldsymbol{K}^{e+N}$, is called a *zero* of the polynomial F or alternatively a *solution* to the polynomial equation $F = 0$. If the various u^0's and x^0's are all numbers of some extension field $\tilde{\boldsymbol{K}}$ of $\boldsymbol{K}$, which still make F vanish when substituted for the variables, then the $(e + N)$-tuple of numbers, considered as a point of the linear space $\tilde{\boldsymbol{K}}^{e+N}$ over $\tilde{\boldsymbol{K}}$, will be called an *extended zero* of F or an *extended solution* of $F = 0$. In order to make the involved field $\tilde{\boldsymbol{K}}$ explicit, it will also be called a $\tilde{\boldsymbol{K}}$-*zero* of F or a $\tilde{\boldsymbol{K}}$-*solution* of $F = 0$.

Let Σ be a set of polynomials. If an $(e + N)$-tuple of numbers as given above is a zero (or an extended zero, or a $\tilde{\boldsymbol{K}}$-zero) of every polynomial in Σ, it will be called a *zero* (respectively, an *extended zero* or a $\tilde{\boldsymbol{K}}$-*zero*) of Σ, or a *solution* (respectively, an *extended solution* or a $\tilde{\boldsymbol{K}}$-*solution*) of the system of equations $F = 0$, $F \in \Sigma$ or simply of $\Sigma = 0$.

Now, let $\Sigma = \Sigma_1$ be a finite set of non-zero polynomials. By Lemma 2, Σ_1 will have some basic set, say Φ_1. If Φ_1 is contradictory, then it consists of a single polynomial A_1 that is of class 0 and belongs to Σ_1. Suppose on the contrary that Φ_1 is not contradictory so that the first polynomial in Φ_1 has its class greater than 0. For all polynomials B, belonging to Σ_1 but not to Φ_1, we form the remainders R_B of B with respect to Φ_1, supposed not all 0. Adjoin all such remainders R_B, whenever non-zero, to the set Σ_1 to get an enlarged set Σ_2 of non-zero polynomials. From the formula about remainders, each R_B will be a linear sum of polynomials in Φ_1 and the polynomial B, with polynomial coefficients. It follows that the set Σ_2 will have the same set of zeros (or extended zeros, or $\tilde{\boldsymbol{K}}$ zeros for any extension field $\tilde{\boldsymbol{K}}$) as that of the original set Σ_1. By Lemma 3, the basic set Φ_2 of Σ_2 will have its rank lower than that of Φ_1. If Φ_2 is contradictory, then it consists of a single polynomial A_2 of class 0. In this case, A_2 belongs to Σ_2 and can be expressed as a linear sum of polynomials in Σ_1, in which all coefficients are polynomials. If Φ_2 is a non-contradictory ascending set, then we can proceed as before. In this way we shall get either a contradictory ascending set after a finite number of steps or a sequence of finite sets

$$\Sigma_1 \subset \Sigma_2 \subset \cdots \subset \Sigma_q \subset \cdots$$

of polynomials, where all Σ_i have the same set of zeros (or extended zeros or $\tilde{\boldsymbol{K}}$ zeros for any extended field $\tilde{\boldsymbol{K}}$) with

$$\Phi_1, \Phi_2, \ldots, \Phi_q, \ldots$$

as their corresponding non-contradictory basic sets, having steadily decreasing ranks. Now by Lemma 1, such a sequence can have only a finite number of terms. In other words, if the last set of such a sequence is Σ_q, with Φ_q as its basic set, then the remainder of any polynomial in Σ_q, not in Φ_q, with respect to Φ_q will be equal to 0.

Let Φ_q be

$$\Phi_q: \quad F_1, F_2, \ldots, F_r,$$

in which each F_i either is contained in Φ_{q-1} or is the non-zero remainder of some polynomial in Σ_{q-1}. Thus, by the remainder formula each F_i is a linear sum of polynomials in Φ_{q-1} with polynomial coefficients. It follows that any zero of Σ_{q-1} and thus any zero of Σ is also a zero of Φ_q.

On the other hand, let the initials of the polynomials in Φ_q be $I_1, I_2, \ldots, I_r$. From the construction we know that for any polynomial G in Σ_q, there should be non-negative integers $s_i \geq 0$ such that

$$I_1^{s_1} \cdots I_r^{s_r} G = Q_1 F_1 + \cdots + Q_r F_r.$$

It follows that any zero of Φ_q, if not a zero of any of the initials $I_1, \ldots, I_r$, is necessarily also a zero of Σ_q and thus a zero of $\Sigma = \Sigma_1$. The same is clearly true for extended zeros or $\tilde{\boldsymbol{K}}$-zeros for any extended field $\tilde{\boldsymbol{K}}$ of $\boldsymbol{K}$.

Denote Φ_q simply by Φ. Summing up the above we obtain the following.

Theorem. There is an algorithm which mechanically determines for any finite set Σ of non-zero polynomials, after a finite number of steps, either a non-zero polynomial A of class 0 (i.e., one involving only the variables $u_1, \ldots, u_e$) so that any zero of Σ is also a zero of A; or, a non-contradictory ascending set

$$\Phi: \quad F_1, \ldots, F_r$$

with initials $I_1, I_2, \ldots, I_r$ such that any zero of Σ is also a zero of Φ, and any zero of Φ which is not a zero of any of the initials I_i will be a zero of Σ. The same is true for extended zeros and $\tilde{\boldsymbol{K}}$-zeros.

We shall call the mechanical procedure of determining Φ from Σ the *well-ordering* of Σ, and the above theorem the *well-ordering theorem*.

4.4 A constructive theory of algebraic varieties – irreducible ascending sets and irreducible algebraic varieties

We shall continue using the notations introduced in the preceding section, denote the basic field of characteristic 0 by $\boldsymbol{K}$, fix the order of the variables as

$$x_1 \prec x_2 \prec \cdots \prec x_N$$

and omit the variables $u_1, \ldots, u_e$. By a polynomial, unless further indicated, we always mean in the variables $x_1, \ldots, x_N$ with coefficients in $\boldsymbol{K}$, i.e., an element of $\boldsymbol{K}[x_1, \ldots, x_N]$.

A finite set of non-zero polynomials will be simply called a *polynomial set.* The union of two sets Σ_1 and Σ_2 of polynomials will be denoted as $\Sigma_1+\Sigma_2$. For polynomials F, G, etc., we shall simply denote $\Sigma+\{F\}$ by $\Sigma+F$, $\Sigma+\{F,G\}$ by $\Sigma+F+G$, and so forth.

We say that a polynomial set Σ defines an *algebraic variety*, denoted by $|\Sigma|$, with Σ as its *defining set.* Let Σ_1 and Σ_2 be two polynomial sets. If any extended zero of Σ_1 is also an extended zero of Σ_2, then the algebraic variety defined by Σ_1 is called a *subvariety* of that defined by Σ_2, denoted as

$$\Sigma_2 = 0|\Sigma_1 \quad \text{or} \quad |\Sigma_1| \subset |\Sigma_2|.$$

If, further, we have $|\Sigma_2| \subset |\Sigma_1|$ so that Σ_1 and Σ_2 have one and the same set of extended zeros, then we say that Σ_1 and Σ_2 are *equivalent*, denoted as

$$\Sigma_1 \approx \Sigma_2.$$

If $|\Sigma_1| \subset |\Sigma_2|$ but $|\Sigma_1| \neq |\Sigma_2|$, i.e.,

$$|\Sigma_1| \subsetneqq |\Sigma_2|,$$

then the algebraic variety defined by Σ_1 is said to be a *true* subvariety of that defined by Σ_2.

Given a polynomial F, if any extended zero of a polynomial set Σ is also one of F, i.e.,

$$\{F\} = 0|\Sigma \quad \text{or} \quad |\Sigma| \subset |\{F\}|,$$

we say that F *vanishes on* Σ, denoted as

$$F = 0|\Sigma.$$

Otherwise we write

$$F \neq 0|\Sigma.$$

Given $k+1$ polynomial sets $\Sigma, \Sigma_1, \ldots, \Sigma_k$ $(k > 1)$ having the following property: Any extended zero of Σ is also an extended zero of at least one Σ_i of the sets $\Sigma_1, \ldots, \Sigma_k$, and conversely, any extended zero of any Σ_i is also one of Σ, then we say that $\Sigma_1, \ldots, \Sigma_k$ are a *decomposition* of Σ, or the corresponding algebraic varieties $|\Sigma_1|, \ldots, |\Sigma_k|$ are a *decomposition* of $|\Sigma|$, denoted as

$$|\Sigma| = |\Sigma_1| \cup \cdots \cup |\Sigma_k|.$$

If all $|\Sigma_i|$ and $|\Sigma_j|$ $(j \neq i)$ in the above union have no $\subset$ relations, then such a decomposition is said to be *non-contractible*. In this case, the variety defined by every Σ_i is a *true* subvariety of $|\Sigma|$, but not a subvariety of any other $|\Sigma_j|$ in the union.

We say that a polynomial set Σ is *reducible* if it has a non-contractible decomposition, while the variety defined by Σ is also said to be *reducible*. In the contrary case we say that Σ as well as the variety defined by it is *irreducible*. If in a certain decomposition of Σ each Σ_i is irreducible, then we say that this decomposition is an *irreducible decomposition* of Σ, and the same for the variety $|\Sigma|$. In this case, each Σ_i or $|\Sigma_i|$ is called an *irreducible component* of Σ or $|\Sigma|$.

In this section we consider the problem of reducibility of a polynomial set or its defining algebraic variety and in the next section we shall consider the problem of irreducible decomposition of an algebraic variety. The following lemma provides a simple criterion for the irreducibility of an algebraic variety.

Lemma 1. A necessary and sufficient condition for a polynomial set Σ to be irreducible is that there cannot exist two non-zero polynomials G and H such that

$$GH = 0|\Sigma$$

while

$$G \neq 0|\Sigma, \quad H \neq 0|\Sigma.$$

Proof. If there do exist G and H satisfying all the above conditions, then we clearly have a decomposition

$$|\Sigma| = |\Sigma + G| \cup |\Sigma + H|,$$

and this decomposition is non-contractible, so Σ is reducible.

On the other hand, let Σ be reducible and

$$|\Sigma| = |\Sigma_1| \cup |\Sigma_2|$$

be a non-contractible decomposition of Σ. Then there must exist a polynomial F_1 in Σ_1 such that $F_1 \neq 0|\Sigma_2$, so $F_1 \neq 0|\Sigma$. Similarly, in Σ_2 there exists a polynomial F_2 such that $F_2 \neq 0|\Sigma_1$, so $F_2 \neq 0|\Sigma$. But any extended zero of Σ should also be an extended zero of Σ_1 or Σ_2, thus also an extended zero of F_1 or F_2. Hence any extended zero of Σ is also one of the product F_1F_2, i.e., $F_1F_2 = 0|\Sigma$. So $G = F_1$ and $H = F_2$ satisfy the condition in the lemma. This completes the proof. □

We shall give another criterion for the irreducibility of a polynomial set Σ

or the algebraic variety $|\Sigma|$ defined by it. For this purpose let us first introduce some notions as follows.

Consider two extension fields $\tilde{\boldsymbol{K}}$ and $\boldsymbol{K}'$ of $\boldsymbol{K}$ and two points $\tilde{\xi} = (\tilde{x}_1, \ldots, \tilde{x}_N)$, $\tilde{x}_i \in \tilde{\boldsymbol{K}}$ and $\xi' = (x'_1, \ldots, x'_N)$, $x'_i \in \boldsymbol{K}'$ in the N-dimensional linear spaces $\tilde{\boldsymbol{K}}^N$ and $\boldsymbol{K}'^N$ over $\tilde{\boldsymbol{K}}$ and $\boldsymbol{K}'$ respectively. Suppose these two points possess the following property:

For any polynomial $F(x_1, \ldots, x_N)$ in $\boldsymbol{K}[x_1, \ldots, x_N]$, if $\tilde{\xi}$ is an extended zero of F then ξ' is also an extended zero of F. In other words, $F(x'_1, \ldots, x'_N) = 0$ as long as $F(\tilde{x}_1, \ldots, \tilde{x}_N) = 0$.

In this case we shall call $\tilde{\xi}$ a *generic point* of ξ', or ξ' a *special point* of $\tilde{\xi}$, or ξ' a *specialization* of $\tilde{\xi}$ with respect to $\boldsymbol{K}$. In case no misunderstanding can occur, the words "with respect to $\boldsymbol{K}$" will be omitted.

Suppose the polynomial set Σ has an extended zero $\tilde{\xi}$ such that any extended zero of Σ is a specialization of $\tilde{\xi}$ with respect to $\boldsymbol{K}$. Then we say that $\tilde{\xi}$ is a *generic point* of the polynomial set Σ or the algebraic variety $|\Sigma|$.

The following lemma gives the second irreducibility criterion of polynomial sets or algebraic varieties.

Lemma 2. A necessary and sufficient condition for a polynomial set Σ or its variety to be irreducible is that Σ has generic points.

Proof. Suppose first that Σ has a generic point $\tilde{\xi} = (\tilde{x}_1, \ldots, \tilde{x}_N)$, in which $\tilde{x}_i$ belongs to some extension field $\tilde{\boldsymbol{K}}$ of $\boldsymbol{K}$. Let G and H be two polynomials in $\boldsymbol{K}[x_1, \ldots, x_N]$ such that $GH = 0|\Sigma$. Then $G(\tilde{\xi})H(\tilde{\xi}) = 0$, so either $G(\tilde{\xi}) = 0$ or $H(\tilde{\xi}) = 0$. To fix the idea let $G(\tilde{\xi}) = 0$. As $\tilde{\xi}$ is a generic point of Σ, for any extended zero ξ' of Σ we have $G(\xi') = 0$, i.e., $G = 0|\Sigma$. Similarly, in the case $H(\tilde{\xi}) = 0$ we have $H = 0|\Sigma$. By Lemma 1, Σ is irreducible.

Suppose secondly that Σ is irreducible. We shall construct a generic point of Σ as follows.

For two arbitrary polynomials F and G, if

$$F - G = 0|\Sigma,$$

then F and G are said to be *equivalent* for Σ, denoted as

$$F \sim G(\Sigma).$$

Thus polynomials can be divided into *equivalence classes*. Clearly, if

$$F_1 \sim G_1, \ F_2 \sim G_2(\Sigma),$$

then

$$F_1 \pm F_2 \sim G_1 \pm G_2(\Sigma),$$

$$F_1 F_2 \sim G_1 G_2(\Sigma).$$

Therefore, on the equivalence classes of Σ we can perform the operations of addition, subtraction and multiplication. We shall call the equivalence class containing the polynomial 0 the *zero class*, denoted as θ, and denote the equivalence class containing the polynomial 1 by ω. Any element a in $\boldsymbol{K}$ may be identified with the equivalence class $a\omega$. Since Σ has been assumed to be irreducible, when two arbitrary equivalence classes α and β are both not equal to the zero class θ, by Lemma 1, $\alpha\beta$ is not equal to θ either. In fact, these equivalence classes form a domain, denoted by R_Σ, with ω as its unit element.

Consider now the quotient field of R_Σ which, in other words, is the totality of all the *pairs* (α, β) of equivalence classes with $\beta \neq \theta$ under the equivalence relation

$$(\alpha, \beta) \sim (\gamma, \delta) \Longleftrightarrow \alpha\delta = \beta\gamma,$$

in which the operations of addition, subtraction and multiplication can be naturally defined. In particular, when $(\gamma, \delta) \nsim (\theta, \omega)$,

$$(\alpha, \beta)/(\gamma, \delta) \sim (\alpha\delta, \beta\gamma).$$

If we identify each element a in $\boldsymbol{K}$ with the equivalence class of the pair $(a\omega, \omega)$, then this quotient field may be considered as an extension field of $\boldsymbol{K}$. We denote this field by $\tilde{\boldsymbol{K}}_\Sigma$. In addition, any element α in R_Σ will be identified with the equivalence class of the pair (α, ω) in $\tilde{\boldsymbol{K}}_\Sigma$.

Let the equivalence class defined by the polynomial x_i in R_Σ be ξ_i and that defined by the pair (ξ_i, ω) in $\tilde{\boldsymbol{K}}_\Sigma$ still be denoted ξ_i. Then $\tilde{\xi} = (\xi_1, \ldots, \xi_N)$ is a point in the linear space $\tilde{\boldsymbol{K}}^N$ over $\tilde{\boldsymbol{K}} = \tilde{\boldsymbol{K}}_\Sigma$. We prove below that $\tilde{\xi}$ is a generic point of Σ.

First, for an arbitrary polynomial $F = F(x_1, \ldots, x_N)$ in Σ we have $F = 0|\Sigma$, so the equivalence class ϕ of F in R_Σ is θ. Obviously, in R_Σ and thus in $\tilde{\boldsymbol{K}}_\Sigma$ too, we have $\phi = F(\xi_1, \ldots, \xi_N)$. Hence $\tilde{\xi}$ is an extended zero of $F \in \tilde{\boldsymbol{K}}[x_1, \ldots, x_N]$. Because F was arbitrary in Σ, $\tilde{\xi}$ is an extended zero of Σ.

Now, let $G = G(x_1, \ldots, x_N)$ be an arbitrary polynomial in $\boldsymbol{K}[x_1, \ldots, x_N]$ with $\tilde{\xi}$ as its extended zero, i.e., $G(\xi_1, \ldots, \xi_N) = \theta$. Then $G(x_1, \ldots, x_N)$ belongs to the zero class θ, i.e., $G = 0|\Sigma$. Therefore, for any extended zero ξ' of Σ we have $G(\xi') = 0$, i.e.,

$$G(\tilde{\xi}) = 0 \Longrightarrow G(\xi') = 0.$$

As G was arbitrary, any extended zero ξ' of Σ is a special point of $\tilde{\xi}$. It follows that $\tilde{\xi}$ is a generic point of Σ and the lemma is proved. □

Although the above two lemmas give necessary and sufficient conditions for a polynomial set to be irreducible, these criteria are merely existential and not constructive. Given a polynomial set Σ, these two lemmas do not provide methods for deciding in a finite number of steps whether or not there exist polynomials G and H to satisfy the conditions of Lemma 1, or whether there exists an extension field $\tilde{\boldsymbol{K}}$ and a point $\tilde{\xi}$ in $\tilde{\boldsymbol{K}}^N$ as a generic point of Σ. To meet the requirement of the gist of this book – mechanical theorem proving – we have to devise a mechanical method which decides in a finite number of steps whether or not a given polynomial set is irreducible, and if not, finds in a finite number of steps every irreducible component of the irreducible decomposition. In this and the next few sections we shall give such a mechanical method in order to set up a *constructive theory* of algebraic geometry.

Consider an ascending set

$$\Phi:\quad A_1, A_2, \ldots, A_n,$$

in which the class of A_i is p_i with

$$0 < p_1 < p_2 < \cdots < p_n.$$

We shall rename the variables x's to y's and u's by setting

$$x_{p_1} = y_1, \ldots, x_{p_n} = y_n$$

and all the other x's in the original order to $u_1, \ldots, u_d$. Here $d = N - n$ is called the *dimension* of the ascending set Φ and denoted as

$$d = \dim \Phi.$$

Now the polynomials A_i in Φ can be written in the form:

$$\Phi:\quad \begin{cases} A_1 = C_{10}y_1^{m_1} + C_{11}y_1^{m_1-1} + \cdots + C_{1m_1}, \\ A_2 = C_{20}y_2^{m_2} + C_{21}y_2^{m_2-1} + \cdots + C_{2m_2}, \\ \quad\cdots \\ A_n = C_{n0}y_n^{m_n} + C_{n1}y_n^{m_n-1} + \cdots + C_{nm_n}, \end{cases}$$

where each $C_{i0} \neq 0$ is the initial of A_i. In addition, each C_{ij} is a polynomial in $u_1, \ldots, u_d, y_1, \ldots, y_{i-1}$ with coefficients in $\boldsymbol{K}$ and each A_i is already reduced with respect to $A_1, \ldots, A_{i-1}$ so that the degrees of C_{ij} in $y_1, \ldots, y_{i-1}$ are less than $m_1, \ldots, m_{i-1}$ respectively. The first problem to be considered is to give conditions for Φ to be the basic set of a certain irreducible polynomial set.

For this problem let us suppose that the ascending set Φ has the following properties:

Let the transcendental extension field $\boldsymbol{K}(u_1, \ldots, u_d)$ of $\boldsymbol{K}$, acquired by adjoining $u_1, \ldots, u_d$, be denoted as $\boldsymbol{K}_0$; then A_1 as a polynomial in $\boldsymbol{K}_0[y_1]$ is irreducible.

Let the algebraic extension field of $\boldsymbol{K}_0$, acquired by adjoining an extended zero η_1 of $A_1 = 0$, be denoted as $\boldsymbol{K}_0(\eta_1) = \boldsymbol{K}_1$; then the polynomial $\tilde{A}_2$ in $\boldsymbol{K}_1[y_2]$ obtained from A_2 by substituting η_1 for y_1 is irreducible in $\boldsymbol{K}_1[y_2]$.

Let the algebraic extension field of $\boldsymbol{K}_1$, acquired by adjoining an extended zero η_2 of $\tilde{A}_2 = 0$, be denoted as $\boldsymbol{K}_1(\eta_2) = \boldsymbol{K}_2$; then the polynomial $\tilde{A}_3$ in

$\boldsymbol{K}_2[y_3]$ obtained from A_3 by substituting η_1 for y_1 and η_2 for y_2 is irreducible in $\boldsymbol{K}_2[y_3]$.

Suppose that proceeding in the same manner we get algebraic extensions $\boldsymbol{K}_i = \boldsymbol{K}_{i-1}(\eta_i)$, polynomials $\tilde{A}_i$ obtained from A_i by substituting $\eta_1, \ldots, \eta_{i-1}$ for $y_1, \ldots, y_{i-1}$ and some extended zeros η_i of $\tilde{A}_i = 0$, where each $\tilde{A}_i$ is irreducible in $\boldsymbol{K}_{i-1}[y_i]$ for $i = 1, 2, \ldots, n$.

Under the above conditions we say that the ascending set Φ is *irreducible*. From Sect. 4.2 we know that there is a mechanical method which decides in a finite number of steps whether or not Φ is irreducible.

Let Φ be irreducible and thus satisfy the above conditions. Then u_i and η_j are all elements of $\tilde{\boldsymbol{K}} = \boldsymbol{K}_n$, so $\tilde{\eta} = (u_1, \ldots, u_d, \eta_1, \ldots, \eta_n)$ is a point of the linear space $\tilde{\boldsymbol{K}}^{d+n} = \tilde{\boldsymbol{K}}^N$. We shall call $\tilde{\eta}$ a *generic point* of Φ and $\tilde{\boldsymbol{K}}$ a *generating field* of Φ.

The following lemma is quite important.

Lemma 3. If an ascending set Φ is irreducible with

$$\tilde{\eta} = (u_1, \ldots, u_d, \eta_1, \ldots, \eta_n)$$

as a generic point, then a necessary and sufficient condition for a polynomial $F \in \boldsymbol{K}[u_1, \ldots, u_d, y_1, \ldots, y_n]$ to have remainder R equal to 0 with respect to Φ is: $\tilde{\eta}$ is an extended zero of F.

Proof. Let the ascending set formed by the first k polynomials in Φ be

$$\Phi_k: \quad A_1, A_2, \ldots, A_k \quad (1 \leq k \leq n).$$

Denote by $\boldsymbol{K}^{d+k}$ the $(d+k)$-dimensional linear space over $\boldsymbol{K}$ with base $u_1, \ldots, u_d, y_1, \ldots, y_k$, and similarly for the others. Then Φ_k is clearly irreducible, and

$$\tilde{\eta}_k = (u_1, \ldots, u_d, \eta_1, \ldots, \eta_k),$$

when considered as a point in $\boldsymbol{K}_k^{d+k}$, is a generic point of Φ_k while $\boldsymbol{K}_k$ is the generating field of Φ_k.

We shall prove by induction on k the following two assertions:

1_k°. $\tilde{\eta}_{k-1}$ is not an extended zero of C_{k0};
2_k°. If $R_k \in \boldsymbol{K}[u_1, \ldots, u_d, y_1, \ldots, y_k]$ is already reduced with respect to Φ_k and $\tilde{\eta}_k$ is an extended zero of R_k, then R_k is identically 0.

As $C_{k+1,0} \in \boldsymbol{K}[u_1, \ldots, u_d, y_1, \ldots, y_k]$ is known to be reduced with respect to Φ_k and is not 0, we see that 1_{k+1}° is a consequence of 2_k°.

Suppose now 2_{k-1}° has already been proved. Consider any R_k satisfying the conditions in 2_k°. Write R_k as a polynomial in y_k:

$$R_k = S_0 y_k^r + S_1 y_k^{r-1} + \cdots + S_r,$$

in which $S_i \in \boldsymbol{K}[u_1, \ldots, u_d, y_1, \ldots, y_{k-1}]$ with $r < m_k$. Substitute $y_1, \ldots, y_{k-1}$ in S_i with $\eta_1, \ldots, \eta_{k-1}$, the resulting S_i being denoted as $\tilde{S}_i \in \boldsymbol{K}_{k-1}$. Set

$$\tilde{R}_k = \tilde{S}_0 y_k^r + \tilde{S}_1 y_k^{r-1} + \cdots + \tilde{S}_r \in \boldsymbol{K}_{k-1}[y_k].$$

By hypothesis, η_k is an extended zero of $\tilde{R}_k$. As $r < m_k$ and η_k is an extended zero of the irreducible $\tilde{A}_k$ in $\boldsymbol{K}_{k-1}$, $\tilde{R}_k$ should be identically 0, so $\tilde{S}_0 = 0, \ldots, \tilde{S}_r = 0$. Since R_k is reduced with respect to Φ_k, each S_i is reduced with respect to Φ_{k-1}, so by the induction hypothesis 2°_{k-1} we have $S_i = 0$ and hence $R_k = 0$, i.e., 2°_k holds true. It follows that 1°_{k+1} is also true. The above proof is clearly valid for 2°_1 while 1°_1 is quite evident. Consequently, 1°_k and 2°_k are true for $k = 1, 2, \ldots, n$.

It is now easy to complete the proof of Lemma 3 as follows. Let the remainder of F with respect to $\Phi_n = \Phi$ be R. Then there are integers $s_i \geq 0$ and polynomials Q_i such that

$$C_{10}^{s_1} \cdots C_{n0}^{s_n} F = Q_1 A_1 + \cdots + Q_n A_n + R.$$

Suppose $R = 0$. Since $\tilde{\eta}$ is an extended zero of the A_k's, but by 1°_k it is not an extended zero of any C_{k0}, by the remainder formula above it is an extended zero of F. Conversely, if $\tilde{\eta}$ is an extended zero of F, then by the same formula $\tilde{\eta}$ is also an extended zero of R. By 2°_n we necessarily have $R = 0$. This completes the proof. □

Lemma 4. Let an ascending set

$$\Phi: \quad A_1, A_2, \ldots, A_n$$

be irreducible with a generic point

$$\tilde{\eta} = (u_1, \ldots, u_d, \eta_1, \ldots, \eta_n)$$

as before. If a polynomial $F \in \boldsymbol{K}[u_1, \ldots, u_d, y_1, \ldots, y_n]$ has non-zero remainder with respect to Φ, then in $\boldsymbol{K}[u_1, \ldots, u_d, y_1, \ldots, y_n]$ there are polynomials G and Q_i, $i = 1, \ldots, n$, such that

$$GF - (Q_1 A_1 + \cdots + Q_n A_n) \in \boldsymbol{K}[u_1, \ldots, u_d]$$

and

$$G(\tilde{\eta}) \neq 0.$$

Proof. We shall prove the following assertion (T_k) by induction on k, while the lemma itself is equivalent to the case $k = n$.

(T_k) Suppose the remainder of

$$F_k \in \boldsymbol{K}[u_1, \ldots, u_d, y_1, \ldots, y_k]$$

with respect to

$$\Phi_k: \quad A_1, A_2, \ldots, A_k$$

is non-zero. Then there must be G_k and Q_{ki} in $\boldsymbol{K}[u_1, \ldots, u_d, y_1, \ldots, y_k]$, $i = 1, \ldots, k$, such that

$$G_k F_k - (Q_{k1} A_1 + \cdots + Q_{kk} A_k) \in \boldsymbol{K}[u_1, \ldots, u_d]$$

and

$$G_k(\tilde{\eta}_k) \neq 0.$$

The case $k = 0$ is trivial. Suppose (T_k) holds true for $k < n$. Let us prove (T_{k+1}) as follows.

Since the initial I_{k+1} of A_{k+1} obviously satisfies the conditions of (T_k), by the induction hypothesis there are polynomials $H_k, R_{k1}, \ldots, R_{kk} \in \boldsymbol{K}[u_1, \ldots, u_d, y_1, \ldots, y_k]$ such that

$$H_k I_{k+1} - (R_{k1} A_1 + \cdots + R_{kk} A_k) \in \boldsymbol{K}[u_1, \ldots, u_d]$$

and

$$H_k(\tilde{\eta}_k) = H_k(\tilde{\eta}_{k+1}) \neq 0.$$

Suppose $F_{k+1} \in \boldsymbol{K}[u_1, \ldots, u_d, y_1, \ldots, y_{k+1}]$ has a non-zero remainder with respect to Φ_{k+1}. By Lemma 3 we have $F_{k+1}(\tilde{\eta}_{k+1}) \neq 0$. Now let us form the resultant of F_{k+1} and A_{k+1} as polynomials in y_{k+1}. Then there are polynomials $U_{k+1}, V_{k+1} \in \boldsymbol{K}[u_1, \ldots, u_d, y_1, \ldots, y_{k+1}]$ and $F_k \in \boldsymbol{K}[u_1, \ldots, u_d, y_1, \ldots, y_k]$ such that

$$U_{k+1} F_{k+1} + V_{k+1} A_{k+1} = F_k \in \boldsymbol{K}[u_1, \ldots, u_d, y_1, \ldots, y_k],$$

in which the degree of U_{k+1} in y_{k+1} is less than m_{k+1} (the degree of A_{k+1} in y_{k+1}) and the degree of V_{k+1} is less than that of F_{k+1} in y_{k+1}. Let $\tilde{U}_{k+1}, \tilde{V}_{k+1}$, etc. be the polynomials in $\boldsymbol{K}_k[y_{k+1}]$ obtained from U_{k+1}, V_{k+1}, etc. by substituting $\eta_1, \ldots, \eta_k$ for $y_1, \ldots, y_k$ respectively. Then the above expression leads to

$$\tilde{U}_{k+1} \tilde{F}_{k+1} + \tilde{V}_{k+1} \tilde{A}_{k+1} = \tilde{F}_k.$$

As $\tilde{A}_{k+1}$ is an irreducible polynomial in $\boldsymbol{K}_k[y_{k+1}]$ and $\tilde{A}_{k+1}(\eta_{k+1}) = 0$, $\tilde{F}_{k+1}(\eta_{k+1}) = F_{k+1}(\tilde{\eta}_{k+1}) \neq 0$. Hence, $\tilde{F}_{k+1}$ and $\tilde{A}_{k+1}$ cannot have common divisors of y_{k+1}, and thus $\tilde{F}_k \neq 0$ or $F_k(\tilde{\eta}_k) = F_k(\tilde{\eta}_{k+1}) \neq 0$. Since $A_{k+1}(\tilde{\eta}_{k+1}) = 0$, we have $U_{k+1}(\tilde{\eta}_{k+1}) \neq 0$.

From $F_k(\tilde{\eta}_k) \neq 0$, by Lemma 3 we know that the remainder of F_k with respect to Φ_k is non-zero. Thus, by the induction hypothesis there are polynomials $G_k, Q_{k1}, \ldots, Q_{kk} \in \boldsymbol{K}[u_1, \ldots, u_k, y_1, \ldots, y_k]$ satisfying the formulas in (T_k):

$$G_k F_k - (Q_{k1} A_1 + \cdots + Q_{kk} A_k) \in \boldsymbol{K}[u_1, \ldots, u_k]$$

and

$$G_k(\tilde{\eta}_k) = G_k(\tilde{\eta}_{k+1}) \neq 0.$$

Now set

$$G_{k+1} = G_k U_{k+1},$$
$$Q_{k+1,1} = Q_{k1}, \ldots, Q_{k+1,k} = Q_{k,k},$$
$$Q_{k+1,k+1} = -G_k V_{k+1}.$$

Then we have

$$G_{k+1}, Q_{k+1,1}, \ldots, Q_{k+1,k+1} \in \boldsymbol{K}[u_1, \ldots, u_d, y_1, \ldots, y_{k+1}],$$
$$G_{k+1}F_{k+1} - (Q_{k+1,1}A_1 + \cdots + Q_{k+1,k+1}A_{k+1}) \in \boldsymbol{K}[u_1, \ldots, u_d],$$

and

$$G_{k+1}(\tilde{\eta}_{k+1}) = G_k(\tilde{\eta}_k)U_{k+1}(\tilde{\eta}_{k+1}) \neq 0.$$

So (T_{k+1}) holds true and the lemma is proved. □

Given an irreducible ascending set Φ as above, let Ω be the set of all those polynomials in $\boldsymbol{K}[u_1, \ldots, u_d, y_1, \ldots, y_n]$ whose remainders with respect to Φ are 0. By Lemma 3, this set will clearly form a *module*. By Hilbert's basis theorem, there will be a finite set of polynomials in Ω such that any polynomial of Ω is a linear combination of these polynomials with polynomial coefficients. We may add $A_1, \ldots, A_n$ of Φ to this finite set and denote the enlarged set by Ω_Φ. By Lemma 3, this polynomial set will clearly have Φ as its basic set and $\tilde{\eta}$ as an extended zero. Let G be any polynomial with $\tilde{\eta}$ as an extended zero. By Lemma 3 again, G has remainder 0 with respect to Φ. According to the construction of Ω_Φ, G can be expressed as a linear sum of polynomials in Ω_Φ, so $G = 0|\Omega_\Phi$. It follows that any extended zero of Ω_Φ is a specialization of $\tilde{\eta}$ or that Ω_Φ is an irreducible polynomial set with $\tilde{\eta}$ as a generic point.

The above proof of the possibility of constructing an irreducible set Ω_Φ of polynomials from an irreducible ascending set Φ is based on Hilbert's finite basis theorem. Since Ω is an infinite set, the proof depends on the axiom of choice and only the existence of Ω_Φ can be proved. Actually, we can construct such a finite set Ω_Φ of polynomials in a finite number of steps by using a mechanical method. In other words, we have the following theorem.

Theorem. There is a mechanical method which determines for any irreducible ascending set Φ a finite number of polynomials including those of Φ that form an irreducible polynomial set Ω_Φ with Φ as its basic set and any generic point of Φ as its generic point in a finite number of steps.

The proof of this theorem is not simple. As for application, the mere existence of such an irreducible polynomial set Ω_Φ as guaranteed by Hilbert's basis theorem will be more than sufficient. We shall satisfy ourselves in merely stating the theorem while putting aside the proof.

4.5 A constructive theory of algebraic varieties – irreducible decomposition of algebraic varieties

In this section we propose a mechanical method to decompose a polynomial set into irreducible polynomial sets in a finite number of steps.

Continuing the use of notations in the preceding section, we shall denote the basic field by $\boldsymbol{K}$ and fix the order of the variables as $x_1 \prec \cdots \prec x_N$. By a polynomial, unless further indicated otherwise, we always mean an element in $\boldsymbol{K}[x_1, \ldots, x_N]$. Let

$$\Phi: \quad A_1, A_2, \ldots, A_n$$

be an ascending set with the class p_i of A_i satisfying

$$0 < p_1 < \cdots < p_n.$$

As in the last section we denote the variables x's by u's and y's, and change their order to

$$u_1 \prec \cdots \prec u_d \prec y_1 \prec \cdots \prec y_n,$$

in which $y_i = x_{p_i}$, $d = N - n$, $u_j = x_k$ $(k \neq p_i)$. As before, set

$$A_i = C_{i0} y_i^{m_i} + C_{i1} y_i^{m_i - 1} + \cdots + C_{im_i},$$

in which $C_{ij} \in \boldsymbol{K}[u_1, \ldots, u_d, y_1, \ldots, y_{i-1}]$, and $C_{i0} \neq 0$ is the initial of A_i, denoted as

$$I_i = C_{i0} \in \boldsymbol{K}[u_1, \ldots, u_d, y_1, \ldots, y_{i-1}].$$

Let Φ be irreducible with

$$\tilde{\eta} = (u_1, \ldots, u_d, \eta_1, \ldots, \eta_n)$$

as a generic point, and let $\tilde{\eta}_i$ have the same meaning as that at the end of the last section. By the theorem in the last section, we know there is an irreducible polynomial set Ω_Φ with Φ as its basic set and $\tilde{\eta}$ as a generic point and that Ω_Φ can be determined from Φ in a mechanical manner. Therefore, the irreducibility of Φ is a sufficient condition for Φ to be the basic set of some irreducible polynomial set Ω_Φ. To this we add the following.

Lemma 1. Let the basic set Φ of a polynomial set Λ be irreducible with the class of each polynomial A_i in Φ being > 0. Denote the initial of A_i by I_i for $i = 1, \ldots, n$. If every polynomial in Λ has remainder 0 with respect to Φ, then Λ has a decomposition

$$|\Lambda| = |\Omega_\Phi| \cup |\Lambda + I_1| \cup \cdots \cup |\Lambda + I_n|,$$

in which Ω_Φ or the corresponding algebraic variety $|\Omega_\Phi|$ is irreducible.

Proof. For any polynomial whose remainder with respect to Φ is 0, in particular, any polynomial G in Λ, there are $s_i \geq 0$ and $Q_i \in \boldsymbol{K}[u_1, \ldots, u_d, y_1, \ldots, y_n]$ such that

$$I_1^{s_1} \cdots I_n^{s_n} G = Q_1 A_1 + \cdots Q_n A_n.$$

From our discussion in the last section, G is a linear sum of polynomials in Ω_Φ so that any extended zero of Ω_Φ is an extended zero of G and hence an extended zero of Λ. Conversely, any extended zero of Λ may be considered as an extended zero of the A_i's. Thus by the above formula it should be an extended zero of either G or some I_i. That is, it should be an extended zero of Ω_Φ or some $\Lambda + I_i$. Hence we have the decomposition claimed in the lemma. □

Lemma 2. Let Λ, Φ be as in Lemma 1 with Λ irreducible. Then

$$\Lambda \approx \Omega_\Phi \quad \text{or} \quad |\Lambda| = |\Omega_\Phi|.$$

Proof. Let the initials of the polynomials in Φ be I_i for $i = 1, \ldots, n$. Then it is clear by definition that

$$|\Lambda + I_1| \cup \cdots \cup |\Lambda + I_n| = |\Lambda + I_1 \cdots I_n|.$$

The decomposition given in Lemma 1 can therefore be written as

$$|\Lambda| = |\Omega_\Phi| \cup |\Lambda + I_1 \cdots I_n|.$$

Any generic point of Φ is also a generic point of Ω_Φ, but not an extended zero of $I_1 \cdots I_n$, so

$$|\Omega_\Phi| \not\subset |\Lambda + I_1 \cdots I_n|.$$

If Λ has some extended zero which is not an extended zero of Ω_Φ, it should be an extended zero of $\Lambda + I_1 \cdots I_n$, so that

$$|\Lambda + I_1 \cdots I_n| \not\subset |\Omega_\Phi|.$$

In this way $|\Lambda|$ would have a non-contractible decomposition contrary to the irreducibility hypothesis on Λ. Hence we have $|\Lambda| \subset |\Omega_\Phi|$. On the other hand, we have $|\Omega_\Phi| \subset |\Lambda|$, and thus $|\Lambda| = |\Omega_\Phi|$. The lemma is proved. □

Consider now an ascending set Φ as before but with Φ reducible. Then there will be some k such that

$$\Phi_{k-1}: \quad A_1, A_2, \ldots, A_{k-1}$$

is irreducible with

$$\tilde{\eta}_{k-1} = (u_1, \ldots, u_d, \eta_1, \ldots, \eta_{k-1})$$

as a generic point, and the polynomial $\tilde{A}_k$ obtained from A_k by substituting $\eta_1, \ldots, \eta_{k-1}$ for $y_1, \ldots, y_{k-1}$ is reducible in $\boldsymbol{K}_{k-1}[y_k]$, where $\boldsymbol{K}_{k-1} = \boldsymbol{K}_0(\eta_1, \ldots, \eta_{k-1})$. Let the irreducible factorization of $\tilde{A}_k$ in $\boldsymbol{K}_{k-1}[y_k]$ be given by

$$\tilde{A}_k = g_1 \cdots g_h,$$

in which each $g_i \in \boldsymbol{K}_{k-1}[y_k]$ is irreducible and $h \geq 2$. As in g_i the coefficients of the powers of y_k are all elements of $\boldsymbol{K}_{k-1}$ and can thus be expressed as the quotients of two polynomials in $u_1, \ldots, u_d, \eta_1, \ldots, \eta_{k-1}$. By reducing fractions to a common denominator, we get an expression of the form

$$\tilde{D}\tilde{A}_k = \tilde{G}_1 \cdots \tilde{G}_h,$$

where $D \in \boldsymbol{K}[u_1, \ldots, u_d, y_1, \ldots, y_{k-1}]$, $G_i \in \boldsymbol{K}[u_1, \ldots, u_d, y_1, \ldots, y_k]$, while $\tilde{D}$, $\tilde{G}_i$ are polynomials in $\boldsymbol{K}_{k-1}[y_k]$, acquired from D, G_i by substituting $\eta_1, \ldots, \eta_{k-1}$ for $y_1, \ldots, y_{k-1}$. We may also consider D to already be reduced with respect to Φ_{k-1} and similarly G_j already reduced with respect to Φ_k.

Write the polynomial $G_1 \cdots G_h - DA_k$ according to the powers of y_k, say

$$G_1 \cdots G_h - DA_k = \sum_j B_j y_k^j,$$

in which $B_j \in \boldsymbol{K}[u_1, \ldots, u_d, y_1, \ldots, y_{k-1}]$. Denote by b_j the element in $\boldsymbol{K}_{k-1} = \boldsymbol{K}_0(\eta_1, \ldots, \eta_{k-1})$ acquired from B_j by substituting $\eta_1, \ldots, \eta_{k-1}$ for $y_1, \ldots, y_{k-1}$. Then we have $b_j = 0$ since $\tilde{D}\tilde{A}_k = \tilde{G}_1 \cdots \tilde{G}_h$. In other words, each B_j will have $\tilde{\eta}_{k-1}$ as an extended zero. It follows from the proof of Lemma 3 in Sect. 4.4 that each B_j will have remainder 0 with respect to the irreducible ascending set Φ_{k-1}, so there are non-negative integers $s_{j,1}, \ldots, s_{j,k-1}$ and polynomials $Q_{ji} \in \boldsymbol{K}[u_1, \ldots, u_d, y_1, \ldots, y_{k-1}]$ satisfying the relation

$$I_1^{s_{j,1}} \cdots I_{k-1}^{s_{j,k-1}} B_j = \sum_{i=1}^{k-1} Q_{ji} A_i.$$

Set $s_i = \max_j(s_{ji})$; then we get

$$I_1^{s_1} \cdots I_{k-1}^{s_{k-1}} (G_1 \cdots G_h - DA_k) = \sum_{i=1}^{k-1} Q_i A_i,$$

or

$$I_1^{s_1} \cdots I_{k-1}^{s_{k-1}} G_1 \cdots G_h = \sum_{i=1}^{k} Q_i A_i,$$

in which the Q_i are all polynomials in $u_1, \ldots, u_d, y_1, \ldots, y_k$.

From the above it is easy to obtain the following.

Lemma 3. Let a polynomial set Λ have Φ as its basic set, and let the class of A_i in Φ be > 0 and the initial of A_i be I_i for $i = 1, \ldots, n$. Suppose Φ is reducible, so that there is some k for which the ascending set Φ_{k-1} formed by the first $k-1$ terms of Φ is irreducible with $\tilde{\eta}_{k-1} \in \boldsymbol{K}_{k-1}$ as a generic point, while the polynomial gotten from A_k by substituting $\tilde{\eta}_{k-1}$ for the corresponding variables is reducible with an irreducible factorization into polynomials $G_1, \ldots, G_h$. Then there is a decomposition of $|\Lambda|$ of the form

$$|\Lambda| = |\Lambda + I_1| \cup \cdots \cup |\Lambda + I_{k-1}| \cup |\Lambda + G_1| \cup \cdots \cup |\Lambda + G_h|.$$

Proof. Any extended zero of either $\Lambda + I_i$ or $\Lambda + G_j$ on the right-hand side of the above expression is clearly an extended zero of Λ. Conversely, any extended zero of Λ is also an extended zero of the A_i's. By the expression just preceding this lemma, an extended zero of Λ is an extended zero of some I_i or some G_j, i.e., one of some $\Lambda + I_i$ or some $\Lambda + G_j$. This proves the lemma. □

Lemma 4. Let Λ be a polynomial set with Φ as its basic set as in Lemma 1 or 3. Then the basic set of any polynomial set $\Lambda + I_i$ or $\Lambda + G_j$ appearing in the right-hand side of the decomposition in Lemma 1 or 3 will have a rank lower than that of Φ.

Proof. As each I_i is already reduced with respect to Φ and each G_j is assumed to be reduced with respect to Φ_k and hence also reduced with respect to Φ, the present lemma is an immediate consequence of Lemma 3 of Sect. 4.3. □

Lemma 5. Let a polynomial set Λ be irreducible with an irreducible ascending set Φ as its basic set. Suppose also that any polynomial in the polynomial sets Λ, Λ' has its remainder 0 with respect to Φ. Then

$$|\Lambda| \cup |\Lambda'| = |\Lambda'|$$

or, in other words, the decomposition $|\Lambda| \cup |\Lambda'|$ is contractible.

Proof. By Lemma 2 we have $|\Omega_\Phi| = |\Lambda|$. According to the hypothesis any polynomial G' in Λ' has its remainder 0 with respect to Φ. By Lemma 3 of Sect. 4.4, it follows that any generic point of Φ or of Ω_Φ is an extended zero of G', whence $G' = 0|\Omega_\Phi$. Consequently, $\Lambda' = 0|\Omega_\Phi$ or $|\Omega_\Phi| \subset |\Lambda'|$, i.e. $|\Lambda| \subset |\Lambda'|$. This proves the lemma. □

From the above lemmas and Sects. 4.3 and 4.4 we get a mechanical method for computing a non-contractible irreducible decomposition of any polynomial set as follows.

Let the given polynomial set be Σ. By the well-ordering theorem given in Sect. 4.3, we can successively enlarge the given set Σ, by following a mechanical procedure, to get a sequence of polynomial sets steadily increasing as shown below:

$$\Sigma = \Sigma_1 \subset \Sigma_2 \subset \cdots \Sigma_q = \Lambda.$$

These polynomial sets are actually mutually equivalent, i.e.,

$$\Sigma = \Sigma_1 \approx \Sigma_2 \approx \cdots \approx \Sigma_q = \Lambda.$$

Two cases may occur. In one case, Λ turns out, in a certain step, to be a contradictory set, i.e., its basic set consists of a single term which is a non-zero element of $\boldsymbol{K}$. In that case, Σ itself is a contradictory set having no extended zeros. Hence it is only necessary to consider the second case, in which Λ has a basic set

$$\Phi: \quad A_1, A_2, \ldots, A_n$$

with $I_1, \ldots, I_n$ as initials and with the class of A_1 greater than 0. Moreover, Λ will possess the following properties: Any polynomial in Λ will have remainder 0 with respect to Φ, any extended zero of Σ is also one of Φ, and any extended zero of Φ, if not one of any initial I_i, is also an extended zero of Σ.

Now, by the mechanical method presented in Sect. 4.2 we can verify whether Φ is reducible, or whether the A_i's are reducible in the successive extension fields $\boldsymbol{K}_{i-1}$. Again we have two subcases.

In the first subcase, Φ is irreducible. By Lemma 1, there is a decomposition

$$|\Lambda| = |\Omega_\Phi| \cup |\Lambda + I_1| \cup \cdots \cup |\Lambda + I_n|,$$

in which Ω_Φ is irreducible while all $\Lambda + I_i$ have basic sets of ranks lower than that of Λ. We may then consider each $\Lambda + I_i$ as a new polynomial set Σ and proceed again in the same way.

In the second subcase, Φ is reducible. Then by Lemma 3 we have a decomposition

$$|\Lambda| = |\Lambda + I_1| \cup \cdots \cup |\Lambda + I_{k-1}| \cup |\Lambda + G_1| \cup \cdots \cup |\Lambda + G_h|,$$

in which each $\Lambda + I_i$ or $\Lambda + G_j$ has a basic set of rank lower than that of Λ. We may then consider each $\Lambda + I_i$ or $\Lambda + G_j$ as a new polynomial set and proceed again as before.

Whatever the subcase we are in, we may take each $\Lambda + I_i$ or $\Lambda + G_i$ as a new polynomial set Σ' in succession and proceed as before to get a sequence

$$\Sigma' = \Sigma'_1 \approx \Sigma'_2 \approx \cdots \approx \Sigma'_{q'} = \Lambda'.$$

If Λ' has a basic set consisting of a single term which is a non-zero element of the field $\boldsymbol{K}$, we may remove $|\Lambda'|$ or the original $|\Lambda + I_i|$ or $|\Lambda + G_j|$ from the decomposition. In the contrary case, $|\Lambda'|$ will be decomposed further into several algebraic varieties with basic sets of ranks lower than that of the corresponding preceding polynomial set, plus possibly one with the corresponding irreducible polynomial set $\Omega_{\Phi'}$ having an irreducible ascending set Φ' as its basic set. By Lemma 5 we can check whether $\Omega_{\Phi'}$ can be removed from the decomposition or remain in the coming decomposition. In this way we will get a further decomposition of $|\Lambda|$ or $|\Sigma|$ itself. In the decomposition, there will appear irreducible polynomial sets of the form $\Omega_\Phi, \Omega_{\Phi'}$ as well as those of the form $\Lambda' + I'$ or $\Lambda' + G'$. For the latter ones we may decompose them further as before.

Since in each step the polynomial sets $\Lambda' + I'$ or $\Lambda' + G'$ to be further decomposed have basic sets of lower rank than those of the preceding ones, by Lemma 1 or 1′ in Sect. 4.3 the decomposition terminates in a finite number of steps. Consequently, after such a finite number of steps we shall arrive at a decomposition of the following form

$$|\Sigma| = |\Omega_{\phi_1}| \cup |\Omega_{\Phi_2}| \cup \cdots \cup |\Omega_{\Phi_s}|,$$

in which each Φ_i is an irreducible ascending set, and Ω_{Φ_i} is the irreducible polynomial set gotten from Φ_i as described in the theorem of Sect. 4.4.

According to the above construction, each $|\Omega_{\Phi_i}|$ is not contained in any $|\Omega_{\Phi_j}|$ for $j > i$, but there is no guarantee of this for $j < i$. This is because we have applied only that part of the theorem in Sect. 4.4 which asserts the existence of Ω_{Φ_i} for Φ_i. If we take into account the assertion of a mechanical procedure for the determination of Ω_{Φ_i} from Φ_i, then we may use Lemma 5 to examine whether $|\Omega_{\Phi_i}|$ is contained in a preceding $|\Omega_{\Phi_j}|$ for $j < i$ and can thus be removed from the decomposition. It follows that, on the basis of the theorem in Sect. 4.4 with the determination procedure given, we can get a non-contractible irreducible decomposition of $|\Sigma|$ in a mechanical manner. In summary, we get the following theorem.

Theorem. There is a mechanical procedure which determines in a finite number of steps, for any polynomial set Σ, a non-contractible irreducible decomposition of the form

$$|\Sigma| = |\Omega_{\Psi_1}| \cup \cdots \cup |\Omega_{\Psi_t}|,$$

in which each Ψ_i is an irreducible basic set of Ω_{Ψ_i}.

However, for the application of mechanical theorem proving it is not necessary to carry out the decomposition to the end and reach a non-contractible one. In fact, it is usually sufficient to have an irreducible decomposition, contractible or not. Hence, for applications the existential part of the theorem in Sect. 4.4 will be quite sufficient.

4.6 A constructive theory of algebraic varieties – the notion of dimension and the dimension theorem

Denote the basic field of characteristic 0 by $\boldsymbol{K}$ and fix the order of the variables

$$x_1 \prec x_2 \prec \cdots \prec x_N$$

as before. Let Σ be an irreducible polynomial set. Then, after having changed the name of x's and reordered them as

$$u_1 \prec \cdots \prec u_d \prec y_1 \prec \cdots \prec y_n \quad (n+d=N),$$

the basic set

$$\Phi: \quad A_1, A_2, \ldots, A_n$$

of Σ possesses the following property:

$$A_i = C_{i0}y_i^{m_i} + C_{i1}y_i^{m_{i-1}} + \cdots + C_{im_i},$$

in which $C_{ij} \in \boldsymbol{K}[u_1, \ldots, u_d, y_1, \ldots, y_{i-1}]$, $\boldsymbol{K}_0 = \boldsymbol{K}(u_1, \ldots, u_d)$, $\boldsymbol{K}_i = \boldsymbol{K}_{i-1}(\eta_i)$, the polynomial $\tilde{A}_i$ obtained from A_i by substituting $\eta_1, \ldots, \eta_{i-1}$ for $y_1, \ldots, y_{i-1}$ is irreducible in $\boldsymbol{K}_{i-1}[y_i]$ and η_i is an extended solution of $\tilde{A}_i = 0$. As before,

$$\tilde{\eta} = (u_1, \ldots, u_d, \eta_1, \ldots, \eta_n)$$

is called a generic point of the irreducible polynomial set Φ, and d the dimension of Φ, denoted as

$$d = \dim \Phi.$$

There are some relations between Σ and Φ as follows. Φ is a basic set of Σ and Σ is equivalent to an irreducible polynomial set Ω_Φ determined from Φ by a mechanical procedure: $\Sigma \approx \Omega_\Phi$ or $|\Sigma| = |\Omega_\Phi|$. Recovering the original name x of u and y and changing $\tilde{\eta}$ to $\tilde{\xi} = (\xi_1, \ldots, \xi_N)$ according to the original order (the order of the variables within the parentheses under the two kinds of names is different, and should not be mixed with each other), $\tilde{\xi}$ is then a generic point of the irreducible algebraic variety $|\Sigma|$. Moreover, for an arbitrary polynomial $F \in \boldsymbol{K}[x_1, \ldots, x_N]$ the following conditions are equivalent:

1. $F = 0|\Sigma$;
2. $F = 0$ has $\tilde{\xi}$ as an extended zero;
3. The remainder of F with respect to Φ is 0;
4. F is a linear sum of polynomials in Ω_Φ with coefficients polynomials in $\boldsymbol{K}[x_1, \ldots, x_N]$.

The properties above were proved in the previous sections. We shall supply one more property below. For this purpose, we say that the *transcendence degree* of an extension field $\tilde{\boldsymbol{K}}$ of $\boldsymbol{K}$ over $\boldsymbol{K}$ is d if there are d elements in $\tilde{\boldsymbol{K}}$ which are algebraically independent, while any $d+1$ elements in $\tilde{\boldsymbol{K}}$ are algebraically dependent over $\boldsymbol{K}$, i.e., d is the largest integer such that $\tilde{\boldsymbol{K}}$ contains d numbers algebraically independent over $\boldsymbol{K}$. Here, *algebraic independency* is defined as follows: e numbers $\tau_1, \ldots, \tau_e$ in $\tilde{\boldsymbol{K}}$ are called *algebraically independent* over $\boldsymbol{K}$ if for any non-zero polynomial $f(t_1, \ldots, t_e) \in \boldsymbol{K}[t_1, \ldots, t_e]$ we have $f(\tau_1, \ldots, \tau_e) \neq 0$.

As before, we say that the extension field

$$\tilde{\boldsymbol{K}}=\boldsymbol{K}(u_1,\ldots,u_d,\eta_1,\ldots,\eta_n)=\boldsymbol{K}(\xi_1,\ldots,\xi_N)$$

obtained by adjoining the generic point $\tilde{\xi}=(\xi_1,\ldots,\xi_N)$ or $\tilde{\eta}=(u_1,\ldots,u_d,\eta_1,\ldots,\eta_n)$ of Φ to $\boldsymbol{K}$ is a *generating field* of Φ. Then, since $\tilde{\boldsymbol{K}}$ is an extension field of $\boldsymbol{K}_0=\boldsymbol{K}(u_1,\ldots,u_d)$ and thus the transcendence degree of $\boldsymbol{K}_0$ over $\boldsymbol{K}$ is d, by the general theory of field extensions we have the following.

Lemma 1. Let Φ be an irreducible ascending set with a generic point $\tilde{\eta}$ or $\tilde{\xi}$ and a generating field $\tilde{\boldsymbol{K}}=\boldsymbol{K}(\tilde{\xi})=\boldsymbol{K}(\tilde{\eta})$. Then $\dim\Phi$ is the transcendence degree of $\tilde{\boldsymbol{K}}$ over $\boldsymbol{K}$.

Lemma 2. Let the linear space $\boldsymbol{K}^N$ over $\boldsymbol{K}$ with base corresponding to $(x_1,\ldots,x_N)$ have two extended points $\tilde{\xi}=(\xi_1,\ldots,\xi_N)$ and $\tilde{\xi}'=(\xi_1',\ldots,\xi_N')$, in which $\tilde{\xi}'$ is a special point of $\tilde{\xi}$. Then the transcendence degree of the extension field $\tilde{\boldsymbol{K}}'=\boldsymbol{K}(\xi_1',\ldots,\xi_N')$ is less than or equal to that of $\tilde{\boldsymbol{K}}=\boldsymbol{K}(\xi_1,\ldots,\xi_N)$ over $\boldsymbol{K}$.

Proof. Let the transcendence degree of $\tilde{\boldsymbol{K}}$ over $\boldsymbol{K}$ be d. Consider a subset, say $\xi_{i_1}',\ldots,\xi_{i_{d+1}}'$, of $\xi_1',\ldots,\xi_N'$. As the transcendence degree of $\tilde{\boldsymbol{K}}$ over $\boldsymbol{K}$ is d, we see that $\xi_{i_1},\ldots,\xi_{i_{d+1}}$ are algebraically dependent over $\boldsymbol{K}$, i.e., there is a non-zero polynomial $f(z_1,\ldots,z_{d+1})\in\boldsymbol{K}[z_1,\ldots,z_{d+1}]$ such that $f(\xi_{i_1},\ldots,\xi_{i_{d+1}})=0$. On the other hand, $\tilde{\xi}'$ is a special point of $\tilde{\xi}$, so $f(\xi_{i_1}',\ldots,\xi_{i_{d+1}}')=0$ and thus $\xi_{i_1}',\ldots,\xi_{i_{d+1}}'$ are algebraically dependent over $\boldsymbol{K}$. From the general theory of algebraic extensions, we know that any $d+1$ elements of $\tilde{\boldsymbol{K}}'$ are algebraically dependent, i.e., the transcendence degree of $\tilde{\boldsymbol{K}}'$ over $\boldsymbol{K}$ is less than or equal to d. This completes the proof of the lemma. □

Theorem. Let two irreducible polynomial sets Σ and Σ' be equivalent with irreducible basic sets Φ and Φ' respectively. Then

$$\dim\Phi=\dim\Phi'.$$

Proof. Let $\tilde{\xi}$ and $\tilde{\xi}'$ be the generic points of Φ and Φ' respectively. Then $\tilde{\xi}$ and $\tilde{\xi}'$ are also the respective generic points of the irreducible algebraic varieties $|\Sigma|$ and $|\Sigma'|$. As $\Sigma\approx\Sigma'$, both $|\Sigma|$ and $|\Sigma'|$ have the same set of extended zeros. Thus $\tilde{\xi}'$ is also an extended zero of $|\Sigma|$ and a special point of the generic point $\tilde{\xi}$ of $|\Sigma|$. Similarly, $\tilde{\xi}$ is also a special point of $\tilde{\xi}'$. By Lemma 2, the extension fields $\boldsymbol{K}(\tilde{\xi})$ and $\boldsymbol{K}(\tilde{\xi}')$ have the same transcendence degree over $\boldsymbol{K}$, i.e., $\dim\Phi=\dim\Phi'$. The proof is complete. □

Owing to this theorem the following definition is proper.

Definition. Let an irreducible polynomial set Σ have Φ as its basic set. Then the dimension of Φ will also be called the *dimension* of Σ or the *dimension* of the irreducible algebraic variety $|\Sigma|$, denoted by $\dim \Sigma$ or $\dim |\Sigma|$. For an arbitrary polynomial set Σ, we can construct a non-contractible irreducible decomposition of $|\Sigma|$ according to the preceding section. In this decomposition, the biggest dimension of the irreducible components will be called the *dimension* of Σ or $|\Sigma|$, denoted by $\dim \Sigma$ or $\dim |\Sigma|$ still.

The main purpose of this section is to prove the following.

Dimension theorem. Let Σ be an irreducible polynomial set with Φ as its irreducible basic set and let B be a polynomial with non-zero remainder with respect to Φ. Then

$$\dim(\Sigma + B) < \dim \Sigma.$$

Proof. Remark first that under a non-singular linear transformation T in $x_1, \ldots, x_N$ the new Σ and $\Sigma + B$ will have their dimension unchanged and the new B will still have non-zero remainder with respect to the new Φ. All these follow easily from the general theory, in particular, the equivalence of properties 1–4 listed above. We may therefore prove below the theorem with some such linear transformation already done.

Let $\Phi : A_1, A_2, \ldots, A_n$ have a generic point $\tilde{\eta} = (u_1, \ldots, u_d, \eta_1, \ldots, \eta_n)$ as before. Then

$$\dim \Sigma = \dim \Phi = d.$$

By Lemma 4 in Sect. 4.4, there are polynomials

$$C, Q_i \in \boldsymbol{K}[u_1, \ldots, u_d, y_1, \ldots, y_n], \quad i = 1, \ldots, n$$

such that

$$D = CB - (Q_1 A_1 + \cdots + Q_n A_n) \in \boldsymbol{K}[u_1, \ldots, u_d]$$

and

$$C(\tilde{\eta}) \neq 0.$$

Thus we have

$$D(u_1, \ldots, u_d) \neq 0.$$

Now suppose the polynomial set Λ is an arbitrary irreducible component of $\Sigma + B$ and

$$\tilde{\zeta} = (v_1, \ldots, v_d, \zeta_1, \ldots, \zeta_n)$$

is a generic point of Λ. As $\tilde{\zeta}$ is also an extended zero of Σ, it is a special point of the generic point $\tilde{\eta}$ of Σ, and especially

$$A_i(\tilde{\zeta}) = 0, \quad i = 1, \ldots, n.$$

In addition, we have

$$B(\tilde{\zeta}) = 0.$$

This implies that $D(\tilde{\zeta}) = 0$, i.e.,

$$D(v_1, \ldots, v_d) = 0.$$

Hence $v_1, \ldots, v_d$ are algebraically dependent.

As the transcendence degree of $\tilde{\eta}$ is d and $u_1, \ldots, u_d$ are algebraically independent, for each i $(1 \le i \le n)$ there is a non-zero polynomial $P_i \in \boldsymbol{K}[u_1, \ldots, u_d, t]$ such that

$$P_i(u_1, \ldots, u_d, \eta_i) = 0,$$

while the coefficient of the highest power of t in $P_i(u_1, \ldots, u_d, t)$ is a non-zero polynomial in $u_1, \ldots, u_d$. Since $\tilde{\zeta}$ is a special point of $\tilde{\eta}$, we have

$$P_i(v_1, \ldots, v_d, \zeta_i) = 0.$$

By a preliminary suitable linear transformation T in $x_1, \ldots, x_N$ supposed already done, we may assume that the coefficient of the highest power of t in each $P_i(u_1, \ldots, u_d, t)$ is some non-zero constant in $\boldsymbol{K}$. Then η_i will actually appear in $P_i(u_1, \ldots, u_d, \eta_i)$ and so will ζ_i in $P_i(v_1, \ldots, v_d, \zeta_i)$. It follows that ζ_i is algebraically dependent on $v_1, \ldots, v_d$, i.e., the number of algebraically independent elements among $v_1, \ldots, v_d, \zeta_1, \ldots, \zeta_n$ is at most $d - 1$. Hence we have

$$\dim \Lambda = \dim \tilde{\zeta} \le d - 1.$$

As this is so for any irreducible component Λ of $\Sigma + B$, the theorem is proved. □

The above dimension theorem is intuitively quite evident, but its proof is not simple at all. For this we may refer to some popular books of algebraic geometry and see that the proofs therein need rather profound tools. For instance, the proof in Hodge and Pedoe (1952) uses Chow coordinates, the proof in van der Waerden (1945) uses the principle of algebraic correspondence, and the proof in Gröbner (1949) uses Hilbert's theory of characteristic polynomials. The proof here seems much simpler and is not only elementary but also constructive.

4.7 Proof of the mechanization theorem of unordered geometry

In this section we give a proof of the mechanization theorem stated in Sect. 4.1. For this let us first make some preparations.

Suppose we are given a set of variables $x_1, \ldots, x_N$ with a fixed order

$$x_1 \prec x_2 \prec \cdots \prec x_N,$$

a basic field $\boldsymbol{K}$ of characteristic 0 and an ascending set

$$\Phi: \quad A_1, A_2, \ldots, A_n$$

of polynomials in $\boldsymbol{K}[x_1, \ldots, x_N]$ for which the classes p_i satisfy the relation

$$0 < p_1 < p_2 < \cdots < p_n.$$

Let us rewrite each x_{p_i} as y_i and all the other x's as $u_1, \ldots, u_d$ with $d = N - n$. Then A_i can be put in the form

$$A_i = c_{i0} y_i^{m_i} + c_{i1} y_i^{m_i - 1} + \cdots + c_{im_i},$$

in which

$$c_{ij} \in \boldsymbol{K}[u_1, \ldots, u_d, y_1, \ldots, y_{i-1}] \quad (i = 1, \ldots, n;\ j = 0, 1, \ldots, m_i).$$

Set

$$I_i = c_{i0} \in \boldsymbol{K}[u_1, \ldots, u_d, y_1, \ldots, y_{i-1}].$$

Thus I_i is the initial of A_i for $i = 1, \ldots, n$. We call each inequation

$$I_i \neq 0$$

a *non-degeneracy condition.*

Let a polynomial $G \in \boldsymbol{K}[u_1, \ldots, u_d, y_1, \ldots, y_n]$ be given and form its remainder R with respect to Φ. By the remainder formula we have

$$I_1^{s_1} \cdots I_n^{s_n} G = Q_1 A_1 + \cdots + Q_n A_n + R$$

for some non-negative integers $s_i \geq 0$ with each $Q_i \in \boldsymbol{K}[u_1, \ldots, u_d, y_1, \ldots, y_n]$.

We shall investigate the necessary and sufficient conditions such that

$$G = 0$$

may be deduced as a consequence of the equations $A_i = 0$, $i = 1, \ldots, n$. In fact we shall prove that, under the non-degeneracy conditions $I_i \neq 0$ and under the hypothesis that Φ is irreducible, a necessary and sufficient condition is just that $R = 0$.

No matter whether the ascending set Φ is irreducible or not, the sufficiency of this condition is quite evident from the above remainder formula.

Theorem 1. Let Φ, A_i, I_i, G be as above and $R = 0$. Then under the non-degeneracy conditions

$$I_i \neq 0, \quad i = 1, \ldots, n,$$

$G = 0$ is a consequence of $A_i = 0, i = 1, \ldots, n$.

If Φ is irreducible, then under the non-degeneracy conditions for $G = 0$ to be a consequence of $A_i = 0$, $i = 1, \ldots, n$, the condition $R = 0$ is also necessary. But in this case the proof is much more difficult.

Theorem 2. Let Φ, A_i, I_i, G be as above and Φ be irreducible. If under the non-degeneracy conditions $I_i \neq 0$ the equation $G = 0$ is a consequence of the equations $A_i = 0$, $i = 1, \ldots, n$ (over a certain extension field of $\boldsymbol{K}$), then the remainder R of G with respect to Φ is 0.

Proof. Suppose R is not 0. We want to derive a contradiction. For this purpose, apply Lemma 4 of Sect. 4.4: there are polynomials $S, Q_i \in \boldsymbol{K}[u_1, \ldots, u_d, y_1, \ldots, y_n]$ such that

$$T = SR - (Q_1A_1 + \cdots + Q_nA_n) \in \boldsymbol{K}[u_1, \ldots, u_d],$$

and T is not 0. From Lemma 3 in Sect. 4.4 together with its proof, we know that the initials I_i of A_i have non-zero remainders with respect to Φ, so there are $J_i, Q_{ij} \in \boldsymbol{K}[u_1, \ldots, u_d, y_1, \ldots, y_{i-1}]$ such that

$$H_i = J_iI_i - (Q_{i1}A_1 + \cdots + Q_{i,i-1}A_{i-1}) \neq 0 \quad \text{and} \quad \in \boldsymbol{K}[u_1, \ldots, u_d].$$

Hence there is a point $\bar{u} = (\bar{u}_1, \ldots, \bar{u}_d)$ in $\boldsymbol{K}^d(u_1, \ldots, u_d)$ such that

$$TH_1 \cdots H_n \neq 0$$

at $\bar{u}$, where $\bar{u}$ may be taken as a rational point from the interior of an arbitrary d-dimensional cube.

We shall prove by induction on i that there is a number $\bar{\eta}_i$ in some appropriate extension field of $\boldsymbol{K}$ such that the point

$$\zeta_i = (\bar{u}_1, \ldots, \bar{u}_d, \bar{\eta}_1, \ldots, \bar{\eta}_i) \in \boldsymbol{K}^{d+i}(u_1, \ldots, u_d, y_1, \ldots, y_i)$$

satisfies the relations

$$A_i(\zeta_i) = 0, \quad I_{i+1}(\zeta_i) \neq 0.$$

Consider first the case $i = 1$. Denote, by $\bar{A}_1 \in \boldsymbol{K}[y_1]$ and $\bar{I}_1 \in \boldsymbol{K}$, the polynomial and the number gotten from A_1 and I_1 by substituting $\bar{u}_1, \ldots, \bar{u}_d$ for $u_1, \ldots, u_d$ respectively. Since

$$H_1 = J_1I_1$$

is not equal to 0 at $\bar{u}$, $\bar{I}_1 \neq 0$ and $\bar{A}_1$ is a polynomial of degree $m_1 \geq 1$ in y_1, so we can take $\bar{\eta}_1$ from some extension field of $\boldsymbol{K}$ such that $\bar{A}_1(\bar{\eta}_1) = 0$ or

$A_1(\zeta_1) = 0$. Furthermore,

$$H_2 = J_2 I_2 - Q_{21} A_1,$$
$$H_2(\zeta_1) = H_2(\bar{u}) \neq 0,$$

so we have

$$I_2(\zeta_1) \neq 0,$$

i.e., the induction hypothesis holds for $i = 1$.

Suppose $\bar{\eta}_1, \ldots, \bar{\eta}_i, \zeta_i$ have already been found to satisfy the induction hypothesis. Let us find $\bar{\eta}_{i+1}$ and ζ_{i+1} as follows.

Denote by $\bar{A}_{i+1} \in \tilde{\boldsymbol{K}}[y_{i+1}]$ the polynomial obtained from A_{i+1} by substituting $\bar{u}_1, \ldots, \bar{u}_d, \bar{\eta}_1, \ldots, \bar{\eta}_i$ for $u_1, \ldots, u_d, y_1, \ldots, y_i$, where $\tilde{\boldsymbol{K}}$ is some extension field of $\boldsymbol{K}$ containing $\bar{\eta}_1, \ldots, \bar{\eta}_i$. Then the coefficient $I_{i+1}(\zeta_i)$ of the highest power $y_{i+1}^{m_i+1}$ of $\bar{A}_{i+1}$ in y_{i+1} is non-zero. Hence we can take $\bar{\eta}_{i+1}$ from $\tilde{\boldsymbol{K}}$, an extension field of $\boldsymbol{K}$, such that $\bar{A}_{i+1}(\bar{\eta}_{i+1}) = 0$ or

$$A_{i+1}(\zeta_{i+1}) = 0.$$

Now we have

$$H_{i+2} = J_{i+2} I_{i+2} - (Q_{i+2,1} A_1 + \cdots + Q_{i+2,i+1} A_{i+1}),$$
$$H_{i+2}(\zeta_{i+1}) = H_{i+2}(\bar{u}) \neq 0$$

and

$$A_1(\zeta_{i+1}) = A_1(\zeta_1) = 0,$$
$$\cdots$$
$$A_{i+1}(\zeta_{i+1}) = 0.$$

Plunging ζ_{i+1} into the formula of H_{i+2}, we get

$$I_{i+2}(\zeta_{i+1}) \neq 0.$$

Now the induction is proved.

Since R is the remainder of G with respect to Φ, there are integers $s_i \geq 0$ and polynomials $B_i \in \boldsymbol{K}[u_1, \ldots, u_d, y_1, \ldots, y_n]$ such that

$$I_1^{s_1} \cdots I_n^{s_n} G = B_1 A_1 + \cdots + B_n A_n + R.$$

From the above induction proof we have

$$I_i(\zeta_n) \neq 0, \quad A_i(\zeta_n) = 0.$$

Since $G = 0$ is a formal consequence of $A_i = 0$ under the non-degeneracy conditions $I_i \neq 0$, the above formulas lead to

$$G(\zeta_n) = 0.$$

Hence, by substituting ζ_n into the remainder formula for G we get

$$R(\zeta_n) = 0.$$

But, according to the choice of $\bar{u}$ we have

$$T(\zeta_n) = T(\bar{u}) \neq 0.$$

Thus, by substituting ζ_n into

$$T = SR - \sum Q_i A_i,$$

we have derived a contradiction.

Therefore, the assumption that $R \neq 0$ cannot be true and the theorem is proved. □

Below we give the proof of the mechanization theorem of unordered geometry.

Let a geometric statement (S) in a certain unordered geometry be given. Our objective is to provide a mechanical method to decide whether (S) is true or not. For this we first choose a coordinate system, represent the points involved by means of coordinates, denote these coordinates by x_i, and arrange them in a definite order

$$x_1 \prec x_2 \prec \cdots \prec x_N.$$

Next we express the various geometric relations in the statement (S) by algebraic relations among these coordinates. Then the hypothesis in the statement (S) will be translated into a system of equations

$$F_1 = 0, \ldots, F_s = 0,$$

in which F_i are polynomials in $\boldsymbol{K}[x_1, \ldots, x_N]$, with $\boldsymbol{K}$ being the basic field of characteristic 0 associated with the geometry in question. Actually all these polynomials are with rational or even integer coefficients. The conclusion of the statement (S) will then be turned into another system of equations

$$G_1 = 0, \ldots, G_t = 0,$$

where all G_j are polynomials in $\boldsymbol{K}[x_1, \ldots, x_N]$, actually with rational or integer coefficients, too. Without loss of generality we may suppose that there is only one such polynomial G_j, denoted simply by G henceforth. The polynomials F_i are then called the *hypothesis polynomials* of the statement (S), and the G_j's or G the *conclusion polynomial(s)* of (S).

The proof of the mechanization theorem consists of exhibiting a mechanical procedure which determines first, in a finite number of steps, a set of polynomials $D_1, \ldots, D_r$ for the *non-degeneracy conditions*, with all D_k in $\boldsymbol{K}[x_1, \ldots, x_N]$ and actually all with rational or even integer coefficients. Secondly, the same mechanical procedure will also decide in a finite number of steps whether under the non-degeneracy conditions

$$D_1 \neq 0, \ldots, D_r \neq 0,$$

the equation $G = 0$ is a consequence of

$$F_1 = 0, \ldots, F_s = 0.$$

Using the language of algebraic geometry developed in the previous sections, the proof of the mechanization theorem can be stated more precisely as follows.

Denote the set of hypothesis polynomials F_i by

$$\Sigma = \{F_i\}.$$

Then Σ defines an algebraic variety $|\Sigma|$, having a fixed dimension d. We need to find a mechanical procedure[1] which determines a set of polynomials $D_1, \ldots, D_r$ such that by adjoining each D_i to Σ, the resulting polynomial set $\Sigma + D_i$ will define an algebraic variety $|\Sigma + D_i|$ of dimension $< d$. Furthermore, the same procedure will decide, under the non-degeneracy conditions $D_1 \neq 0, \ldots, D_r \neq 0$, whether or not $G = 0$; in other words, whether or not G will vanish on the remaining part of the algebraic variety $|\Sigma|$ after removal of the true subvarieties $|\Sigma + D_i|$.

The theory of constructive algebraic varieties in the previous sections of this chapter and the two theorems in this section already provide such a mechanical method.

According to Sect. 4.5, we can decompose the algebraic variety $|\Sigma|$ into irreducible components, of which each has an irreducible basic set Φ_i, which determines in turn that irreducible component in question, denoted by $|\Omega_{\Phi_i}|$ as before. Furthermore, if the dimension d_i of $|\Omega_{\Phi_i}|$ is less than the dimension d

1 The argument based on dimensionality of algebraic varieties needs some clarification. When $x_1, \ldots, x_N$ are considered as corresponding to the base of the N-dimensional linear space $\tilde{\boldsymbol{K}}^N$, it may occur that some irreducible component $|\Omega_{\Phi_i}|$ corresponds to a degenerate case of the geometric theorem while $\dim |\Omega_{\Phi_i}| = d_i = d = \dim |\Sigma|$, and that $|\Omega_{\Phi_i}|$ corresponds to a non-degenerate case of the theorem yet $\dim |\Omega_{\Phi_i}| = d_i < d$. This situation can be settled roughly as follows. Let the variables $x_1, \ldots, x_N$ be properly distinguished into parameters $u_1, \ldots, u_d$ and geometric dependents $y_1, \ldots, y_n$ $(d+n = N)$ as before. In the irreducible decomposition, we remove those components $|\Omega_{\Phi_i}|$ for which Φ_i contains a polynomial D_i involving the parameters $u_1, \ldots, u_d$ only and take each D_i as a non-degeneracy polynomial. That is, only those components $|\Omega_{\Phi_i}|$ for which Φ_i contains no polynomial involving $u_1, \ldots, u_d$ only are considered. They should all be of dimension d. Otherwise, the theorem is not well-formulated. [Transl.]

of $|\Sigma|$, then this true subvariety should be a subvariety of some $|\Phi_j + D_i|$ obtained from a certain previous $|\Omega_{\Phi_j}|$ of dimension d by adjoining to Φ_j a polynomial D_i (which is either an initial I_k or some G_l in the previous notations). In this case, we take each D_i as a non-degeneracy polynomial. Suppose after removal of all these true subvarieties, the remaining irreducible components of dimension d are

$$|\Omega_{\Phi_1}|, \ldots, |\Omega_{\Phi_t}|.$$

Denote the initials of polynomials in each Φ_j by $I_{j1}, \ldots, I_{ju}$ and consider them as non-degeneracy polynomials D_{jk}. By Theorems 1 and 2, whether $G = 0$ is a consequence of $F_1 = 0, \ldots, F_s = 0$ under the non-degeneracy conditions $D_i \neq 0$, $D_{jk} \neq 0$, is just the same as whether $G = 0$ on the remaining part of $|\Omega_{\Phi_1}|, \ldots, |\Omega_{\Phi_t}|$ after removal of the components $|\Phi_j + D_i|$ and those defined by $D_{jk} = 0$, which can be decided by verifying whether the remainders of G with respect to Φ_j are all 0. This furnishes the mechanical procedure required and thus gives the proof of the mechanization theorem in question.

The above mechanical procedure of theorem proving is quite simple in appearance. However, it would be rather difficult to apply this method to the proof of concrete theorems. The reason is that the irreducible decomposition of algebraic varieties depends on factorization of polynomials which, though theoretically almost self-evident, is a rather difficult problem in practice and for which no method of high efficiency is now available. Consequently, the above method, if followed according to that procedure, is entirely impracticable, i.e., non-feasible in the terminology of mathematical logic. Fortunately, for theorem proving in geometries, we may hope that the theorem under consideration is really a true theorem and we want to prove its truth in an affirmative manner. For this purpose it is enough to verify, by using Theorem 1, that the remainder of the conclusion polynomial G is 0 with respect to some ascending set, no matter whether it is irreducible or not. Therefore, to each concrete theorem whose truth is to be tested and to be proved in the case it is really true, we may apply Theorem 1 directly. If we know from computation that G has its remainder 0 with respect to the ascending set, then the theorem under consideration is true and the computation furnishes a proof of this theorem, and so in this case everything is done. Only in the case that the remainder is not zero should we ask further whether the corresponding ascending set is irreducible or not. For this reason we shall modify the above mechanical procedure of theorem proving to the following form which has proved to be very efficient in practice.

The modified mechanical procedure runs as follows.

Consider a set ρ of polynomial sets and a set Δ of polynomials, where Δ is called the *non-degeneracy set*. In the outset, ρ will consist of a single polynomial set, viz. the set

$$\Sigma = \{F_1, \ldots, F_s\}$$

of hypothesis polynomials, and the non-degeneracy set will be an empty one, viz.

$$\Delta = \phi.$$

During the procedure we shall increase or decrease the number of polynomial sets in ρ and also adjoin non-degeneracy polynomials into Δ to get the final

$$\Delta = \{D_1, \ldots, D_r\}$$

as required.

Step 1. Let ρ be non-empty. Then take an arbitrary polynomial set Σ from ρ, and remove it from ρ to get a new ρ. Use the well-ordering theorem in Sect. 4.3 to enlarge Σ to successive polynomial sets as shown below:

$$\Sigma = \Sigma_1 \to \Sigma_2 \to \cdots \to \Sigma_q = \Lambda.$$

If Λ has an element which is a non-zero number in $\boldsymbol{K}$, then Λ is contradictory. In this case, the hypothesis in the statement (S) is contradictory in itself and the procedure terminates. In the contrary case, let the basic set of Λ be

$$\Phi: \quad A_1, A_2, \ldots, A_n.$$

The initials of A_i will be denoted as I_i. By construction, any polynomial in Λ will have its remainder 0 with respect to Φ. In that case we also have

$$\dim|\Sigma| = \dim\Phi = N - n = d.$$

If Step 1 is just the first step of the whole procedure, then the dimension d will be recorded for future reference.

If Step 1 is a later step during the procedure, then we compare the new dimension d with the previous d recorded at the beginning.

If this new d is equal to the previously recorded d, then adjoin the initials I_i to Δ to get an enlarged non-degeneracy set Δ, and proceed to Step 2.

If this new d is less than the previously recorded d, and the present Σ is obtained as some $\Lambda + I_i$ or $\Lambda + G_j$ from Step 3 below, then adjoin this I_i or G_j to Δ to get a new Δ. Then go back to Step 1 and proceed as before.

Step 2. Find the remainder R of G with respect to Φ.

Suppose first $R = 0$. If in ρ there are no more polynomial sets, then the statement (S) is true under the non-degeneracy conditions

$$D_k \neq 0 \quad (D_k \in \Delta),$$

and the procedure terminates. In this case, the theorem is true and is proved under the non-degeneracy conditions. Otherwise go back to Step 1 and proceed as before.

If $R \neq 0$, then proceed to Step 3.

Step 3. Check the irreducibility of the basic set Φ according to Sect. 4.2.

If Φ is irreducible, then, as G has remainder $R \neq 0$ with respect to Φ, by Theorem 2 under the non-degeneracy conditions

$$D_k \neq 0 \quad (D_k \in \Delta)$$

the statement (S) is not true and the procedure terminates. In this case, the theorem is not true under the non-degeneracy conditions.

If Φ is reducible, then according to Sect. 4.5 there will be some decomposition

$$|\Lambda| = |\Lambda + I_1| \cup \cdots \cup |\Lambda + I_{l-1}| \cup |\Lambda + G_1| \cup \cdots \cup |\Lambda + G_h|.$$

Consider such $\Lambda + I_i$ and $\Lambda + G_j$ as new polynomial sets Σ, and adjoin all these to ρ to get an enlarged set ρ. Then go back to Step 1 and proceed as before.

According to the previous sections, the above procedure will terminate in a finite number of steps. In this way we shall get a final non-degeneracy set

$$\Delta = \{D_k\}$$

and one of the following three conclusions should be true:

1. Under the non-degeneracy conditions

$$D_k \neq 0 \quad (D_k \in \Delta)$$

   the hypotheses in the statement (S) are contradictory;
2. under the above non-degeneracy conditions, or under the additional hypotheses $D_k \neq 0$, the statement (S) is true, or what is the same, the theorem under consideration is true;
3. under the above non-degeneracy conditions, or under the additional hypotheses $D_k \neq 0$, the statement (S) is not true, or what is the same, the theorem is not true.

Generally speaking, the degeneracy conditions

$$D_k = 0$$

are not worthy of any more consideration. If there is necessity to consider some degeneracy condition $D_k = 0$, we may simply take it as a new hypothesis to adjoin to the statement (S), i.e., we consider $\{F_1, \ldots, F_s, D_k\}$ instead of $\{F_1, \ldots, F_s\}$ and then proceed as above.

The above simplified mechanical procedure is very feasible. The next section will give a few illustrative examples (mainly with hand calculation). More examples by computer may be seen in relevant books by the author to be published in the near future.

4.8 Examples for the mechanical method of unordered geometry

The following examples illustrate the main idea of the mechanical procedure presented in the preceding section. Since the calculations in this book are done only by hand, we shall take examples which are as simple as possible and choose suitable coordinate systems in order to further simplify the calculations.

In the book “Theory, Method and Practice of Mechanical Theorem Proving in Geometries” under preparation by the author, the mechanized proofs of more complicated theorems will be given, but the principle is the same as what is shown in this section.

Example 1 (Theorem of orthocenter). The three altitudes of any triangle are concurrent.

In orthogonal geometry, this theorem is introduced as an axiom (cf. Sect. 2.2). Here we take it as a theorem to be proved as an explanation of the method of mechanical proving. This is independent of the problem of circular proof.

According to Sect. 2.2, the three sides of $\triangle ABC$ cannot all be isotropic. We may suppose AB is a non-isotropic line. In this case, AB and the altitude on AB should intersect, say at a point O. We take an orthogonal coordinate system with O as its origin, AB and CO as its first and second axes respectively. Let D be the intersection point of the two altitudes on the sides BC and AC. Then we need to prove that D lies on the altitude CO (see Fig. 4.1).

For this purpose, let the orthogonal rate of the orthogonal coordinate system be k and the coordinates of points be

$$A=(x_1,0),\quad B=(x_2,0),\quad C=(0,x_3),\quad D=(x_4,x_5).$$

According to Sect. 2.3, the hypothesis of the theorem consists of

$$AD\perp BC \Longleftrightarrow kx_3x_5-x_2(x_4-x_1)=0,$$
$$BD\perp AC \Longleftrightarrow kx_3x_5-x_1(x_4-x_2)=0.$$

Denote the two polynomials on the left-hand side by F_1 and F_2 respectively, and well-order the set $\Sigma=\{F_1,F_2\}$ of polynomials according to Sect. 4.3 to get a polynomial set Λ with basic set

$$\Phi:\quad A_1,A_2,$$

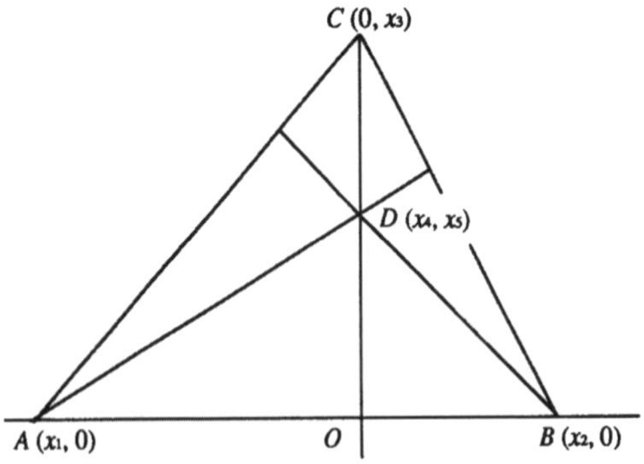

Fig. 4.1

in which

$$A_1 = (x_1 - x_2)x_4,$$

$$A_2 = kx_3x_5 - x_2(x_4 - x_1).$$

Now the non-degeneracy condition

$$D = kx_3(x_1 - x_2) \neq 0$$

is introduced. The conclusion of the theorem is

$$CO \text{ passes through } D \Longleftrightarrow x_4 = 0.$$

Under the non-degeneracy conditions

$$x_3 \neq 0, \quad x_1 \neq x_2,$$

the theorem is clearly true.

The geometric meanings of the non-degeneracy conditions are also evident:

$$x_3 \neq 0 \Longleftrightarrow C \text{ does not lie on } AB;$$

$$x_1 \neq x_2 \Longleftrightarrow A \text{ and } B \text{ do not coincide.}$$

For these simple degeneracy cases it is obviously unnecessary to make a further investigation and the following conclusion may be drawn:

(In orthogonal geometry) the three altitudes of any non-degenerate triangle are concurrent.

In fact, this is an axiom in orthogonal geometry. Although the "proof" here is simple, it is representative: The non-degeneracy conditions appear successively in the course of mechanical proving and need not be considered in advance. As for the occurring non-degeneracy conditions, we can take them successively as degeneracy ones to be adjoined to the hypothesis in order to further verify whether or not the theorem holds, if necessary. It all proceeds systematically and mechanically, without need of further consideration.

Example 2 (Theorem about incenter and excenters). The bisectors of the three angles of any triangle, three-to-three, intersect at four points.

In unordered metric geometry, there is only the concept of total angles but not the concept of general angles. The two bisectors of a total angle cannot be distinguished either. Therefore, the four intersection points in the theorem cannot be distinguished into incenter and excenters as in ordinary geometry. This fact is reflected in the following mechanized proof.

To simplify the calculation, we shall take one side (non-isotropic line) AB of $\triangle ABC$ as the first axis of a Descartes coordinate system and the line passing through C perpendicular to AB, i.e., the altitude on side AB, as the second axis. Let one of the bisectors of the total angle A and one of those of the total angle

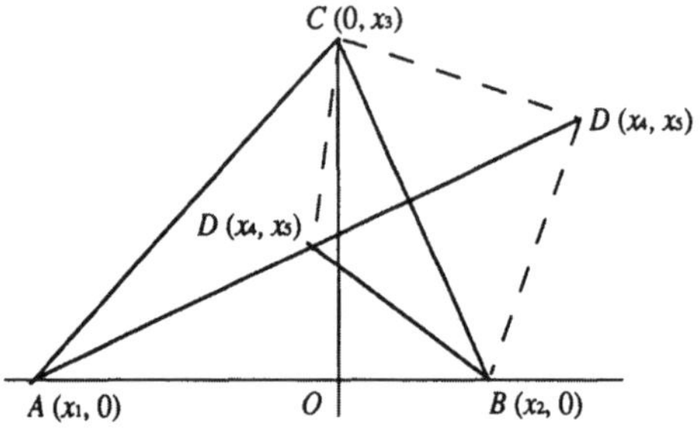

Fig. 4.2

B intersect at D. We want to prove that CD is a bisector of the total angle C (see Fig. 4.2).

Let the coordinates of points be

$$A=(x_1,0),\quad B=(x_2,0),\quad C=(0,x_3),\quad D=(x_4,x_5).$$

According to Sect. 2.4, the hypothesis of the theorem consists of

$$AD=\text{the bisector of the total angle } A \Longleftrightarrow$$
$$F_1\equiv x_3[x_5^2-(x_4-x_1)^2]-2x_1x_5(x_4-x_1)=0,$$
$$BD=\text{the bisector of the total angle} \Longleftrightarrow$$
$$F_2\equiv x_3[x_5^2-(x_4-x_2)^2]-2x_2x_5(x_4-x_2)=0.$$

The conclusion of the theorem is

$$CD=\text{the bisector of the total angle } C \Longleftrightarrow$$
$$G\equiv[x_1(x_5-x_3)+x_3x_4]\cdot[x_3(x_5-x_3)-x_2x_4]$$
$$+[x_2(x_5-x_3)+x_3x_4]\cdot[x_3(x_5-x_3)-x_1x_4]=0.$$

Starting from the hypothesis-polynomial set

$$\Sigma=\{F_1,F_2\},$$

we can get a polynomial set Λ having basic set

$$\Phi:\quad A_1,A_2,$$

where

$$A_1=4x_4(x_4-x_1)(x_4-x_2)(x_4-x_1-x_2)-x_3^2(2x_4-x_1-x_2)^2,$$
$$A_2=2(x_4-x_1-x_2)x_5-x_3(2x_4-x_1-x_2).$$

The non-degeneracy conditions introduced are

$$x_3 \neq 0, \quad x_1 \neq x_2, \quad x_4 \neq x_1 + x_2.$$

Reducing G with respect to Φ, we can easily find that $4(x_4 - x_1 - x_2)^2 G$ is a linear sum of A_1 and A_2 under the above non-degeneracy conditions, and the theorem thus holds true under these non-degeneracy conditions.

The geometric meanings of the non-degeneracy conditions

$$x_3 \neq 0 \quad \text{and} \quad x_1 \neq x_2$$

are rather evident. To verify whether the theorem still holds in the degeneracy case

$$x_4 = x_1 + x_2,$$

we may adjoin

$$F_3 = x_4 - x_1 - x_2$$

to the original Σ and proceed in the same way for the new set

$$\Sigma = \{F_1, F_2, F_3\}.$$

We then get

$$x_1 + x_2 = 0, \quad x_4 = 0$$

under the same non-degeneracy conditions

$$x_3 \neq 0, \quad x_1 \neq x_2.$$

So the theorem holds true as well. Actually, in this case the triangle is symmetric with respect to CO.

In the original basic set Φ: A_1, A_2, the polynomial A_1 is of degree 4 in x_4 and A_2 is linear in x_5. This indicates a geometric fact: The common intersection points of the angular bisectors are exactly four in number.

Example 3. The theorem is the same as the preceding one.

Since the bisectors of a total angle in (unordered) orthogonal geometry may not necessarily exist, in the above example the theorem has been proved as one in (unordered) metric geometry. In the course of its proof, we have chosen a Descartes coordinate system. Such a system generally does not exist in orthogonal geometry. But in fact, the same theorem also holds in orthogonal geometry, as long as we modify its statement a little bit.

In orthogonal geometry, though angular bisectors may not necessarily exist, we can always define the symmetric line of any straight line with respect to another non-isotropic line. In this case, the non-isotropic line is just a bisector of the total angle constituted by the original line and its reflection. Hence, we may consider the assertion having the following meaning in orthogonal geometry.

Let the three sides of $\triangle ABD$ be all non-isotropic and the symmetric lines

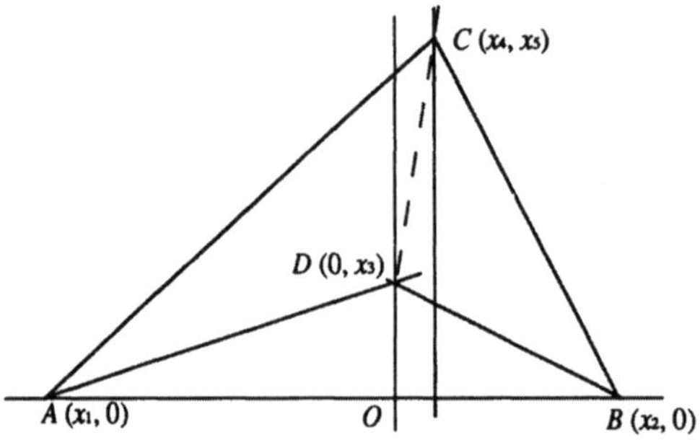

Fig. 4.3

of AB with respect to AD and with respect to BD intersect at a point C. Then AC and BC are symmetric with respect to CD (cf. Fig. 4.3).

To prove this, choose an orthogonal coordinate system with AB as its first axis and the line passing through D and perpendicular to AB as its second axis. Let the orthogonal rate of the coordinate system be k and represent the coordinates of points as

$$A = (x_1, 0), \quad B = (x_2, 0), \quad D = (0, x_3), \quad C = (x_4, x_5).$$

Then the hypothesis of the theorem consists of

$$AC = \text{the reflection of } AB \text{ with respect to } AD \Longleftrightarrow$$

$$F_1 \equiv (x_1^2 - kx_3^2)x_5 + 2x_1x_3(x_4 - x_1) = 0,$$

$$BC = \text{the reflection of } AB \text{ with respect to } BD \Longleftrightarrow$$

$$F_2 \equiv (x_2^2 - kx_3^2)x_5 + 2x_2x_3(x_4 - x_2) = 0,$$

and the conclusion is

$$AC = \text{the reflection of } BC \text{ with respect to } CD \Longleftrightarrow$$

$$G \equiv [(x_4 - x_1)(x_5 - x_3) - x_4x_5] \cdot [kx_5(x_5 - x_3) + x_4(x_4 - x_2)]$$
$$+ [(x_4 - x_2)(x_5 - x_3) - x_4x_5] \cdot [kx_5(x_5 - x_3) + x_4(x_4 - x_1)] = 0.$$

Starting from the polynomial set $\Sigma = \{F_1, F_2\}$, by the well-ordering procedure we get a polynomial set Λ having basic set

$$\Phi: \quad A_1, A_2,$$

where

$$A_1 = (kx_3^2 + x_1x_2)x_4 - k(x_1 + x_2)x_3^2,$$

$$A_2 = (x_1^2 - kx_3^2)x_5 + 2x_1x_3(x_4 - x_1).$$

In the course of well-ordering, the non-degeneracy conditions

$$x_3 \neq 0, \quad x_1 \neq x_2,$$

have already been introduced.

Reducing G by the two polynomials in Φ, we see that G has its remainder 0 with respect to Φ under the above non-degeneracy conditions. Therefore, the theorem is true under the non-degeneracy conditions

$$x_3 \neq 0, \quad x_1 \neq x_2,$$
$$kx_3^2 + x_1x_2 \neq 0, \quad x_1^2 - kx_3^2 \neq 0.$$

The geometric meanings of these non-degeneracy conditions can be easily explained. For instance,

$$kx_3^2 + x_1x_2 = 0$$

means $AD \perp BD$. If we adjoin it to the hypothesis and well-order the obtained set in the same way, we shall get

$$x_1 = x_2, \quad kx_3^2 + x_1^2 = 0,$$

under the non-degeneracy conditions

$$x_1 \neq 0, \quad x_2 \neq 0, \quad x_3 \neq 0.$$

In other words, A and B coincide while $AD = BD$ is an isotropic line. So the hypothesis of the theorem itself is contradictory.

Let us compare this example with the preceding one. This example, being a theorem in orthogonal geometry, is obviously more general than the preceding one, which is meaningful only in metric geometry. Therefore, if the statement is proper, we can get a more general theorem while the calculation used for its proof becomes simpler. For example, both A_1 and A_2 here are polynomials of degree 1 in x_4 and x_5, whereas A_1 in the preceding example is of degree 4 in x_4. The reason is that if point D is fixed, then point C can be uniquely determined. The contrary is not so, however.

For this example, as a theorem in an arbitrary orthogonal geometry, the proof by applying the metric concepts such as distance and congruence is clearly not suitable. Although it is not difficult to get a direct proof by following the Euclidean fashion (for instance, the direct proof in Sect. 2.2), the proof method would depend upon the theorem and is not in accordance with the teachings of Descartes.

Example 4 (Feuerbach's theorem). The nine-point circle of a triangle is tangent to any circle that is tangent to all the three sides of the triangle.

This is one of the most famous theorems in elementary geometry. There is an interesting discussion of this theorem in Davis (1927). To prove the theorem

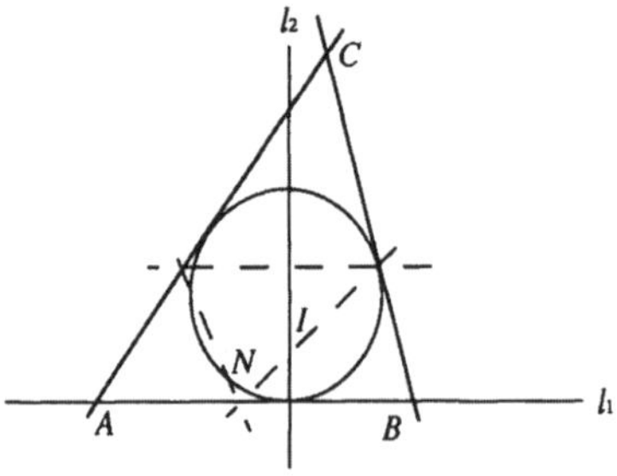

Fig. 4.4

using the classical method of Euclid, one certainly needs some techniques. Even using analytic geometry, some ingenious ideas are still necessary: see the proof in Salmon (1879a: p. 127), in which trigonometric functions are used and the theorem was thus proved as one in ordinary geometry. In fact, the statement of the above theorem does not concern the internal tangency and external tangency but only the tangency of circles; hence, the theorem is one in unordered metric geometry. By applying the mechanical method of this chapter, the proof is very easy – not much different from the previous three examples. The only difference is the number of calculations. It is both complicated and error-prone to use hand-calculation, so we should use a computer to perform the proof. Let us state the proof process as follows.

According to Sect. 2.4, there are four circles that are tangent to the three sides of a triangle. Take any one of them and denote its center by I. Choose a Descartes coordinate system with AB as its first axis l_1 and the line passing through I and perpendicular to AB as its second axis. Let the coordinates of points be as follows

$$I = (0, x_1), \quad A = (x_2, 0), \quad B = (x_3, 0), \quad C = (x_4, x_5).$$

Let the midpoints of the pairs (BC), (AC), (AB) of $\triangle ABC$ be M_A, M_B, M_C respectively. According to the definition, the nine-point circle of $\triangle ABC$ is the circle passing through M_A, M_B and M_C. It is easy to prove that this circle also passes through six other well-known points. Let the center of this nine-point circle be N with coordinates

$$N = (x_6, x_7).$$

Suppose

$$x_8^2 = \text{the square of radius of the nine-point circle} = \overline{\overline{NM_A}}, \text{ etc.},$$

$$x_9^2 = \text{the square of distance of the pair } (NI), \text{ i.e., } \overline{\overline{NI}}.$$

Since we do not assume any order relation, radius and distance have no meaning in our geometry. But according to Sect. 2.4, we can define the square of radius and the square of distance of pair of points. So the hypothesis of the theorem reads as follows:

AI is one of the bisectors of $\measuredangle A(BC)$, or
$AC =$ the symmetric line of l_1 with respect to AI

$$\Longleftrightarrow (x_1^2 - x_2^2)x_5 + 2x_1x_2(x_2 - x_4) = 0,$$

BI is one of the bisectors of $\measuredangle B(AC)$, or
$BC =$ the symmetric line of l_1 with respect to BI

$$\Longleftrightarrow (x_1^2 - x_3^2)x_5 + 2x_1x_3(x_3 - x_4) = 0,$$

$N =$ the center of the nine-point circle $M_AM_BM_C$, or
the square of radius of the nine-point circle $= \overline{\overline{NM_A}} = \overline{\overline{NM_B}} = \overline{\overline{NM_C}}$

$$\Longleftrightarrow \begin{cases} x_8^2 = \left(x_6 - \dfrac{x_2 + x_3}{2}\right)^2 + x_7^2 \\ \quad = \left(x_6 - \dfrac{x_2 + x_4}{2}\right)^2 + \left(x_7 - \dfrac{x_5}{2}\right)^2 \\ \quad = \left(x_6 - \dfrac{x_3 + x_4}{2}\right)^2 + \left(x_7 - \dfrac{x_5}{2}\right)^2. \end{cases}$$

Now fix the order of the x's as

$$x_1 \prec x_2 \prec \cdots \prec x_8 \prec x_9,$$

and well-order the set of polynomials appearing in the above hypothesis. Then we may introduce the non-degeneracy conditions

$$x_1 \neq 0, \quad x_2 \neq \pm x_3, \quad x_1 \neq \pm x_2,$$

and get, under these conditions, a *quasi-basic set*

$$A_1 \equiv (x_1^2 + x_2x_3)x_4 - x_1^2(x_2 + x_3) = 0,$$

$$A_2 \equiv (x_1^2 + x_2x_3)x_5 - 2x_1x_2x_3 = 0,$$

$$A_3 \equiv 4x_6 - (x_2 + x_3 + 2x_4) = 0,$$

$$A_4 \equiv 4x_5x_7 - x_5^2 - (x_3 - x_4)(4x_6 - 2x_2 - x_3 - x_4) = 0,$$

$$A_5 \equiv 4x_8^2 - (2x_6 - x_2 - x_3)^2 - 4x_7^2 = 0,$$

$$A_6 \equiv x_9^2 - x_6^2 - (x_7 - x_1)^2 = 0,$$

where *quasi*-basic set means that the initial of each polynomial in the set is reduced, though the polynomial itself may not necessarily be reduced, with respect to the preceding polynomials. This is enough for application of Theorem 1 in Sect. 4.7.

The conclusion of the theorem becomes

$$G \equiv x_9^4 + x_8^4 + x_1^4 - 2x_9^2x_8^2 - 2x_9^2x_1^2 - 2x_8^2x_1^2 = 0.$$

Introduce further non-degeneracy conditions

$$x_1^2 + x_2x_3 \neq 0, \quad x_5 \neq 0$$

for the initial of each A_i to be non-zero. Then the successive remainders of G with respect to A_i are $R_5, R_4, \ldots, R_1, R_0$ listed as follows:

$$\begin{aligned}
R_5 = {} & x_8^4 - 4x_1^2x_8^2 - 2x_7^2x_8^2 + 4x_1^2x_7^2 - 2x_6^2x_8^2 \\
& + 4x_1x_7x_8^2 + x_7^4 + 2x_6^2x_7^2 - 4x_1x_7^3 + x_6^4 \\
& - 4x_1x_6^2x_7,
\end{aligned}$$

$$\begin{aligned}
R_4 = {} & -64x_1^2x_6^2 + 64x_1^2x_2 - 64x_1x_2x_6x_7 + 64x_1^2x_3x_6 \\
& - 64x_1x_3x_6x_7 - 32x_1^2x_2x_3 + 32x_2x_3x_6^2 + 32x_1x_2x_3x_7 \\
& - 16x_1^2x_2^2 + 16x_2^2x_6^2 + 16x_1x_2^2x_7 - 16x_1^2x_3^2 + 16x_3^2x_6^2 \\
& + 16x_1x_3^2x_7 - 24x_2^2x_3x_6 - 8x_2^3x_6 - 24x_2x_3^2x_6 - 8x_3^3x_6 \\
& + 6x_2^2x_3^2 + 4x_2^3x_3 + 4x_2x_3^3 + x_2^4 + x_3^4,
\end{aligned}$$

$$\begin{aligned}
R_3 = {} & -64x_1^2x_5x_6^2 + 64x_1^2x_2x_5x_6 + 64x_1^2x_3x_5x_6 \\
& - 32x_1^2x_2x_3x_5 + 32x_2x_3x_5x_6^2 - 16x_1^2x_2^2x_5 \\
& + 16x_2^2x_5x_6^2 - 16x_1^2x_3^2x_5 + 16x_3^2x_5x_6^2 - 24x_2^2x_3x_5x_6 \\
& - 8x_2^3x_5x_6 - 24x_2x_3^2x_5x_6 - 8x_3^3x_5x_6 + 6x_2^2x_3^2x_5 \\
& + 4x_2^3x_3x_5 + 4x_2x_3^3x_5 + x_2^4x_5 + x_3^4x_5 - 16x_1x_2x_5^2x_6 \\
& - 16x_1x_3x_5^2x_6 + 8x_1x_2x_3x_5^2 + 4x_1x_2^2x_5^2 + 4x_1x_3^2x_5^2 \\
& - 64x_1x_2x_3x_6^2 - 64x_1x_3^2x_6^2 + 80x_1x_2x_3^2x_6 + 49x_1x_2^2x_3x_6 \\
& + 32x_1x_3^3x_6 + 64x_1x_2x_4x_6^2 + 64x_1x_3x_4x_6^2 - 64x_1x_2x_3x_4x_6 \\
& - 48x_1x_2^2x_4x_6 - 16x_1x_3^2x_4x_6 - 20x_1x_2^2x_3^2 - 8x_1x_2^3x_3 \\
& - 16x_1x_2x_3^3 + 16x_1x_2^2x_3x_4 + 8x_1x_2^3x_4 + 8x_1x_2x_3^2x_4 \\
& - 4x_1x_3^4 - 16x_1x_2x_4^2x_6 - 16x_1x_3x_4^2x_6 + 8x_1x_2x_3x_4^2 \\
& + 4x_1x_2^2x_4^2 + 4x_1x_3^2x_4^2,
\end{aligned}$$

$$\begin{aligned}
R_2 = {} & 2x_1x_2x_3^2x_4 + 2x_1x_2^2x_3x_4 + 4x_1^2x_2x_4x_5 \\
& + 4x_1^2x_3x_4x_5 - 2x_1x_2x_4x_5^2 - 2x_1x_3x_4x_5^2 \\
& - 4x_1x_2x_3x_4^2 - 2x_1x_2^2x_4^2 + 2x_1x_2x_4^3 - 2x_1x_3x_4^3 \\
& - x_1^2x_2^2x_5 - 2x_1^2x_2x_3x_5 - x_1^2x_3^2x_5 - 2x_1x_3^2x_4^2 \\
& - 4x_1^2x_4^2x_5 + 2x_2x_3x_4^2x_5 + x_2^2x_4^2x_5 + x_3^2x_4^2x_5,
\end{aligned}$$

$$\begin{aligned}
R_1 = {} & 5x_1^4x_2x_3^2x_4 + 5x_1^4x_2^2x_3x_4 - 6x_1^4x_2x_3x_4^2 \\
& - x_1^4x_2^2x_4^2 + x_1^4x_2x_4^3 + x_1^4x_3x_4^3 - x_1^4x_3^2x_4^2 \\
& + 2x_1^2x_2^2x_3^3x_4 + 2x_1^2x_2^3x_3^2x_4 + 2x_1^2x_2^2x_3x_4^3 \\
& + 2x_1^2x_2x_3^2x_4^3 - x_1^4x_2^3x_3 - 2x_1^4x_2^2x_3^2 - x_1^4x_2x_3^3 \\
& - 6x_1^2x_2^2x_3^2x_4^2 - x_1^2x_2^3x_3x_4^2 - x_1^2x_2x_3^3x_4^2 \\
& - x_2^3x_3^4x_4 + x_2^4x_3^3x_4 + x_2^3x_3^2x_4^3 + x_2^2x_3^3x_4^3 \\
& - x_1^2x_2^4x_3^2 - 2x_1^2x_2^3x_3^3 - x_1^2x_2^2x_3^4,
\end{aligned}$$

$$R_0 = 0.$$

The number of terms of the remainders $R_5, \ldots, R_0$ is

$$11, \quad 23, \quad 45, \quad 18, \quad 24, \quad 0,$$

respectively. Since the remainder R_0 is equal to 0, the remainder of G with respect to the quasi-basic set is 0 under the previous non-degeneracy conditions. Therefore, by Theorem 1 of the last section (which also holds in the case of quasi-basic sets) Feuerbach's theorem holds true under these non-degeneracy conditions.

The geometric meanings of the non-degeneracy conditions are clear and it is unnecessary to check further every degeneracy case. We conclude that Feuerbach's theorem is generically true, or simply say in accordance with the convention that Feuerbach's theorem is true, while the above process of calculations provides a proof of this theorem.

The above examples show that the mechanical method is sufficient to prove some rather difficult theorems with the aid of a computer or by hand. For the proof of these theorems, not only is the classical method of Euclid difficult to use but also the usual method of analytic geometry is unable to give a solution owing to the overly complicated calculations. However, by using computers and within the limitation of their speed, all kinds of theorems, as long as they belong to the class considered in this chapter, can be easily proved.

Our contention is not restricted to proving known theorems. We have already shown a means to make theorem proving easy, and consequently such a means also makes it possible to concentrate our creativity on the discovery of new theorems. We can make various kinds of conjectures and verify them on a computer. If a conjecture is verified to be true, we then get a new theorem. Of course, here the computer is employed only as a tool for computation. As for this role, it is not much different from the subsidiary apparatus of pen and paper used for mathematical calculations and proofs, but as for the efficiency, there is a big difference. The manner for discovering new theorems with the aid of computers will be simply called *mechanical discovering*. We shall give two such examples as follows.

Example 5 (The configuration of Pappusian lines). The so-called Pappus's theorem states that, if there are two sets of distinct points A, B, C and A', B', C' respectively on two distinct lines l and l', and these points are also distinct from the possible intersection point of l and l', then the following intersection points

$$P = AB' \wedge BA', \quad Q = AC' \wedge CA', \quad R = BC' \wedge CB',$$

if they exist (i.e., the corresponding lines are not parallel to each other), are collinear. The line determined by P, Q, R will be called a Pappusian line and denoted by

$$\begin{bmatrix} A & B & C \\ A' & B' & C' \end{bmatrix}.$$

Making various permutations of A, B, C; A', B', C', if all the corresponding intersection points exist, we get six Pappusian lines by applying the above method. In addition to the one above, the other five lines are

$$\begin{bmatrix} A & B & C \\ B' & C' & A' \end{bmatrix}, \quad \begin{bmatrix} A & B & C \\ C' & A' & B' \end{bmatrix}, \quad \begin{bmatrix} A & B & C \\ C' & B' & A' \end{bmatrix},$$

$$\begin{bmatrix} A & B & C \\ B' & A' & C' \end{bmatrix}, \quad \begin{bmatrix} A & B & C \\ A' & C' & B' \end{bmatrix}.$$

Let us investigate the geometric property of the configuration of these six Pappusian lines.

In unordered Pascalian geometry, we may naturally introduce some derived concepts such as algebraic curves, the degree of algebraic curves and tangent lines. It is easy to know that generally five lines may be tangent to a curve of the second degree (conic), but for six lines this is possible only if these lines satisfy some condition (the so-called Brianchon theorem). Hence we may ask the following question: Are the above six Pappusian lines all tangent to a conic?

Drawing a figure, one finds that these six Pappusian lines are indeed tangent to a conic, but this conic has degenerated into two points. In other words, each of the six Pappusian lines passes through one of the two points. Formulating this observation as a geometric sentence, we make the following:

Conjecture. The three Pappusian lines

$$\begin{bmatrix} A & B & C \\ A' & B' & C' \end{bmatrix}, \quad \begin{bmatrix} A & B & C \\ B' & C' & A' \end{bmatrix}, \quad \begin{bmatrix} A & B & C \\ C' & A' & B' \end{bmatrix}$$

in the above-mentioned configuration, if not parallel to each other, are concurrent, and so are the other three.

Applying our mechanical procedure and performing the computation on a computer, it is easy to verify the truth of this conjecture, from which we thus get a new theorem.[2]

Example 6 (Pascal-conic theorem). As another example of mechanical discovering, let us consider the following problem.

Suppose we are given 6 points $A_1, \ldots, A_6$ on the same conic. Take 4 distinct ones A_i, A_j, A_k, A_l from them and construct the intersection point

$$P_{ij,kl} = A_i A_j \wedge A_k A_l.$$

There are 45 such intersection points in all (assume the corresponding lines not parallel to each other), which are called Pascalian *points*.

Let us fix the six points A_i in an arbitrary order $A_{i_1} A_{i_2} \cdots A_{i_6}$ to get a hexagon. Then the intersection points

$$P_{i_1 i_2, i_4 i_5} = A_{i_1} A_{i_2} \wedge A_{i_4} A_{i_5},$$

$$P_{i_2 i_3, i_5 i_6} = A_{i_2} A_{i_3} \wedge A_{i_5} A_{i_6},$$

$$P_{i_3 i_4, i_6 i_1} = A_{i_3} A_{i_4} \wedge A_{i_6} A_{i_1}$$

of the three opposite sides of the hexagon are collinear, called a Pascalian *line* of the hexagon $A_{i_1} A_{i_2} \cdots A_{i_6}$. There are 60 such Pascalian lines in all and they constitute a complicated configuration. Among these 45 Pascalian points and 60 Pascalian lines, there are many interesting geometric properties which have been much studied by numerous geometers including Steiner, Cayley, Kirkmann, Veronese (cf. the remark at the end of Salmon's book (Salmon 1879a)). However, most of the interesting theorems found by them are of a *linear* character: Collinearity of certain points and concurrency of certain lines. Now we pose the question: What relations of quadratic or even higher degree character can be found among these 45 Pascalian points and 60 Pascalian lines?

Of course, such relations should exist. For instance, six Pascalian points on two arbitrary Pascalian lines have a quadratic relation: They lie on a *degenerate* conic constituted by these two Pascalian lines. Such a quadratic relation is clearly trivial. Therefore, for the simplest case we may ask: Do there exist six points among the 45 Pascalian points lying on the same *non-degenerate* conic?

---

2 This theorem is now known as a special case of Steiner's theorem. [Transl.]

We shall consider this question in the following way.

In the symmetric group S_6 generated by $1, 2, \ldots, 6$, let us consider a subgroup G of order 6, for example, the subgroup

$$G_6 = \{1, g, g^2, \ldots, g^5\}$$

generated by a permutation

$$g = (123456).$$

Consider now an arbitrary Pascalian point

$$P = P_{ij,kl}$$

and apply an element σ of G_6 to P; we get a point

$$\sigma P = P_{\sigma(i)\sigma(j),\sigma(k)\sigma(l)}.$$

Let us ask: Do the six points

$$P, gP, g^2P, \ldots, g^5P$$

generated by P and G_6 lie on the same conic?

According to the relation between $P_{ij,kl} = A_iAj \wedge A_kA_l$ and $g = (123456)$, we may classify the 45 Pascalian points into 10 types listed in Table 1, in which the second and the last actually belong to the same type.

Table 1

Point $P = A_iA_j \wedge A_kA_l$	Quadratic relation among the six points g^sP
$A_1A_2 \wedge A_3A_4$	No quadratic relation exists
$A_1A_2 \wedge A_3A_5$	No quadratic relation exists
$A_1A_2 \wedge A_3A_6$	On a degenerate conic constituted by two Pascalian lines
$A_1A_2 \wedge A_4A_5$	On a degenerate conic constituted by two coincident Pascalian lines
$A_1A_3 \wedge A_2A_4$	No quadratic relation exists
$A_1A_3 \wedge A_2A_5$	?
$A_1A_3 \wedge A_2A_6$	No quadratic relation exists
$A_1A_3 \wedge A_4A_5$	No quadratic relation exists
$A_1A_3 \wedge A_4A_6$	No quadratic relation exists or meaningless
$A_1A_3 \wedge A_5A_6$	No quadratic relation exists.

The only case that is not easy to determine in Table 1 is the case

$$P_{13,25} = A_1A_3 \wedge A_2A_5.$$

By applying the mechanical procedure of this chapter with an implementation of it on a computer, we verified that the six points are indeed on the same conic, from which the following theorem is obtained.

Theorem. Let $g = (123456)$ and $P = A_1A_3 \wedge A_2A_5$. Then the six points

$$P, gP, \ldots, g^5P$$

lie on the same non-degenerate conic. The conics of this type obtained from different g and P are 60 in all.

The computer we used is one with a speed of 3000 additions per second. The computing time required for proving this theorem is about sixty hours. If we use a microcomputer with a speed of 200,000–300,000 additions per second, the proof can be completed within one hour. This demonstrates the efficiency of our mechanical method. By combining and permuting the six points $A_1, \ldots, A_6$, we should be able to find *all* possible non-degenerate quadratic relations among the 45 Pascalian points and give their proof without much difficulty. We can even consider non-degenerate cubic relations and other complicated relations. This is almost impossible to be accomplished by using the classical methods (whether they are synthetic or analytic).

This section has given several examples to explain the role of the mechanical method. Naturally, both mechanical proving and mechanical discovering may be replaced by using the usual, traditional method. The following is a direct proof of the theorem in Example 6.

Denote $P_{13,25}$ by A_7 and $gA_7, g^2A_7, \ldots, g^5A_7$ by $A_8, A_9, \ldots, A_{12}$, respectively. To prove that the six points $A_7, A_8, \ldots, A_{12}$ lie on the same conic, it suffices to prove that $A_7A_8A_{11}A_{12}A_9A_{10}$ is a Pascalian hexagon, while the latter is equivalent to proving that

$$A_{13} = A_7A_8 \wedge A_9A_{12}, \quad A_{14} = A_8A_{11} \wedge A_9A_{10}, \quad A_{15} = A_{11}A_{12} \wedge A_7A_{10}$$

are collinear.

By the hypothesis that $A_1, A_2, \ldots, A_6$ lie on the same conic, $A_1A_3A_6A_5A_2A_4$ is a Pascalian hexagon, so the three points

$$A_1A_3 \wedge A_2A_5 = A_7, \quad A_3A_6 \wedge A_2A_4 = A_8, \quad A_5A_6 \wedge A_1A_4$$

are collinear. In other words, $A_5A_6 \wedge A_1A_4$ is identical to A_{13}.

Similarly, the fact that $A_1A_2A_5A_3A_6A_4$ is a Pascalian hexagon implies that $A_1A_2 \wedge A_3A_6 = A_{14}$, and that $A_1A_5A_2A_6A_3A_4$ is a Pascalian hexagon implies that $A_2A_5 \wedge A_3A_4 = A_{15}$.

Finally, from the hypothesis that $A_1A_2A_5A_6A_3A_4$ is a Pascalian hexagon, one knows that A_{13}, A_{14}, A_{15} are collinear, so that $A_7, A_8, \ldots, A_{12}$ lie on the same conic. This is what we wanted to prove.

The above proof is simple and direct, and it sufficiently reflects the character of the traditional, purely geometric proof. But as pointed out by Descartes, every such proof needs some ingenious idea, an individual theorem needs an individual proof and thus an individual, ingenious idea. For mechanical proving, the situation is just the contrary. So long as one has theoretically proved that some class of theorems or some geometry is mechanizable and has given the corresponding mechanical procedure, the proof of every theorem in that class – no matter which one it is – is the same without any difference between difficulty and ease; and, within the limitation of the computer's memory and speed, the proof can be constructed in practice without requiring any other consideration. Here, the relation between mechanical proving and traditional proving is similar to the relation between algebraic methods and arithmetic methods. Taking the four-arithmetic-operation problem of the so-called chicken-rabbit-co-coop as an example, the arithmetic method supposes that every chicken would have four feet or every rabbit would have only two feet, which may be well considered of exhausting all ingenious ideas. The method is also vivid, beautiful, fascinating and attractive. On the contrary, the algebraic method (i.e., the Thieh Yuan Shu of ancient China) is mechanical and uninteresting, but it is applicable to a large class of problems, not just a special one. At the beginning, the creation of the algebraic method or Thieh Yuan Shu was only directed to a certain class of problems in order to "save labour many times." But as for its development up to this day, the role has been as we all know.

Remark. The author has written a program on a small computer – HP 1000 – according to the mechanical method proposed in this book and made some experiments, including examples for mechanical theorem discovering. We present them as follows.

1. Theorems about the inversion center and the inversion-center line

Fix a line l in the plane. If the two angles formed by the lines l_1, l_2 with l are complementary, we say that l_1, l_2 are *inversional* (with respect to l).

Theorem 1. The three inversion lines of the three sides of an arbitrary triangle $A_1A_2A_3$ drawn from their opposite vertices respectively are concurrent. The intersection point will be called the inversion center of $\triangle A_1A_2A_3$ (with respect to l).

Theorem 2. The four inversion centers of the four triangles determined by an arbitrary complete quadrilateral are collinear.

The proof of these two theorems on our HP 1000 takes about 1 min and 4 min running time respectively.

2. Pappus-point theorem in Example 5 of this section
The running time for the proof of the Pappus-point theorem on HP 1000 is about 33 min.

This theorem was discovered and proved in 1980 on an HP 9835A. At that time the printings were recorded, but we did not record the exact running time which is around 20 h.

3. Pascal-conic theorem in Example 6 of this section
The running time for the proof of this theorem on our HP 1000 is about 1 h and 23 min.

The theorem was discovered and proved also in 1980 on the HP 9835A. At that time, the printings were also recorded and the computing time required for the proof is about 60 h.

We have recorded the detailed printings of the proving process for the above three theorems on the HP 1000.

Further remark. Since the proving process was displayed in detail on the screen during the computation, the above-listed timings are actually much more than the true CPU time. A Chinese student, S. C. Chou, studying now in the University of Texas at Austin, USA for his doctoral degree, has written a program on a Dec-20 computer based on our method. He has proved 150 theorems in ordinary geometry and has also discovered some interesting theorems, of which one is as follows:

Draw a circle cocentered with the circumcircle of a triangle. From an arbitrary point on this circle drop three perpendiculars to the three sides of the triangle respectively. Then the triangle formed by the three perpendicular feet has a constant area. In case the cocentered circle is identical to the circumcircle, this constant becomes 0, i.e., the three perpendicular feet are collinear – this is commonly called Simson's theorem.

Chou's paper "Proving Elementary Geometry Theorems Using Wu's Algorithm" will appear in Contemporary Mathematics, American Mathematical Society, 1984.

5 Mechanization theorems of (ordinary) ordered geometries

5.1 Introduction

The geometries, such as unordered Pascalian geometry and orthogonal geometry, which were considered in Chaps. 3 and 4 do not assume any relation of order. In geometries of this kind the statement of theorems, after fixing the coordinate system, involves only some equality relations. We have proposed a mechanical method which usually is quite efficient and can be used to prove rather difficult theorems. Practice will show the extent of efficiency of our method for unordered geometries. In the book "Theory, Method and Practice of Mechanical Theorem Proving in Geometries" (under preparation), we shall explain in detail the algorithmic procedure, implementation and computer experiments of this method and give more examples. This book is restricted only to the basic principles. If a geometry assumes an order relation and the statement – especially the conclusion part – of a theorem involves such an order relation, then the situation becomes not only much more complicated but also different in essence. In this case, there are methods for mechanical proving in theory, but their efficiency is not high. It still seems difficult to prove non-trivial theorems by using these methods.

In Chaps. 1 and 2 the following kinds of ordered geometries were presented:

1. Ordered Pascalian geometry, i.e., the geometry obtained by adjoining the Pascalian geometry defined in Chap. 1 with the relation of order and assuming Hilbert's axioms of order H II.
2. Ordered orthogonal geometry (Chap. 2).
3. Ordered metric geometry (Chap. 2).
4. Ordinary geometry, i.e., the geometry mentioned in Sect. 1.1 which has such fundamental relations as incidence, order, congruence, and parallelism given in Hilbert's "Grundlagen der Geometrie" and which satisfies all five groups of axioms H I–H V. This geometry is usually called Euclidean geometry in the literature.

For the sake of simplicity, any ordered geometry mentioned above will be called a *usual ordered geometry*. In this chapter, we sometimes omit the word *usual* and simply call it an *ordered geometry*.

The associated field of every ordered geometry is an ordered field. In particular, the associated field of ordinary geometry is just the usual real number field. The problem of mechanical theorem proving in these geometries depends upon the order property of the associated field. For this purpose, in the following we first give a brief review of ordered fields. The reader may refer to van der Waerden (1930a: chap. 10, 1955: chap. 9), or Artin (1950: chap. 1 sect. 9), Jacobson (1974: chap. 5), or the original article by Artin and Schreier (1927) for details. Since the axiom of infinity holds in the considered geometries, the corresponding associated fields are all of characteristic 0, which we shall not explain further.

Definition 1. A number field $\boldsymbol{K}$ is called an *ordered field* if the numbers other than 0 in $\boldsymbol{K}$ can be divided into two parts $\boldsymbol{K}^+$ and $\boldsymbol{K}^-$, satisfying the following properties:

1. If $a \in \boldsymbol{K}^+$, then $-a \in \boldsymbol{K}^-$. Similarly, if $a \in \boldsymbol{K}^-$, then $-a \in \boldsymbol{K}^+$.
2. If $a, b \in \boldsymbol{K}^+$, then $a + b \in \boldsymbol{K}^+$ and $ab \in \boldsymbol{K}^+$.

For an ordered field $\boldsymbol{K}$, the numbers in $\boldsymbol{K}^+$ are called *positive numbers* and the numbers in $\boldsymbol{K}^-$ are called *negative numbers*. If $a - b \in \boldsymbol{K}^+$, it is denoted as $a > b$ or $b < a$. Moreover, we denote the *absolute value* of a by $|a|$ which is 0, a or $-a$, depending on $a = 0$, $a \in \boldsymbol{K}^+$ or $a \in \boldsymbol{K}^-$ respectively.

From the definition one knows that in an ordered field the integers and rational numbers remain their usual order relation. For example, $+1$ is a positive number and -1 is a negative number.

Definition 2. A number field $\boldsymbol{K}$ is called a *formally real number field* if the square sum of any finitely many numbers in $\boldsymbol{K}$ is not equal to -1.

Definition 3. A number field $\boldsymbol{K}$ is called a *real closed field* if $\boldsymbol{K}$ is a formally real number field and any algebraic extension of $\boldsymbol{K}$, so long as it is different from $\boldsymbol{K}$, is not a formally real number field.

Definition 4. A polynomial

$$f = f(x_1, \ldots, x_n) \in \boldsymbol{K}[x_1, \ldots, x_n]$$

over an ordered field $\boldsymbol{K}$ is said to be *positive definite* (or *semi-positive definite*) if for an arbitrary set of numbers $a_1, \ldots, a_n$ in $\boldsymbol{K}$

$$f(a_1, \ldots, a_n) > 0 \quad (\text{or } f(a_1, \ldots, a_n) \geq 0).$$

If $-f$ is positive definite (or semi-positive definite), then we say that f is *negative definite* (or *semi-negative definite*).

Below are some main results about ordered fields which will be used for the mechanization of geometry.

Theorem 1. It is not possible to introduce an order relation in a non-formally real number field so that it becomes an ordered field. On the contrary, one can introduce at least one order relation in an arbitrary formally real number field so that it becomes an ordered field.

Theorem 2. If $\boldsymbol{K}$ is an ordered field, then there is one and only one real closed algebraic extension $\bar{\boldsymbol{K}}$ of $\boldsymbol{K}$, up to equivalence, so that the order in $\boldsymbol{K}$ remains in $\bar{\boldsymbol{K}}$, i.e., for $a, b \in \boldsymbol{K}$, if $a < b$ (or $a > b$, $a = b$), then as elements in $\bar{\boldsymbol{K}}$, $a < b$ (or $a > b$, $a = b$) still holds.

Theorem 3. The rational number field $\boldsymbol{Q}$ has only one uniquely determined order so that it becomes an ordered field with the integers retaining the usual order; and there is one and only one real closed algebraic extension of $\boldsymbol{Q}$, up to isomorphism.

Theorem 4 (Rolle's zero theorem). Let $\boldsymbol{R}$ be a real closed field and $f(x) \in \boldsymbol{R}[x]$ be a polynomial over $\boldsymbol{R}$. If $a, b \in \boldsymbol{R}$ are such that $f(a) < 0$, $f(b) > 0$, then there exists a number c between a and b such that $f(c) = 0$.

As a corollary of Theorem 4, we have the following.

Theorem 5. Let $\boldsymbol{R}$ be a real closed field. Then:

1. An arbitrary positive number in $\boldsymbol{R}$ has exactly two square roots whose absolute values are equal and signs are opposite;
2. An arbitrary polynomial equation $f(x) = 0$ of odd degree in $\boldsymbol{R}$ has at least one root in $\boldsymbol{R}$.

For a real closed field, we have the following.

Sturm's theorem. Let $\boldsymbol{R}$ be a real closed field, $f(x)$ be a polynomial over $\boldsymbol{R}$ of degree greater than 0 and $f'(x)$ be the formal derivative of $f(x)$. Form the Sturm series of $f(x)$

$$S = \{f_0(x), f_1(x), \ldots, f_s(x)\},$$

where the polynomials are obtained by using the division algorithm as follows:

$$\begin{aligned}
f_0(x) &= f(x),\\
f_1(x) &= f'(x),\\
f_2(x) &= q_1(x) f_1(x) - f_0(x), \quad \deg f_2 < \deg f_1,\\
&\cdots\\
f_{i+1}(x) &= q_i(x) f_i(x) - f_{i-1}(x), \quad \deg f_{i+1} < \deg f_i,\\
&\cdots \quad \cdots \quad \cdots \quad \cdots\\
f_{s+1}(x) &= q_s(x) f_s(x) - f_{s-1}(x) = 0.
\end{aligned}$$

For any $c \in \boldsymbol{R}$ with $f(c) \neq 0$, let $V(c)$ denote the number of sign changes in the series

$$\{f_0(c), f_1(c), \ldots, f_s(c)\}.$$

Then, when $f(a) \neq 0$, $f(b) \neq 0$, $a, b \in \boldsymbol{R}$ and $a < b$, the number of roots of $f(x) = 0$ in the interval (a, b) is $V(a) - V(b)$.

Properties 1 and 2 in Theorem 5 are considered as two characteristics of real closed fields. In fact, in some books an ordered field satisfying Properties 1 and 2 is also called a real closed field. Moreover, the proof of Sturm's theorem depends only on these two properties (cf. Jacobson 1974; and Sect. 5.2).

Like the relation between the real number field and the complex number field, we have the following.

Theorem 6. The algebraic extension $\boldsymbol{R}(\sqrt{-1})$ obtained from a real closed field $\boldsymbol{R}$ by adjunction of the algebraic element $\sqrt{-1}$ is algebraically closed. On the contrary, if a formally real number field $\boldsymbol{R}$ becomes an algebraically closed field after adjoining $\sqrt{-1}$ to it, then $\boldsymbol{R}$ is a real closed field.

From this theorem, we know that the roots of any equation $f(x) = 0$ $(f(x) \in \boldsymbol{R}[x])$ over an arbitrary real closed field $\boldsymbol{R}$ are contained in $\boldsymbol{R}(\sqrt{-1})$. If the roots themselves are not in $\boldsymbol{R}$, they appear in the form of a conjugate pair $a \pm b\sqrt{-1}$, where $a, b \in \boldsymbol{R}$. The symmetric functions of the roots can be obtained constructively via operations on the coefficients of $f(x)$, which is analogous to the usual case.

Now, let $\boldsymbol{K}$ be an ordered field. Then the inequality relations among numbers of $\boldsymbol{K}$ may be derived from the existence relations on the *solvability* of some equations, where *solvability* means that the equations have solutions in $\boldsymbol{K}$ (which we shall not further explain). For example, if a, b are numbers in $\boldsymbol{K}$ and x, y are variables, then

$ax^2 = 1$	has a solution (in $\boldsymbol{K}$)	$\Longrightarrow$	$a > 0$,
$ax^2 = -1$	....	$\Longrightarrow$	$a < 0$,
$x^2 = a$	....	$\Longrightarrow$	$a \geq 0$,
$x^2 = -a$	....	$\Longrightarrow$	$a \leq 0$,
$abx^2 = 1$	....	$\Longrightarrow$	a, b have the same sign, i.e., they are non-zero and both positive or both negative,
$abx^2 = -1$	....	$\Longrightarrow$	a, b have different signs,
$ax = 1$	....	$\Longrightarrow$	$a \neq 0$.

If $\boldsymbol{K}$ is not only an ordered field but moreover a real closed field or an ordered field satisfying the Properties 1 and 2 in Theorem 5, then the implication relation in the above expressions can all be replaced by equivalence relation. In this case, the inequality relations among numbers can be completely reduced to existence relations on the solvability of equations.

Consider now an arbitrary ordered geometry as mentioned at the beginning of this section, with which the associated field is an ordered field $\boldsymbol{K}$. After fixing a proper coordinate system, some of the fundamental relations such as incidence, parallelism, and perpendicularity in this geometry can be expressed as equality relations among the coordinates. However, in expressing the fundamental relation of order, inequalities in the coordinates have to be involved. We give some examples according to the discussions in Sect. 3.3 as follows.

Example 1. For three points

$$A_1 = (a_1, a_4), \quad A_2 = (a_2, a_5), \quad A_3 = (a_3, a_6)$$

lying on the same line, the order relation that point A_2 lies between A_1 and A_3 may be expressed as

$$a_1 < a_2 < a_3 \quad \text{or} \quad a_1 > a_2 > a_3 \quad \text{or} \quad a_4 < a_5 < a_6 \quad \text{or} \quad a_4 > a_5 > a_6$$

or

$$(a_2 - a_1)(a_2 - a_3) < 0 \quad \text{or} \quad (a_5 - a_4)(a_5 - a_6) < 0.$$

Thus, according to the discussion before, this order relation can be derived from the solvability of an equation:

$$(a_2 - a_1)(a_2 - a_3)x^2 = -1 \quad \text{is solvable,}$$

or

$$(a_5 - a_4)(a_5 - a_6)x^2 = -1 \quad \text{is solvable,}$$

where the solvability of an equation is meant to be in $\boldsymbol{K}$. If $\boldsymbol{K}$ itself is a real closed field, then the geometric fact that A_2 lies between A_1 and A_3 is equivalent to the solvability of either of the two equations.

If the line is represented by a parametric equation with parameter t, and A_i corresponds respectively to $t = t_i$, then the order relation that A_2 lies between A_1 and A_3 can be expressed as an inequality relation

$$(t_2 - t_1)(t_2 - t_3) < 0,$$

and thus can be derived from or is equivalent to the solvability of the equation

$$(t_2 - t_1)(t_2 - t_3)x^2 = -1.$$

Example 2. Two points $A = (a_1, a_2)$, $A' = (a'_1, a'_2)$ lie on the same side (or different sides) of a line

$$l: \quad L(x_1, x_2) \equiv c_1 x_1 + c_2 x_2 + c_3 = 0.$$

This geometric relation corresponds to the fact that

$$L(a_1, a_2) = c_1a_1 + c_2a_2 + c_3$$

and

$$L(a'_1, a'_2) = c_1a'_1 + c_2a'_2 + c_3$$

have the same sign (or different signs). Hence it can be derived from the solvability of the equation

$$(c_1a_1 + c_2a_2 + c_3)(c_1a'_1 + c_2a'_2 + c_3)x^2 = +1 \quad \text{(same side)}$$

or

$$(c_1a_1 + c_2a_2 + c_3)(c_1a'_1 + c_2a'_2 + c_3) = -1 \quad \text{(different sides)}$$

in $\boldsymbol{K}$. If $\boldsymbol{K}$ is a real closed field, whether A, A' lie on the same side or different sides of l is equivalent to whether the former or the latter equation has a solution.

Now consider an arbitrary theorem in the above-mentioned ordered geometry. Let $\boldsymbol{K}$ be the geometry-associated field and $\bar{\boldsymbol{K}}$ be a real closed algebraic extension of $\boldsymbol{K}$. In a proper coordinate system, the hypothesis and conclusion of this theorem can both be expressed by some equality and/or inequality relations over $\boldsymbol{K}$. If we introduce – besides all the coordinates of points – some new variables, then the inequality relations can be replaced by the solvability relations of some polynomial equations for these new variables over $\bar{\boldsymbol{K}}$, according to the discussion before. Now we may divide the theorems into two classes, according to whether their conclusions can be expressed by equality relations or inequality relations. If the conclusion of a theorem does not involve the order relation and thus can be expressed by the usual equality relations over $\boldsymbol{K}$ while all the variables appearing in the equalities are the original coordinate variables, then this theorem is called a *theorem of equality type*; otherwise, it is called a *theorem of inequality type*. For instance, in ordinary geometry the following theorem is one of equality type.

Theorem 7. Construct *outside* of each side of a triangle an equilateral triangle. Then the three lines connecting, respectively, each vertex of the original triangle to the opposite vertex of the equilateral triangle on its opposite side are concurrent.

In this theorem, the hypothesis involves the order relation, but the conclusion involves only the incidence relation.

Let us come to the next theorem.

Theorem 8. The nine-point circle of any triangle is tangent *internally* to the *inscribed* circle and *externally* to the escribed circles of the triangle.

Since the tangency of circles involves the order relation, this theorem is one of inequality type. However, if we are not required to prove that the nine-point circle is tangent internally to the inscribed circle and externally to the escribed circles, but only to prove that the circles are tangent to each other, then the conclusion no longer involves the order relation. If the radii of the nine-point circle and the inscribed or an escribed circle are denoted by R and r respectively, and the distance between the center of the nine-point circle and the center of the inscribed or escribed circle is denoted by d, then the tangency relation of the two circles can be expressed by the following equality:

$$R^4 + r^4 + d^4 - 2R^2r^2 - 2R^2d^2 - 2r^2d^2 = 0.$$

Hence the theorem now is still one of equality type.

There are too many theorems of this kind in elementary geometry to be mentioned one by one. Generally speaking, since the notion of angles does not exist or at least is indeterminate in unordered geometry and the angular bisectors of two intersecting lines, two in number, cannot be distinguished from each other, the above situation occurs often in the case that the theorem concerns angles and angular bisectors.

For a theorem of equality type, after introducing some new variables the original hypothesis can be expressed by equality relations in the new and old variables over $\bar{\boldsymbol{K}}$ while the conclusion may be expressed by equality relations only in the *original* variables with coefficients not only in $\bar{\boldsymbol{K}}$ but also in the original unextended number field $\boldsymbol{K}$ (actually, the coefficients can all be taken as integers). In this case, the algebraic form of the theorem may be taken as that in Sect. 4.1: Deduce, from a set of polynomial relations in some variables over a number field, another polynomial relation in these variables (actually, some of the variables in the hypothesis relations) over the same number field, while the coefficients of these polynomials may be actually taken as integers. Therefore, by the method of Chap. 4 we have the following.

Mechanization theorem 1. There is a mechanical method for proving theorems of equality type in an arbitrary (usual) ordered geometry.

For theorems of inequality type, the situation is totally different. Let the variables in the hypothesis of a theorem be $x_1, \ldots, x_m$. Then the conclusion of the theorem can be expressed as an inequality relation in the variables $x_1, \ldots, x_m$. After introducing a new variable, say y, whether this inequality relation holds is equivalent to whether an equation of the form

$$g(y, x_1, \ldots, x_m) = 0$$

is solvable for y in $\bar{\boldsymbol{K}}$. This is an *existence problem* of the solution of an equation, which is essentially different from the problem of *formula inference* as for theorems of equality type, where only the inference among equalities is encountered. The manners of dealing with these two kinds of problems are also completely different. After introducing some new variables in order to reduce the hypothesis of a theorem of inequality type to equality relations over $\bar{\boldsymbol{K}}$ and letting the new and old variables altogether be $x_1, \ldots, x_n$, the problem of proving the theorem is reduced to an algebraic problem of the following form:

Let $\boldsymbol{K}$ be an ordered field and

$$\Sigma = \{f_i(x_1, \ldots, x_n)\}$$

be a set of finitely many polynomials in the variables $x_1, \ldots, x_n$ over $\boldsymbol{K}$. Let y be another variable and

$$g(y, x_1, \ldots, x_n) = 0$$

be another polynomial equation over $\boldsymbol{K}$. Find a mechanical method which determines for any numbers $a_1, \ldots, a_n$ in $\boldsymbol{K}$ or $\bar{\boldsymbol{K}}$, satisfying the equations

$$f_i(x_1, \ldots, x_n) = 0,$$

whether or not the equation

$$g(y, a_1, \ldots, a_n) = 0$$

has a solution for y in $\bar{\boldsymbol{K}}$.

A solution to this problem is provided by the so-called Tarski's theorem. The proof of this theorem and the underlying method will be presented in the next section. From it we immediately obtain the following.

Mechanization theorem 2. There is a mechanical method for proving theorems in usual ordered geometries.

It should be stressed that neither the mechanization theorem 1 nor the mechanization theorem 2 here contains the mechanization theorem in Chap. 4 or even in Chap. 3. This is because the mechanization theorems here can only be applied to really ordered geometries, so the associated field $\boldsymbol{K}$ must be an ordered field. However, the geometries to which the mechanization theorems in Chaps. 4 and 3 can be applied do not need to have an order relation and do not even have to have the possibility of introducing an order relation. This is, for instance, the case for orthogonal geometry, in which isotropic lines exist, or for complex geometry. Hence, the mechanization theorems in Chaps. 3 and 4 and those in this chapter do not contain each other, while for the geometries and theorems to which both kinds of mechanical methods can be applied (i.e., their overlapping part), the methods in Chaps. 3 and 4 are completely different from that in this chapter and the former are much more efficient than the latter.

5.2 Tarski's theorem and Seidenberg's method

We have mentioned in the preceding section that in a proper coordinate system the problem of proving theorems of inequality type in ordered geometries can be transformed into a purely algebraic problem, which can then be solved by using Tarski's theorem. However, both the original method of Tarski and its later simplification by Seidenberg are all rather complicated. These methods are not only less efficient, but the provided solution also does not well satisfy the requirements of geometry. In what follows we reduce this algebraic problem to another form and treat it by an improved version of Tarski–Seidenberg's method. For this purpose, let us first restate below the algebraic problem formulated at the end of the last section. When polynomials are mentioned, they will be assumed to have integer coefficients, unless indicated otherwise. The polynomial sets will all be assumed to consist of only a finite number of polynomials.

Algebraic problem 1. Let $\boldsymbol{K}$ be a real closed field,

$$\Sigma = \{f_i(x_1, \ldots, x_n)\}$$

be a finite set of polynomials and $g(y, x_1, \ldots, x_n)$ be a polynomial in which y actually appears. Find a mechanical method which decides from

$$\text{Hypothesis:} \quad f_i(x_1, \ldots, x_n) = 0, \quad f_i \in \Sigma$$

whether or not

$$\text{Conclusion:} \quad g(y, x_1, \ldots, x_n) = 0$$

has a solution in $\bar{\boldsymbol{K}}$.

In detail, when for any $a_k \in \bar{\boldsymbol{K}}$, $k = 1, \ldots, n$, that satisfy the hypothesis, the equation $g(y, a_1, \ldots, a_n) = 0$ has a solution for y, the corresponding geometric theorem is true. If there are a_k's such that the above equation has no solution, then the corresponding geometric theorem is not true at least in the case $x_k = a_k$.

As only the non-degenerate cases of a geometric theorem need to be considered, we may leave out of account the degenerate cases. For in the degenerate cases, a theorem is either meaningless or even false. Hence the algebraic problem we really need to solve does not completely agree with the problem 1 above. The problem in fact is to find a set

$$\Delta = \{D_1, \ldots, D_r\}$$

of non-degeneracy conditions, where each D_i is also a polynomial in the variables $x_1, \ldots, x_n$ with integer coefficients, so as to determine whether or not the equation $g = 0$ has a solution under the subsidiary conditions

$$D_i \neq 0 \quad (D_i \in \Delta).$$

The resolution of the algebraic problem in this form is not only much easier than that of the problem 1, but also more in accordance with the real situation of geometry.

On the other hand, the above conclusion polynomial $g(y, x_1, \ldots, x_n)$ is not arbitrary but is obtained from a polynomial inequality

$$h(x_1, \ldots, x_n) > 0 \quad \text{or} \quad < 0$$

according to the original meaning of the geometric theorem by introducing a new variable y, i.e.,

$$g = y^2 h \mp 1,$$

in which the sign is taken to be negative when $h < 0$, to be positive when $h > 0$. We will not directly prove the theorem to hold (under some non-degeneracy conditions) but prove its negation to fall into fallacies, i.e., prove that the set of following equations and inequality

$$f_i(x_1, \ldots, x_n) = 0 \quad (f_i \in \Sigma),$$
$$h(x_1, \ldots, x_n) \leq 0 \quad \text{or} \quad \geq 0,$$

has no solution in $\bar{\boldsymbol{K}}$. In other words, if we introduce a new variable z and set

$$g(z, x_1, \ldots, x_n) = z^2 \pm h(x_1, \ldots, x_n),$$

in which the sign is taken to be $+$ or $-$ depending on $h \leq 0$ or ≥ 0, then proving the original theorem to hold is equivalent to proving that the set of equations

$$f_i(x_1, \ldots, x_n) = 0 \quad (f_i \in \Sigma),$$
$$g(z, x_1, \ldots, x_n) = 0,$$

has no solution in $\bar{\boldsymbol{K}}$ under some subsidiary conditions.

Now according to Seidenberg, set

$$F(z, x_1, \ldots, x_n) = \sum f_i^2 + g^2.$$

As $\bar{\boldsymbol{K}}$ is a real closed field, whether or not the set of equations in $\bar{\boldsymbol{K}}$ has a solution is equivalent to whether or not $F = 0$ has a solution. Furthermore, by taking into account the geometric reality as before, we need only to consider the non-degenerate cases. Therefore we arrive finally at the following.

Algebraic problem 2. Let $\bar{\boldsymbol{K}}$ be a real closed field, $z, x_1, \ldots, x_n$ be variables and

$$F(z, x_1, \ldots, x_n)$$

be a polynomial. Find a mechanical method which determines a set

$$\Delta = \{D_1, \ldots, D_r\}$$

of polynomials in the variables $x_1, \ldots, x_n$ and decides whether or not the set of equation and inequations

$$F(z, x_1, \ldots, x_n) = 0,$$
$$D_i(x_1, \ldots, x_n) \neq 0 \quad (D_i \in \Delta)$$

has a solution in $\bar{\boldsymbol{K}}$ in a finite number of steps.

If it is decided that the set has no solution, then the original geometric theorem is true under the non-degeneracy conditions $D_i \neq 0$. Otherwise, the theorem is false under the same non-degeneracy conditions. As for D_i, they should be so restricted that every $D_i = 0$ is not a formal consequence of $F = 0$ (cf. Sect. 4.1).

The method of Tarski–Seidenberg may be used to deal with the more general algebraic problem 1, but it is not very appropriate for theorem proving in geometries. If we restrict ourselves to algebraic problem 2, then the method can be much simplified to satisfy the requirements of geometry and to have the possibility for proving some really complicated theorems. Let us give a solution to algebraic problem 2 as follows.

The method will make induction on the number n of variables x. For this purpose, we first introduce some notations. Denote simply the polynomial with integer coefficients of the form

$$F(z, x_1, \ldots, x_n)$$

by

$$F_n = F_n(z, x^{(n)}),$$

where

$$x^{(n)} = (x_1, \ldots, x_n).$$

Similarly, an n-tuple $(a_1, \ldots, a_n)$ of values in $\bar{\boldsymbol{K}}$ is simply denoted by $a^{(n)}$ and

$$x^{(n)} = a^{(n)}$$

means that

$$x_1 = a_1, \ldots, x_n = a_n.$$

In addition, let

$$\Delta_n = \{D_{n1}, \ldots, D_{nr_n}\}$$

be a set of non-degeneracy conditions, where $D_{ni} = D_{ni}(x_1, \ldots, x_n)$ are all non-zero polynomials in the variables $x_1, \ldots, x_n$ with integer coefficients and each $D_{ni} \neq 0$ is not a formal consequence of $F = 0$. Simply write

$$\{F_n, \Delta_n\} = \Omega_n.$$

We say that Ω_n *has a solution* if for an arbitrary n-tuple $a^{(n)}$ in $\bar{\boldsymbol{K}}$ with

$$D_{ni}(a^{(n)}) \neq 0 \quad (D_{ni} \in \Delta_n),$$

the equation

$$F_n(z, a^{(n)}) = 0$$

has a solution for z in $\bar{\boldsymbol{K}}$. If $z = c$ is a solution of the above equation, we say that $(c, a^{(n)})$ *satisfies* Ω_n. Now we are ready to state the following.

Theorem. There is a mechanical method which finds, for an arbitrary polynomial

$$F_n = F_n(z, x^{(n)}),$$

a polynomial F_{n-1} in the variables z and $x^{(n-1)}$ and a finite set Δ_n of polynomials $D_{n1}, \ldots, D_{nr_n}$ in $x^{(n-1)}$ such that whether or not

$$\Omega_n = \{F_n, \Delta_n\}$$

has a solution depends on whether or not

$$\Omega_{n-1} = \{F_{n-1}, \Delta_n\}$$

has a solution.

By using this theorem we can immediately solve the algebraic problem 2: Starting from the polynomial $F_n = F_n(z, x^{(n)})$, we get F_{n-1} and Δ_n, and from F_{n-1}, we get F_{n-2} and Δ_{n-1}. On the analogy of this, we finally get a polynomial $F_0(z)$. Merge the successively obtained

$$\Delta_n, \Delta_{n-1}, \ldots, \Delta_1$$

into a set Δ of polynomials. Then whether or not

$$\Omega = \{F_n, \Delta\}$$

has a solution depends on whether or not

$$F_0(z) = 0$$

has a solution for z, while the latter can be determined by using Sturm's theorem. Therefore, when the geometry-associated field is a real closed field, there is a mechanical method which finds a set $\{D_i\} = \Delta$ of polynomials for the non-degeneracy conditions from $F_n = F_n(z, x_1, \ldots, x_n)$ and determines whether or not $F_n = 0$ has a solution and thus determines whether or not the geometric theorem holds under the non-degeneracy conditions $D_i \neq 0$. In other words, theorem proving in the corresponding geometry is mechanizable or the mechanization theorem of the corresponding geometry holds.

The above theorem solves algebraic problem 2 and thus the corresponding geometric problem. It is especially worth pointing out that our main interest is in *positively proving* theorems rather than negating them. For the former, we need only to verify that the final $F_0 = 0$ has no solution in $\bar{\boldsymbol{K}}$ so that the original polynomial equation $F_n = 0$ has no solution in $\bar{\boldsymbol{K}}$, at least under some non-degeneracy conditions, and thus has no solution in the original geometry-associated field. Therefore, even if the geometry-associated field is not a real closed but an ordered field, we can take its extended real closed field and apply the above conclusion to get a set of non-degeneracy conditions and *positively prove* the theorem to be true under these non-degeneracy conditions, which is what we aim to do. As for the case in which the final equation $F_0 = 0$ has a solution, we know that the theorem is false only when the solution is considered for a real closed field. When it is considered for a general ordered geometry-associated field we cannot conclude the falsity of the theorem yet, but this is not really a problem we are concerned with and thus less important to us. For our point of interest, the result achieved by using the above method is sufficient.

The proof of the theorem is based on the simplified method of Seidenberg (cf. Seidenberg 1954, Jacobson 1974: chap. 5). We only give an outline of the proof and mainly explain where the proof of Seidenberg can be modified.

Seidenberg's proof of the theorem is based principally on a geometric fact stated in the following.

Seidenberg's lemma. Let the coordinates of a Descartes coordinate system be (x, y) and let

$$C: \quad f(x, y) = 0$$

be a curve, where f is a polynomial over a real closed field $\bar{\boldsymbol{K}}$. If C has points in $\bar{\boldsymbol{K}}$ or $f = 0$ has solutions in $\boldsymbol{K}$, then among these points there is one (a, b) with shortest distance to the origin. Furthermore, if (a, b) is neither the origin nor a singular point of C, then the curve C is tangent at this point to the circle

$$S: \quad x^2 + y^2 = a^2 + b^2$$

which passes through the point (a, b).

Since we also need to consider more general ordered geometries such as ordered Pascalian geometry, in which the concept of orthogonality and other metric concepts do not necessarily exist, and so-called Descartes coordinates, circles and distance, etc. have no meaning, we cannot directly use the proof of Seidenberg. However, as long as some concepts are properly modified, the lemma still holds and the original proof is available as well. Let us give an account of this modification as follows.

Let Γ be a coordinate system with (x, y) as coordinates in an ordered geometry. Here Γ is arbitrary, so there are no concepts of circles and distance, etc. However, some concepts may still retain their usual meaning. For example, let the equation

$$f(x, y) = 0$$

define a curve C, where f is supposed, without loss of generality, to be a polynomial with integer coefficients. If a, b are both in $\bar{K}$ while $f(a, b) = 0$, we say that $P(a, b)$ is a *point on* C. Set

$$\alpha = \frac{\partial f}{\partial x}\Big|_{(a,b)}, \quad \beta = \frac{\partial f}{\partial y}\Big|_{(a,b)}.$$

If $P(a, b)$ is a point on C with

$$\alpha = 0, \quad \beta = 0,$$

then P is called a *singular point* of C. If P is a point on C, not singular, then the equation

$$\alpha(x - a) + \beta(y - b) = 0$$

defines a line, called the *tangent line* of C at point P. These definitions are proper in an arbitrary coordinate system and coincide with the usual concepts. Although there is no concept of orthogonality in the considered geometry, we still say that the line

$$\beta(x - a) = \alpha(y - b)$$

is the *"normal line" of C at point P in the coordinate system* Γ, which is, however, different from the usual case. Of course, the "normal line" here is to be understood with respect to the coordinate system. Similarly, any equation

$$x^2 + y^2 = r^2$$

($r \geq 0$) is said to define a *"circle"* with *"radius"* r and *"center"* at the origin in the *coordinate system* Γ and a point (a, b) is said to be *inside* the "circle," or *on* the "circle" or *outside* the "circle," respectively, depending on

$$a^2 + b^2 < r^2, \quad = r^2 \quad \text{or} \quad > r^2.$$

Furthermore, for any point $P_0(x_0, y_0)$ there is a number r_0 greater than 0 such that $r_0^2 = x_0^2 + y_0^2$. This number r_0 is called the *"distance"* between P_0 and the center $(0, 0)$ of the "circle."

Let $P = (a, b)$ be an intersection point of the curve C and the "circle"

$$S: \quad x^2 + y^2 = r^2,$$

where

$$r^2 = a^2 + b^2$$

and $r \geq 0$ is the "radius" of S. Suppose P is not a singular point of C and set

$$c = b\alpha - a\beta,$$
$$d = -a\alpha - b\beta.$$

Then by transforming the coordinate system into a new one that has P as its origin and the tangent line and "normal line" of C at P as its axes l_1', l_2' respectively, with new coordinates (x', y'), we have

$$x' = \beta(x - a) - \alpha(y - b),$$
$$y' = \alpha(x - a) + \beta(y - b).$$

Thus the equation of the "circle" S in the new coordinate system Γ' becomes

$$(x' - c)^2 + (y' - d)^2 = c^2 + d^2,$$

where

$$c^2 + d^2 = (\alpha^2 + \beta^2)(a^2 + b^2).$$

As $\bar{\boldsymbol{K}}$ is a real closed field, there is a $\gamma \in \bar{\boldsymbol{K}}$ such that $\alpha^2 + \beta^2 = \gamma^2$ and $\gamma \geq 0$. Hence the new equation of S is

$$(x' - c)^2 + (y' - d)^2 = r'^2,$$

where $r' = r\gamma \geq 0$.

We call S a "circle" with "center" (c, d) and "radius" $r' \geq 0$ in the new coordinate system Γ'. Similarly, we can define the notions "on circle," or "inside circle" or "outside circle" and "distance" between a point and the "center" etc.

In this way, with respect to a fixed coordinate system we may still define the notions of circles, radius and distance etc., so that Seidenberg's lemma holds under the corresponding notions and his proof is also available. For the detailed proof, the reader may refer to Jacobson (1974: chap. 5 sect. 5). We do not repeat it here.

Another place where we need to modify the proof of Tarski–Seidenberg is the introduction of non-degeneracy conditions. For instance, in the original proof the division of polynomials was often applied: Let $t_1, \ldots, t_r$ be parameters and

$$F(t, x) = u_n x^n + u_{n-1} x^{n-1} + \cdots + u_0,$$
$$G(t, x) = v_m x^m + v_{m-1} x^{m-1} + \cdots + v_0$$

be two polynomials in x with all coefficients u_i, v_i polynomials in $t_1, \ldots, t_r$. In dividing F by G, the original proof needs to proceed by distinguishing

$$v_m = 0, v_{m-1} = 0, \ldots, v_{k+1} = 0, v_k \neq 0,$$

for all cases $k = m, m-1, \ldots, 0$. But in our modified proof, we need only to consider the case

$$v_m \neq 0$$

by taking v_m as a non-degeneracy polynomial, adjoined to the non-degeneracy set Δ. If one of the parameters $t_1, \ldots, t_r$, say t_r, occupies a special position, we may consider v_m as a polynomial in t_r and take the leading coefficient of t_r

$$\omega_m = \omega_m(t_1, \ldots, t_{r-1})$$

as a non-degeneracy polynomial, adjoined to Δ, i.e., only consider the case

$$\omega_m \neq 0$$

during the proof. In this way, the proof and method of Tarski–Seidenberg can be simplified a great deal.

As the proof and method of Tarski–Seidenberg have been described in detail by Jacobson (1974: chap. 5 sects. 3–6), we are satisfied to only point out the places where the original proof needs to be modified in order to fit our requirements and do not make a cumbersome presentation again. From the previous explanation of mechanical theorem proving, we can summarize the main results of this chapter in a theorem of the following form.

Mechanization theorem. Theorem proving is mechanizable in all such ordered geometries as ordered Pascalian geometry, ordered orthogonal geometry and ordered metric geometry, which are subordinate to ordinary geometry and satisfy the Pascalian axiom for intersecting lines, and ordinary geometry itself.

5.3 Examples for the mechanical method of ordered geometries

The mechanical method presented in the preceding two sections for proving theorems in ordered geometries is not only essentially different from but also much more complicated than the methods in Chaps. 3 and 4 for proving theorems in unordered geometries. Within the capability of hand-calculation, below we give a few examples for this method. Some simplification is made by combining it with the method in Chap. 4.

Example 1 (Pasch's axiom). As an alternative form of Pasch's axiom, let us consider the following proposition.

Let the sides AB, AC, BC of $\triangle ABC$ intersect a line at three points D, E, F respectively. If B lies between A and D, and C between A and E, then F does not lie between B and C.

To simplify the calculation, let us take a coordinate system with origin A

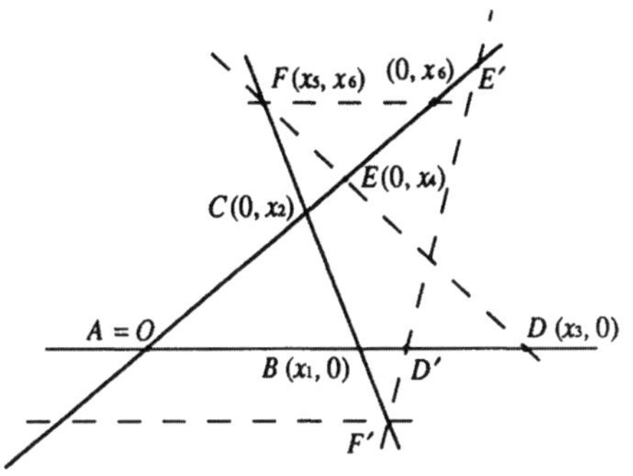

Fig. 5.1

and axes AB and AC. Let the coordinates of points be

$$B = (x_1, 0), \quad C = (0, x_2),$$
$$D = (x_3, 0), \quad E = (0, x_4),$$
$$F = (x_5, x_6).$$

From the hypothesis of the order relations, we know that there are $y_1, y_2 \in \bar{\boldsymbol{K}}$ such that

$$y_1^2 x_1 (x_1 - x_3) = -1,$$
$$y_2^2 x_2 (x_2 - x_4) = -1.$$

Moreover, we have

$$x_2 x_5 + x_1 x_6 = x_1 x_2,$$
$$x_4 x_5 + x_3 x_6 = x_3 x_4.$$

Fix the order of all the variables as

$$y_2 \prec y_1 \prec x_1 \prec x_2 \prec x_3 \prec x_4 \prec x_5 \prec x_6.$$

The hypothesis relations, after being well-ordered, are

$$y_1^2 x_1 x_3 = y_1^2 x_1^2 + 1,$$
$$y_2^2 x_2 x_4 = y_2^2 x_2^2 + 1,$$
$$(x_1 x_4 - x_2 x_3) x_5 = x_1 x_3 (x_4 - x_2),$$
$$(x_1 x_4 - x_2 x_3) x_6 = x_2 x_4 (x_1 - x_3),$$

provided with the non-degeneracy conditions

$$x_1 \neq 0, \quad x_2 \neq 0, \quad x_1 x_4 - x_2 x_3 \neq 0$$

and

$$y_1 \neq 0, \quad y_2 \neq 0.$$

The geometric meanings of these non-degeneracy conditions are obvious and the conditions are all naturally satisfied by the hypothesis of the theorem. Under these non-degeneracy conditions the conclusion of the theorem to be proved corresponds to

$$g_1 \equiv x_5(x_5 - x_1) > 0,$$

$$g_2 \equiv x_6(x_6 - x_2) > 0.$$

We prove that $g_2 > 0$ as follows. Applying the last hypothesis relation to eliminate x_6, we have

$$(x_1x_4 - x_2x_3)^2 g_2 = x_2^2 x_3 x_4 (x_1 - x_3)(x_2 - x_4).$$

Applying then the first two hypothesis relations to eliminate x_3 and x_4, we finally get

$$y_1^4 y_2^4 x_1^2 (x_1x_4 - x_2x_3)^2 g_2 = (y_1^2x_1^2 + 1)(y_2^2x_2^2 + 1) > 0.$$

This implies that $g_2 > 0$.

In the last step, to prove

$$(y_1^2x_1^2 + 1)(y_2^2x_2^2 + 1) > 0,$$

we can of course proceed mechanically by using the method of Tarski–Seidenberg, but that would be very complicated. Also, the theorem can be proved instead by determining whether or not some equations have solutions in $\bar{\boldsymbol{K}}$.

Example 2 (Geometric absurdity). Euclid's "Elements" is a model of books edited according to axiom systems in history. Nevertheless, its axiom system is incomplete. This fact has constantly received critiques with discussions concentrated on the independence of the axiom of parallels. In the later nineteenth century, along with the upsurge of critical tides for mathematics, mathematicians started making a comprehensive analysis of Euclid's axiom system, no longer only restricted to the axiom of parallels. At that time some geometric absurdities appeared. The one presented below is better-known.

Absurdity. Any triangle is an isosceles triangle.

Proof. Let ABC be an arbitrary triangle (see Fig. 5.2). Construct a bisector of the angle C to intersect the perpendicular bisector of the opposite side AB at a point D. Construct further $DE \perp AC$, $DF \perp BC$. Then, by the congruence of triangles we have

$$CE = CF, \quad AE = BF.$$

This implies that $AC = BC$. Hence any triangle ABC is an isosceles triangle. □

For the above absurdity and its proof, the reader may refer to Klein (1939: p. 202) or Kline (1972: p. 1007). The error in the proof occurs because point

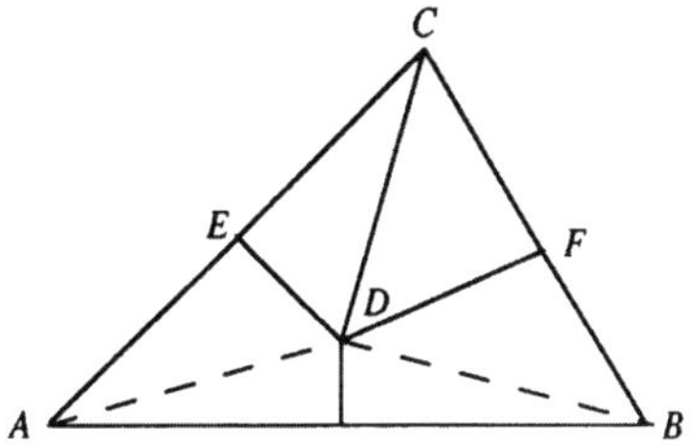

Fig. 5.2

D in the figure should be outside, not inside $\triangle ABC$. This can be proved by introducing the axioms of order, while the order relation was not considered in "Elements" at all. In detail, we have the following proposition.

Let D be the intersection point of a bisector of angle C and the perpendicular bisector of the opposite side AB of $\triangle ABC$. If CD is the *internal* angular bisector, then D and C lie on *different sides* of line AB; if CD is the *external* angular bisector, then D and C lie on the *same side* of AB.

The statement of this proposition, actually the same as that of other geometric theorems, is directed only to the generic cases. For example, if in the proposition $AB = AC$, then the bisector of the angle C coincides with the perpendicular bisector of AB so that the statement of the proposition itself becomes meaningless. Therefore, the case $AB = AC$ should be excluded from the hypothesis and be treated as a degenerate case of the proposition. For the occurrence and treatment of such kinds of non-degeneracy conditions, the traditional Euclidean method is powerless, while our mechanical method can give a systematic, mechanical treatment (see the previous chapters, in particular Sects. 3.1 and 3.2).

As an explanation of the method of this chapter, let us prove the above proposition as follows.

For the sake of simplifying calculations, we take a Descartes coordinate system with C as its origin and the bisector of angle C as its first axis (where it is not necessary to use the theorem that the two bisectors of an angle are perpendicular to each other). Then the point D lies on the first axis. Let the coordinates of points be

$$A = (x_1, x_2), \quad B = (x_3, x_4), \quad D = (x_5, 0).$$

Then the hypothesis of the proposition consists of

$$OD \text{ is a bisector of angle } C \Longleftrightarrow x_1x_4 + x_2x_3 = 0,$$
$$D \text{ lies on the perpendicular bisector of } AB \Longleftrightarrow$$
$$2(x_3 - x_1)x_5 + x_1^2 + x_2^2 - x_3^2 - x_4^2 = 0.$$

As to whether OD is the *internal* or the *external* bisector of angle C, it corresponds to whether A and B lie on *different* sides or the *same* side of the first

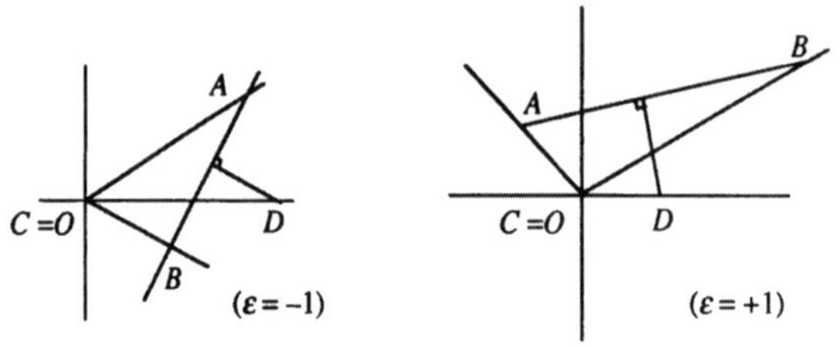

Fig. 5.3

axis, i.e., whether x_2 and x_4 have *different* signs or the *same* sign, or whether or not the equation

$$x_0^2 x_2 x_4 = \varepsilon$$

has a solution, where

$$\varepsilon = \begin{cases} -1 & (OD \text{ is the } \textit{internal} \text{ bisector of angle } C), \\ +1 & (OD \text{ is the } \textit{external} \text{ bisector of angle } C). \end{cases}$$

The conclusion that D and C lie on the same side or different sides of AB corresponds to whether or not the equation

$$y^2(x_2x_5 - x_4x_5 + x_1x_4 - x_2x_3)(x_1x_4 - x_2x_3) = \eta$$

has a solution, where

$$\eta = \begin{cases} +1 & (\text{on the same side}), \\ -1 & (\text{on different sides}). \end{cases}$$

Hence, the proposition is equivalent to saying that under its hypothesis

the equation has a solution for y when $\eta = \varepsilon$, and
the equation has no solution for y when $\eta = -\varepsilon$.

Fix now the order of the variables appearing in the hypothesis as

$$x_0 \prec x_1 \prec x_2 \prec x_3 \prec x_4 \prec x_5.$$

The hypothesis relations, after being well-ordered, are

$$x_0^2 x_2^2 x_3 + \varepsilon x_1 = 0,$$
$$x_0^2 x_2 x_4 - \varepsilon = 0,$$
$$2(x_0^2 x_2^2 + \varepsilon)x_1 x_5 - x_0^2 x_2^2 (x_1^2 + x_2^2 - x_3^2 - x_4^2) = 0,$$

of which the last equation can be reduced to

$$2x_0^2x_2^2 \cdot x_1x_5 - (x_1^2 + x_2^2)(x_0^2x_2^2 - \varepsilon) = 0.$$

The non-degeneracy conditions already used are

$$x_0 \neq 0, \quad x_1 \neq 0, \quad x_2 \neq 0, \quad x_0^2x_2^2 + \varepsilon \neq 0.$$

The equation corresponding to the conclusion is

$$g \equiv y^2h - \eta = 0,$$

in which

$$h = (x_2x_5 - x_4x_5 + x_1x_4 - x_2x_3)(x_1x_4 - x_2x_3).$$

Reducing h and g by the hypothesis polynomials, we have

$$x_0^6x_2^4h = \varepsilon[x_2^2(x_0^2x_2^2 - \varepsilon)^2 + x_1^2(x_0^2x_2^2 + \varepsilon)^2].$$

As $x_2 \neq 0$ and $x_0 \neq 0$, under the above non-degeneracy conditions $g = 0$ is equivalent to the equation

$$\varepsilon y^2[x_2^2(x_0^2x_2^2 - \varepsilon)^2 + x_1^2(x_0^2x_2^2 + \varepsilon)^2] - \eta = 0,$$

where the expression inside the brackets is positive definite under the non-degeneracy conditions

$$x_1 \neq 0, \quad x_2 \neq 0, \quad x_0^2x_2^2 - \varepsilon \neq 0, \quad x_0^2x_2^2 + \varepsilon \neq 0.$$

Hence under these conditions, the equation has a solution when ε and η have the same sign, and it has no solution when ε and η have different signs. This is what we wanted to prove.

The geometric meanings of the non-degeneracy conditions can be easily understood. The condition $x_0^2x_2^2 \pm \varepsilon \neq 0$ means that $x_3 \neq x_1$ when $\varepsilon = -1$ and $x_3 \neq -x_1$ when $\varepsilon = +1$. In any case, this condition means that $AC \neq BC$. As we said at the beginning of this example, it is worth repeating that these non-degeneracy conditions occur naturally during the process of mechanical proving and need not be pre-considered.

In the above two examples, we did not completely follow the mechanical method of Tarski–Seidenberg, but used a simplified one. In detail, let the set of hypothesis polynomials be denoted by $\{f_i\}$ and the conclusion polynomial by g. We did not first form $F = \sum f_i^2 + g^2$ and then repeatedly apply Seidenberg's lemma and theorem. Instead, we have used the method of Chap. 4 by first well-ordering $\{f_i\}$ to get a basic set Φ: $A_1, A_2, \ldots, A_n$, then reducing the polynomial g by Φ (where the existence of solutions of $g = 0$ needs to be determined) so as to successively eliminate the variables $x_n, x_{n-1}, \ldots, x_1$ and to get the remainder polynomials $g_{n-1}, \ldots, g_0$ (where $g_n = g$, $g_i \in \bar{K}[u_1, \ldots, u_e, x_1, \ldots, x_{i-1}]$ and u_i are parameters other than the x's), and finally determining whether or not the last polynomial equation $g_0 = 0$ has a solution. All non-degeneracy conditions automatically occur during the process of well-ordering and reduction. This method is feasible because in the above two examples the degree of g_i in x_i is 1 so that under the corresponding non-degeneracy conditions the question of whether or not $g_{i+1} = 0$ has a solution is equivalent to whether or not $g_i = 0$ has a solution. This situation is similar to the case of applying Hilbert's mechanical method to prove the class of (linear) PIP theorems in Chap. 3. If some $g_i = 0$ has degree in x_i greater than 1, this method, of course, cannot be applied.

In the above two examples, the problem is reduced to finally determining whether a polynomial g is positive definite or semi-positive definite. In fact, the mechanization problem of ordered geometries can be reduced to the problem of determining under the hypothesis conditions

$$f_1 = 0, \ldots, f_n = 0,$$

how to adjoin subsidiary conditions

$$D_1 \neq 0, \ldots, D_s \neq 0,$$

so that a polynomial g becomes positive definite or semi-positive definite. If there is no hypothesis condition and g is known to be positive definite or semi-positive definite, then by the 17th of Hilbert's 23 famous problems, g can be represented as the square sum of some rational functions. This problem was positively solved by Artin (1927) and extended to the general case of having hypothesis conditions $f_1 = 0, \ldots, f_n = 0$ by A. Robinson (1955, 1956). As pointed out by Artin (1927), his proof is indirect and cannot lead to an explicit representation of the square sum decomposition. So it is necessary to have a more complete proof. Such a constructive proof was later given by Kreisel (1958). But for us, the more important problem is the pre-problem of finding an efficient method for determining whether or not a polynomial is positive. Although this method has been provided by Tarski–Seidenberg, a problem that still remains to be solved is how to find more efficient methods.

6 Mechanization theorems of various geometries

6.1 Introduction

We introduced the concept of *geometry* from the viewpoint of axiomatization in Sect. 2.6 and the concepts of mechanical methods and *mechanizability* of theorem proving in geometry in Sect. 3.3. Furthermore, in Chaps. 3–5 we showed the mechanizability and presented the corresponding mechanical methods for some geometries or some classes of theorems. From the axiomatization to the mechanization, we have roughly gone through such a path as

$$\text{Axiomatization} \longrightarrow \text{Algebraization} \longrightarrow \text{Coordinatization} \longrightarrow \text{Mechanization.}$$

For the mechanization problem after having established the coordinate system, the previous three chapters provided three different manners which may be classified into three different types of mechanization problems in algebraic form and correspondingly three mechanical methods for their solutions were given. However, to establish the coordinate system, one first has to pass through a rather lengthy and tedious process which starts from the axiom system via algebraization and ends up with the number system. The steps in this way are often not as simple as usually imagined.

The geometries considered in the previous chapters are actually all subordinate to ordinary geometry, so the process of algebraization may have the aid of the familiar proof methods of ordinary geometry and analytic geometry. For other geometries which are not subordinate to ordinary geometry, we need to proceed to algebraization and coordinatization according to the special forms of the geometric axioms. We give in this chapter a few such examples – projective geometry and two non-Euclidean geometries. In these geometries, after algebraization and coordinatization the theorems can also be formulated into the algebraic form as in the previous three chapters. Therefore, the conclusion that theorem proving is mechanizable drawn from methods in the previous three chapters is also valid for all these theorems, see the mechanization theorems for these geometries in Sects. 6.2–6.5.

For the sake of simplicity, we restrict these geometries to the planar case, so their fundamental objects are points and lines. But as pointed out in the general definition in Sect. 2.6, we may also consider geometries with other elements as their fundamental objects, for instance, Möbiusian geometry with points and circles as its fundamental objects and Laguerrean geometry with points and oriented circles as its fundamental objects – two circle geometries one often encounters. For these two geometries, we may get the corresponding mechanization theorems by following the general approach, i.e., proceeding from the axioms through algebraization and coordinatization, and by using the methods of the previous three chapters. As a full presentation is too long (on this topic the author is going to write a technical article later on), we only give a short introduction in Sect. 6.5. For many other axiomatizable geometries not included in this book the mechanization problem may be considered in the same way.

The last section of this chapter will consider the mechanization problem of proving formulas involving transcendental functions. From this section one will see that, as for theorem proving in geometry, the way of using transcendental functions, which is popularized in geometry books and periodicals, may be avoided or, in other words, is mechanizable, too.

6.2 The mechanization of theorem proving in projective geometry

Projective geometry in what follows is restricted to *plane projective geometry*. In the higher dimensional case the situation is basically the same. The places where there are differences will be pointed out if necessary.

As it is restricted to the planar case, the fundamental objects in this geometry consist of two kinds: *points* and *lines*, and the fundamental relation consists of only one kind: *points lying on lines* or *lines passing through points*. Below are the axioms which the fundamental relation satisfies.

Axioms of incidence P

P_1. There is one and only one line passing through two distinct points.
P_2. Two arbitrary lines have one and only one intersection point.
P_3. There are at least three distinct points on an arbitrary line.
P_4. There are at least three distinct points which do not lie on the same line.

From P_1–P_4 one knows that there are at least three distinct lines passing through an arbitrary point. Moreover, the axioms P_1–P_4 are not completely independent of each other, but this is not important for this book.

In order to exclude the case in which there is only a finite number of points, i.e., the case of *finite geometry*, we introduce the following axiom.

Axiom of infinity I. Take three distinct points O_0, O_1, and O_∞ on a line l, a point Q_∞ not on l and a line m passing through point Q_∞ but distinct from $Q_\infty O_\infty = l_\infty$ and l. Extend $Q_\infty O_0$ and $Q_\infty O_1$ to meet m at points P_0 and P_1 respectively. Connect P_1 and $O_1 P_0 \wedge l_\infty = R_\infty$ to meet l at point O_2, extend $Q_\infty O_2$ to meet m at point P_2, and construct the intersection point O_3 of $P_2 R_\infty$ and l. In this way, we can construct $O_4, O_5, \ldots$. Similarly, connect P_0 and $O_0 P_1 \wedge l_\infty = S_\infty$ to meet l at point O_{-1}, and O_{-1} and Q_∞ to meet m at point P_{-1}. Extend $P_{-1} S_\infty$ to meet l at point O_{-2}. In this way, we can get $O_{-3}, O_{-4}, \ldots$. Then the points in the obtained series

$$\ldots, O_{-n}, \ldots, O_{-1}, O_0, O_1, O_2, \ldots, O_n, \ldots$$

are distinct from each other (see Fig. 6.1).

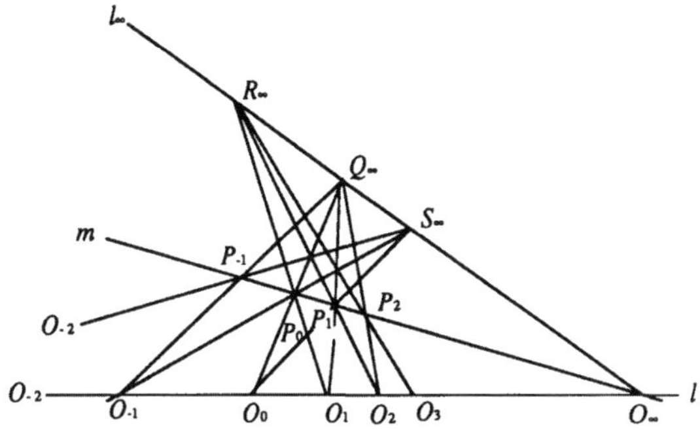

Fig. 6.1

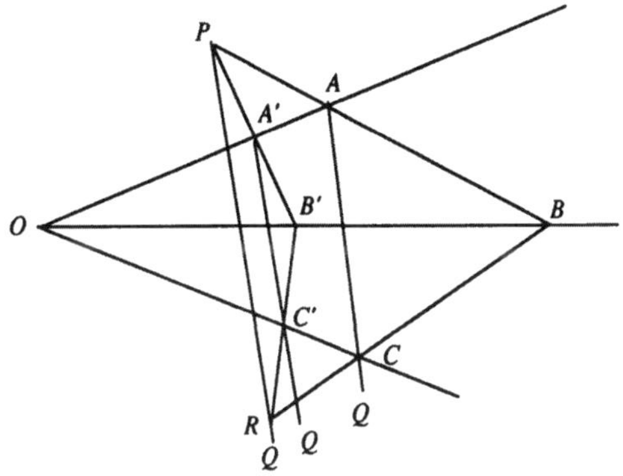

Fig. 6.2

Furthermore, we introduce the following axioms (cf. Fig. 6.2).

Desargues' axiom $\mathbf{D_1}$. Let $\triangle ABC$ and $\triangle A'B'C'$ be two triangles, where A, B, C do not lie on the same line and neither do A', B', C'. Let the lines AB and $A'B'$; AC and $A'C'$; and BC and $B'C'$ respectively be distinct from each other, and the three intersection points

$$P = AB \wedge A'B', \quad Q = AC \wedge A'C', \quad R = BC \wedge B'C'$$

be collinear. Then, when A, A' are distinct, and so are B, B' and C, C', the three lines AA', BB' and CC' are concurrent.

Desargues' axiom $\mathbf{D_2}$. Let $\triangle ABC$ and $\triangle A'B'C'$ be two triangles, where A, B, C respectively are distinct from A', B', C' and the lines AA', BB', CC' are pairwise distinct and concurrent at O. Then, when the lines AB, $A'B'$ are distinct, and so are AC, $A'C'$ and BC, $B'C'$, the three intersection points

$$P = AB \wedge A'B', \quad Q = AC \wedge A'C', \quad R = BC \wedge B'C'$$

are collinear.

As we know, Desargues' axioms can be proved as theorems in space pro-

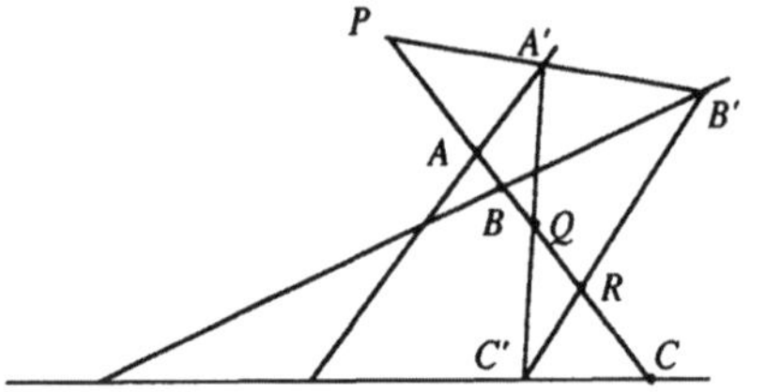

Fig. 6.3

jective geometry. But here the geometry is restricted to the planar case, they cannot be derived from the axioms of incidence and the axiom of infinity, and thus have to be introduced as independent axioms. The statement of Desargues' axioms above is much more complicated than that in a usual book on projective geometry. The reason, as pointed out in Sect. 3.1, is that without adding some *non-degeneracy* conditions the axioms may become meaningless and even fall into fallacies. For example, in Desargues' axiom D_1, if the three points A, B, C lie on the same line, even though the other non-degeneracy conditions are all satisfied and the three points

$$P = AB \wedge A'B', \quad Q = AC \wedge A'C', \quad R = BC \wedge B'C'$$

are collinear (on the line determined by A, B, C), the three lines AA', BB', CC' are not necessarily concurrent (see Fig. 6.3).

It is very important to repeat this point, and the occurrence of similar cases is rather widespread, not limited to projective geometry. So we restate the view expressed in Sect. 3.1 as follows:

Even though every geometry may have a rigorous and non-contradictory axiom system as the starting point of all reasoning and proving, *using the traditional Euclidean method it is impossible to logically reach the necessary rigor, at least in practice*, as the method cannot deal with various degenerate cases.

Under the assumption of the axioms of incidence P, the axiom of infinity I and Desargues' axioms D, it is possible to introduce a number system associated intrinsically with the geometry as follows.

First, take three distinct points, denoted by O_0, O_1, O_∞, on an arbitrary line l and regard a point A distinct from O_∞ as a number, denoted by the corresponding low case a, with $A \leftrightarrow a$ inducing a one-to-one correspondence. All such numbers constitute a set $N = N(O_0, O_1, O_\infty)$. Now we introduce the operations of addition and multiplication among elements of N in the following way.

1. Addition. See Fig. 6.4. Let A, B be two points distinct from O_∞ on l and be regarded respectively as two numbers a and b. We define $a + b$ and the corresponding point C on l as follows.

Suppose first that A, B are distinct from each other and distinct from O_0. Draw two lines distinct from l through A and B and let their intersection point be P. Draw a line distinct from l through O_0 and not passing through P, meeting the two previous lines at points A', B' respectively. Extend $O_\infty B'$ to intersect AA' at point M and $O_\infty A$ to intersect BB' at point N. Since both M

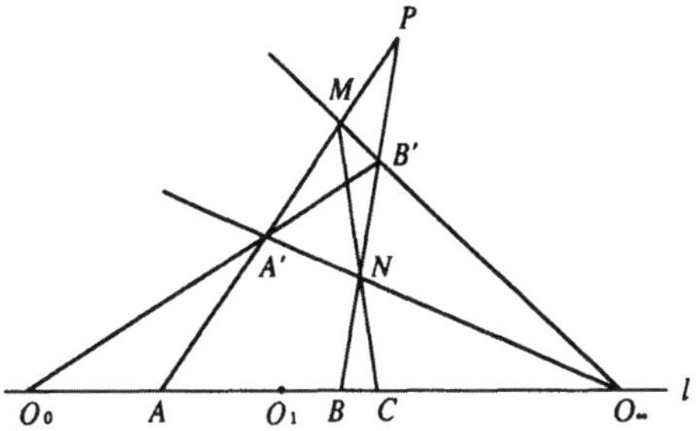

Fig. 6.4

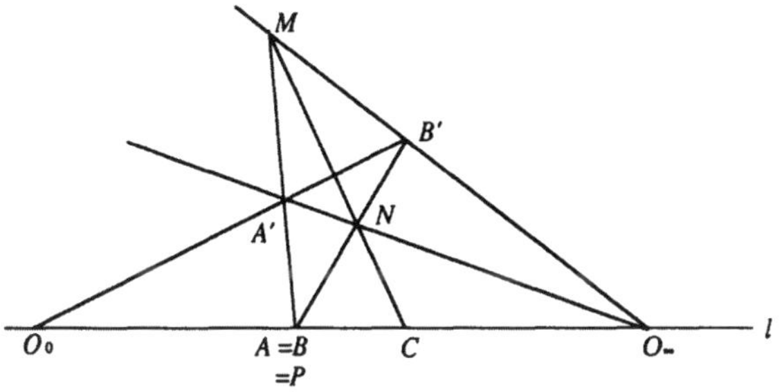

Fig. 6.5

and N are distinct from P, they are distinct from each other, so we can extend MN to meet l at a point C. Obviously C is distinct from O_∞ and, by applying Desargues' axioms, one can prove that C is independent of the choice of all lines in the construction. Hence C corresponds to a unique number in N, defined to be the *sum* of the two numbers a and b and denoted as $a+b$: $C \leftrightarrow a+b$.

When A and B are the same but distinct from O_0 (see Fig. 6.5), we can still define $C \leftrightarrow a+b$ according to the construction above, so long as a slight modification is made. When A and O_0 are the same, the above construction is impossible. But in this case we can directly define C to be B or $a+b=b$. Similarly, when B and O_0 are the same, we directly define C to be A or $a+b=a$.

In any case we have defined an *addition* in $N = N(O_0, O_1, O_\infty)$.

2. Multiplication. Let A, B be distinct from O_∞ with $A \leftrightarrow a$, $B \leftrightarrow b$ as before. As in Fig. 6.6, suppose first that A, B are distinct from each other and distinct from O_0, O_1. Draw two lines distinct from l through A, B respectively, meeting at point P. Draw another line distinct from l passing through O_0 but not through P, meeting the two lines previously drawn at points L and K respectively. Extend O_1L to meet BP at point M, $O_\infty M$ to meet AP at point N and KN to meet l at point C. Then C is distinct from O_∞ and by applying Desargues' axioms one can prove that C is independent of the choice of the lines drawn through A, B, O_0. The number corresponding to C in N is defined to be the *product* of a and b, denoted as ab, i.e., $C \leftrightarrow ab$.

When A, B are the same but distinct from O_0, O_1, O_∞, we can still define C according to the above construction, as long as a slight modification is made

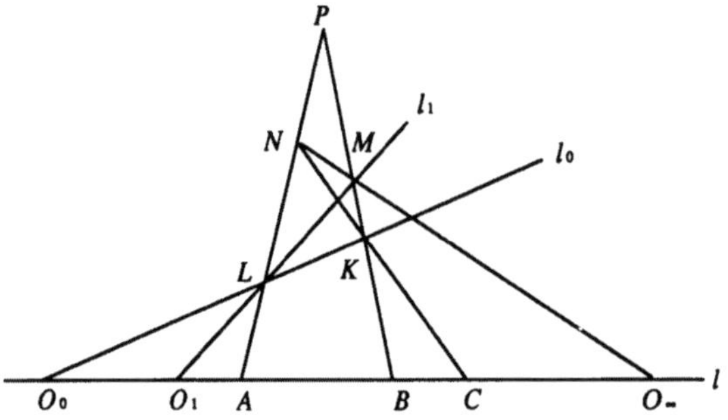

Fig. 6.6

as well. In other cases, the above construction is impossible, but we can directly define ab as follows: $C = O_0$ or $ab = 0$ when $A = O_0$ or $B = O_0$; $C = B$ or $ab = b$ when $A = O_1$; $C = A$ or $ab = a$ when $B = O_1$.

In any case the *multiplication* of a and b is well-defined.

The first and also the crucial step of the algebraization of projective geometry is the following.

Theorem 1. After fixing three distinct points O_0, O_1, O_∞ on an arbitrary line l, the points distinct from O_∞ on l correspond one-to-one to numbers in the number system $N = N(O_0, O_1, O_\infty)$, in particular O_0 corresponds to 0 and O_1 to 1. Under the addition and multiplication defined above, N constitutes a Desarguesian number system, i.e., a number sfield, satisfying the axioms N 1–N 12 in Sect. 1.4. Moreover, when taking three distinct points O'_0, O'_1, O'_∞ on another line l' and establishing a one-to-one correspondence between points distinct from O'_∞ on l' and numbers in the Desarguesian number system $N' = N'(O'_0, O'_1, O'_\infty)$ in the same way, there is an isomorphism F: $N \approx N'$ such that under F, the numbers 0, 1 in N correspond to 0, 1 in N'.

The number sfield N uniquely determined up to isomorphism in the theorem is called a geometry-*associated number sfield.*

The proof of the propriety of the definitions of $a+b$ and ab and many assertions involved in Theorem 1 requires repeated use of Desargues' axioms. Such a method of introducing number sfield in geometry originates from V. Staudt. If one line in the plane is taken as the infinite line and O_∞ is on this infinite line, then we get a Desarguesian plane and the corresponding Desarguesian number system. Actually, Hilbert's method of introducing a Desarguesian number system in Chap. 1 is obtained from the method of V. Staudt.

The geometry-associated number system N introduced in Theorem 1 is a number sfield, in which the commutative law of multiplication generally does not hold. Now we introduce the following

Pappus' axiom. Let A, B, C and A', B', C' be two sets of points on two distinct lines l and l' respectively, which are distinct from each other and distinct from the intersection point of l and l'. Then the three points

$$P = BC' \wedge B'C, \quad Q = CA' \wedge C'A, \quad R = AB' \wedge A'B$$

are collinear.

The significance of this axiom for the foundation of the geometry is the following theorem which was first proposed by V. Staudt.

Theorem 2. Whether or not Pappus' axiom holds is equivalent to whether or not the commutative law of multiplication for the number sfield ***N*** holds. In other words, Pappus' axiom provides a necessary and sufficient condition for ***N*** to become a number field.

We call a geometry, which takes points and lines as its fundamental objects, the incidence relation as its fundamental relation and satisfies the following axioms, an *unordered projective geometry*:

1. Axioms of incidence P_1–P_4;
2. Axiom of infinity I;
3. Pappus' axiom.

By the well-known Hessenberg theorem, Desargues' axioms can be derived from axioms P_1, P_2, and P_3 and can thus be considered as theorems in this geometry.

Pappus' axiom also provides a condition for the so-called fundamental theorem of projective geometry to hold, and under the assumption of other axioms both are equivalent to each other. Assume that Pappus' axiom holds and thus the fundamental theorem of projective geometry holds. Let two lines l and l' and three pairwise distinct points O_0, O_1, O_∞ on l and O_0', O_1', O_∞' on l' be given and map the points on l to the points on l' such that O_0, O_1, O_∞ correspond to O_0', O_1', O_∞' respectively, by a variety of perspective transformations. Then the correspondence between points on l and points on l' can be uniquely determined. From this it follows that the isomorphism

$$F: \quad N(O_0, O_1, O_\infty) \approx N(O_0', O_1', O_\infty')$$

in Theorem 1 can be taken as a *canonical* one realized by a variety of perspective transformations, not depending on the choice of these transformations. Therefore, all number sfields $N(O_0, O_1, O_\infty)$ (which are also number fields in this case) can be identified as a single one. Denoting it by ***N*** as before, ***N*** not only is isomorphic to an arbitrary $N(O_0, O_1, O_\infty)$, but also is uniquely determined. In addition, the concept of *cross ratio* may be introduced as a number in ***N*** in the usual manner.

The above assertions hold only under the assumption of Pappus' axiom, i.e., in unordered projective geometry as defined above. In the "affine" case where

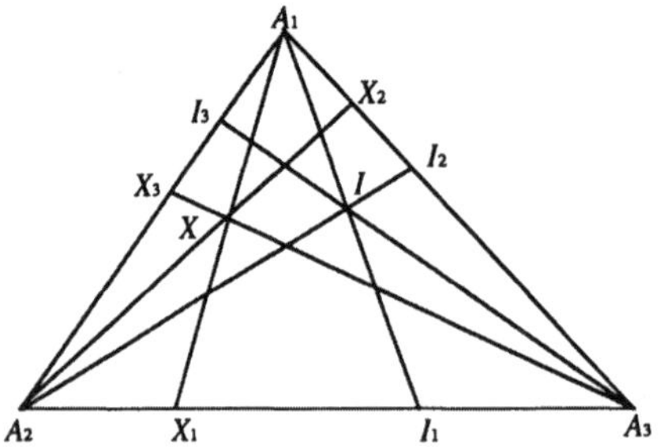

Fig. 6.7

the situation is somewhat different, the so-called Pascalian axiom for intersecting lines corresponding to Pappus' axiom is not necessary (see Sect. 1.6).

We introduce a coordinate system in unordered projective geometry as follows.

As in Fig. 6.7, take four points A_1, A_2, A_3, I in the plane, of which any three are not collinear. Let

$$A_1I \wedge A_2A_3 = I_1, \quad A_2I \wedge A_3A_1 = I_2, \quad A_3I \wedge A_1A_2 = I_3$$

determine the number systems

$$N_1 = N(A_2, I_1, A_3), \quad N_2 = N(A_3, I_2, A_1), \quad N_3 = N(A_1, I_3, A_2)$$

on the lines A_2A_3, A_3A_1, A_1A_2 respectively.

These number systems are all number fields and determine isomorphisms with each other. Denote the geometry-associated number field by $\boldsymbol{N}$; then there are canonical isomorphisms

$$F_i\colon \quad \boldsymbol{N} \approx N_i, \quad i = 1, 2, 3.$$

We identify the numbers in N_i to those in $\boldsymbol{N}$ under these canonical isomorphisms and do not distinguish them in notation.

We say that $A_1A_2A_3$ is a *coordinate tripoint*, I is a *unit point* and $(A_1A_2A_3I)$ constitutes a *coordinate system.*

For an X that is not on any of A_2A_3, A_3A_1 and A_1A_2, set

$$A_1X \wedge A_2A_3 = X_1, \quad A_2X \wedge A_3A_1 = X_2, \quad A_3X \wedge A_1A_2 = X_3.$$

Then X_1, X_2, X_3 correspond to numbers

$$X_1 \leftrightarrow \bar{x}_1, \quad X_2 \leftrightarrow \bar{x}_2, \quad X_3 \leftrightarrow \bar{x}_3$$

in N_1, N_2, N_3 or $\boldsymbol{N}$. It can be proved that

$$\bar{x}_1\bar{x}_2\bar{x}_3 = 1.$$

In general, this relation does not hold for points on A_1A_2, A_2A_3 and A_3A_1. For this reason we rewrite the $\bar{x}$'s as

$$\bar{x}_1 = \frac{x_3}{x_2}, \quad \bar{x}_2 = \frac{x_1}{x_3}, \quad \bar{x}_3 = \frac{x_2}{x_1},$$

where (x_1, x_2, x_3) are uniquely determined, up to a non-zero factor. We call (x_1, x_2, x_3) the *homogeneous coordinates* of X with respect to the coordinate system $(A_1A_2A_3I)$ and write

$$X = (x_1 : x_2 : x_3).$$

If $X = X_1$ lies on A_2A_3 but it is neither A_2 nor A_3 and X_1 corresponds to $\bar{x}_1$ in N_1 or $\boldsymbol{N}$, then the *homogeneous coordinates* of X_1 are uniquely determined, up to a non-zero factor, as $(0, 1, \bar{x}_1)$, i.e.,

$$X_1 = (0 : 1 : \bar{x}_1).$$

For points which lie on A_3A_1 or A_1A_2 but are distinct from A_1, A_2, A_3, it is similar. The *homogeneous coordinates* of A_1, A_2, A_3 are given as

$$A_1 = (1:0:0), \quad A_2 = (0:1:0), \quad A_3 = (0:0:1)$$

respectively.

In the coordinate system $(A_1A_2A_3I)$, any point $P = (x_1 : x_2 : x_3)$ on a line l satisfies the homogeneous linear equation

$$u_1x_1 + u_2x_2 + u_3x_3 = 0,$$

where u_1, u_2, u_3 are not all 0, determined up to a non-zero factor. The triple (u_1, u_2, u_3) of numbers is called the *homogeneous coordinates* of the line l in the coordinate system $(A_1A_2A_3I)$, which we denote as

$$l = (u_1 : u_2 : u_3).$$

Therefore a necessary and sufficient condition for a point $P = (x_1 : x_2 : x_3)$ and a line $l = (u_1 : u_2 : u_3)$ to be incident is given by the above homogeneous equation.

Since in unordered projective geometry the fundamental objects consist of points and lines and the incidence relation between points and lines is the only fundamental relation, and since this relation can be expressed, after fixing a certain coordinate system and representing points and lines by homogeneous coordinates, as a polynomial relation

$$u_1x_1 + u_2x_2 + u_3x_3 = 0$$

in the number field, according to the general principle indicated in Sect. 6.1 we have the following.

Mechanization theorem 1. There is a mechanical method for proving theorems in unordered projective geometry.

The following two theorems are not very relevant to the mechanization problem of this section, but they will be used in the proof of other mechanization theorems, see Sects. 6.4 and 6.5. We first introduce some notions.

Suppose there is a one-to-one correspondence T which maps the set of points and lines in the plane of an unordered projective geometry to itself. If T transforms points to points, lines to lines and preserves the incidence relation (i.e., if a point P lies on a line l then the point $T(P)$ lies on the line $T(l)$), then T is called a *collineatory transformation.* If T transforms points to lines, lines to points and preserves the incidence relation (i.e., if a point P lies on a line l then the point $T(l)$ lies on the line $T(P)$), then T is called an *inverse morphism transformation.*

Theorem 3. In a coordinate system $(A_1A_2A_3I)$, for every collineatory transformation T there are numbers a_{ij}, $i, j = 1, 2, 3$ in $\boldsymbol{N}$ and an automorphism J of $\boldsymbol{N}$ such that $|a_{ij}| \neq 0$ and, if T transforms the point $P = (x_1 : x_2 : x_3)$ to the point $P' = (x'_1 : x'_2 : x'_3)$, we have $\rho \neq 0$ and

$$\rho x'_i = \sum_{j=1}^{3} a_{ij} J(x_j), \quad i = 1, 2, 3.$$

Theorem 4. As above, for any inverse morphism transformation T there are a_{ij} and J such that $|a_{ij}| \neq 0$ and, if T transforms the point $P = (x_1 : x_2 : x_3)$ to the line $l = (u_1 : u_2 : u_3)$, we have $\rho \neq 0$ and

$$\rho u_i = \sum_{j=1}^{3} a_{ij} J(x_j), \quad i = 1, 2, 3.$$

Up to now we have not yet introduced any concept of order in the discussed projective geometry, so the geometry-associated number field may be an arbitrary unordered number field such as the complex number field. When considering the relation of order, we certainly cannot take the relation that one of three points lies between the two others as a fundamental relation as in ordinary geometry. Instead, we take the relation of separation among four points. In other words, for any four points A, B, C, D we introduce the relation that A, C are *separated by* B, D (without definition) as a fundamental relation, denoted as

$$AC/BD.$$

The following axioms are taken as those satisfied by this fundamental relation.

Axioms of separation S

S_1. If AB/CD, then the points A, B, C, D are pairwise distinct and lie on the same line.
S_2. If AB/CD, then CD/AB and BA/CD.
S_3. If AB/CD, then AC/BD does not hold.
S_4. If four points A, B, C, D are pairwise distinct and lie on the same line, then one of the three cases AB/CD, AC/BD and AD/BC holds.
S_5. If three points A, B, C are pairwise distinct and lie on the same line, then there is a point D such that AB/CD.
S_6. For five points A, B, C, D, E, pairwise distinct and on the same line, if AB/DE, then either AB/CD or AB/CE holds.
S_7 (Perspective axiom). Let A, B, C, D and A', B', C', D' be two sets of points lying respectively on two lines l and l' with $l \neq l'$. Let O be a point such that O, A, A' lie on the same line and so do O, B, B'; O, C, C' and O, D, D'. Then $A'B'/C'D'$ holds if AB/CD, where the case of $A = A'$ does not have to be excluded.

An unordered projective geometry that satisfies the axioms of separation S_1–S_7 is called an *ordered projective geometry*.

For an ordered projective geometry, let us take three distinct points O_0, O_1, O_∞ on a line l to form a number field $N = N(O_0, O_1, O_\infty)$. For any point A distinct from O_∞ on l, if AO_1/O_0O_∞, then the corresponding number a of A is said to be *negative*, denoted as $a < 0$. If a point B is distinct from O_∞ and O_0 and the corresponding number b is not less than 0, then b is said to be *positive*, denoted as $b > 0$. From the perspective axiom S_7, one knows that the positiveness and negativeness of numbers in $\boldsymbol{N}$ are independent of the choice of line l and the points O_0, O_1, O_∞, so that one may define *positiveness* and *negativeness* of numbers in the geometry-associated number field $\boldsymbol{N}$.

Using the usual method one may define the relation of *order* among numbers in the geometry-associated number field $\boldsymbol{N}$ and prove that this relation satisfies the axioms N 1–N 17 in Sect. 1.4, so that $\boldsymbol{N}$ becomes an *ordered number field*.

In a projective coordinate system $(A_1A_2A_3I)$, the relation of separation among four points on a line can clearly be expressed by some linear inequalities in their coordinates. Therefore, according to the general principle of Sect. 6.1 we have the following.

Mechanization theorem 2. There is a mechanical method for proving theorems in ordered projective geometry.

Note that the two mechanization theorems in this section are independent of each other. This is because, on the one hand, ordered projective geometry is only a special case of unordered projective geometry and, on the other hand, the content of theorems in the former is richer than that in the latter. Hence, there is no subordinate relation between these two mechanization theorems and none can be inferred from the other. Also, the mechanical methods for them are quite different. This also holds for similar cases in later sections.

6.3 The mechanization of theorem proving in Bolyai–Lobachevsky's hyperbolic non-Euclidean geometry

The axiom of parallels in Euclid's "Elements" says that, through a point not on a given line there is one and only one line parallel to the given line (i.e., they never intersect).

The long-term dispute about whether or not this axiom is independent of the others in Euclid's whole axiom system led to the discovery of *non-Euclidean geometries* in the 19th century, to a completely logical analysis and investigation of the axiom system of ordinary geometry, and finally to the birth of Hilbert's book "Grundlagen der Geometrie."

In Sect. 1.1, we recalled the axiom system of Hilbert about ordinary geometry, in which the sharper axiom of parallels H IV is Euclid's axiom of parallels mentioned above. If we appropriately modify this and the axioms of incidence but preserve all the others, we may get two usual non-Euclidean geometries, i.e.,

Balyai–Lobachevsky's hyperbolic non-Euclidean geometry, and
Riemann's elliptic non-Euclidean geometry.

These two non-Euclidean geometries will be called BL geometry and R geometry in what follows, of which the former has been more extensively studied. This section focuses on the problem of how to realize the mechanization of BL geometry from axiomatization.

For usual BL geometry, we only modify the axiom of parallels. As the axioms of incidence H I and the axioms of order H II are assumed, we can define the concepts of segments, half-lines, angles and the interior of angles, etc. The substitution axiom for H IV in this geometry is the following:

Axiom H IVBL. For any given line b and a point A not on b, there always exist two half-lines a_1, a_2 emanating from A which neither form one and the same line nor intersect line b, while any half-line emanating from A that lies in the interior of the angle formed by a_1, a_2 does intersect line b.

In this axiom, the lines determined by the two half-lines a_1, a_2 are usually called two *parallel lines* of b passing through A. More precisely, let a half-line emanating from A that lies in the interior of the angle formed by a_1, a_2 intersect

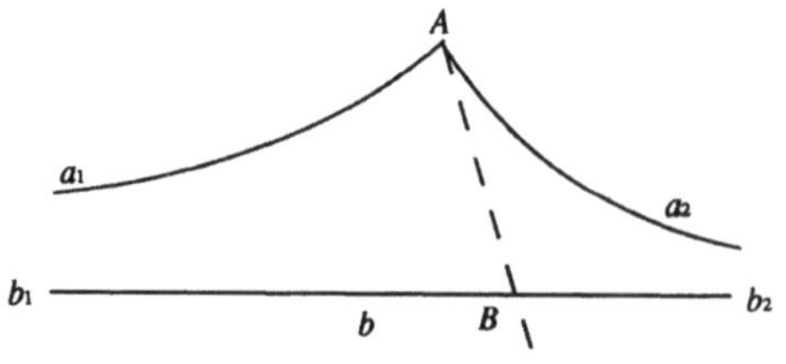

Fig. 6.8

b at a point B, and let B divide b into two half-lines b_1 and b_2 with b_1, a_1 lying on one side and b_2, a_2 on the other side of the line AB. Then we say that the half-line a_1 is *parallel* to b_1 and a_2 is *parallel* to b_2.

We may take the axioms of incidence H I and the new axiom of parallels H IVBL as the basis and adjoin other axioms to it to establish various kinds of geometries (as in Chaps. 1 and 2) so as to explore the dependence relations among axioms. Leaving aside the significance of that attempt, it at least does not agree with the gist of this book – to provide mechanical methods for theorem proving by algebraization and coordinatization starting from axioms.

The representative book in this respect is the one by Hilbert. After the appearance of "Grundlagen der Geometrie" in 1899, Hilbert published an article entitled "A New Development of Bolyai–Lobachevskian Geometry" (see Hilbert 1903) and included it as one of the appendices in later editions of "Grundlagen der Geometrie." In this article, Hilbert described a method for the establishment of this non-Euclidean geometry starting from the axioms H I, H II, H III, and H IVBL. In comparison with a number of other publications about non-Euclidean geometry around that time, this article has the following two characteristics:

1. Hilbert clearly pointed out at the end of his article that the establishment of BL geometry and the derivation of many famous formulas in this geometry do not need the axioms of *continuity* but only the axioms H I–H IV. Here the axiom H IV is taken to be the axiom H IVBL above.
2. Hilbert introduced an algorithm called *end* (Enden) by which he set up the coordinate system and proved that points can be expressed as linear equations. From that he could conclude that the establishment of BL geometry does not have to use the axioms of continuity.

As we pointed out in Chaps. 1 and 2 for the mechanization problem of ordinary geometry, since the assumption of continuity is not necessary, the possibility for the mechanization of BL geometry is guaranteed in theory. The second point above actually indicates a way of mechanization via algebraization and coordinatization from axiomatization.

The article by Hilbert is brief. Afterwards Gerretsen (1942a, b), Szász (1958, 1959a, b) and Szmielew (1959, 1961) elaborated further on the basis of this article. In what follows we give a brief introduction by following the works of Hilbert and Szász and show the reason for BL geometry to be mechanizable.

We say that every half-line determines an *end* and all parallel half-lines determine the same end. Ends are denoted by letters from the Greek alphabet. Denote the half-line emanating from A with end α by $A\alpha$ or (A, α). Each line has exactly two ends, say α and β; this line is also called the *connecting line* of α and β, denoted by $\alpha\beta$ or $\beta\alpha$ and sometimes by (α, β) or (β, α).

By the axioms H I–H IV in which H IV means H IVBL, one may prove that for any two half-lines, with α and β as their ends, there must be a line which has α, β as its two ends. Similarly, for any point A there is a half-line $A\alpha$. One may also prove that two lines which are neither parallel nor intersecting must have a common perpendicular and the relevant propositions of the concept of reflection etc. (which we will not list here).

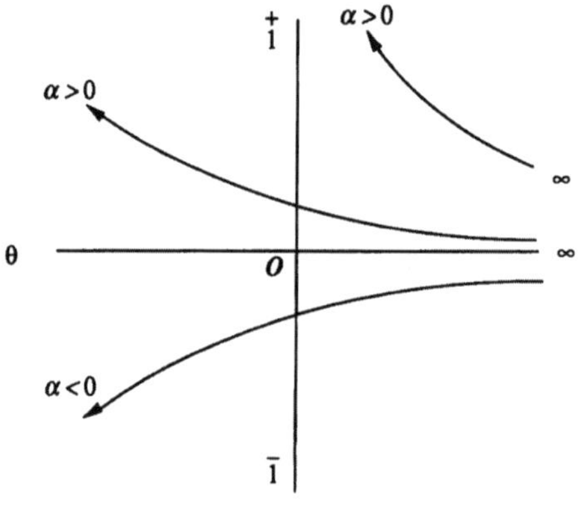

Fig. 6.9

We take an arbitrary point O as the *origin* and two half-lines emanating from O, perpendicular to each other, as the *axes* in the plane of BL geometry to form a coordinate system, called a Hilbert *coordinate system*. Denote the ends of these two half-lines by ∞ and $\overset{+}{1}$ and the ends of the two half-lines which have an opposite direction to $O\infty$ and $O\overset{+}{1}$ by θ and $\overset{-}{1}$ respectively.

Now we introduce the notions of positiveness and negativeness and define addition and multiplication for ends other than ∞ in the plane as follows.

1. Positiveness and negativeness of ends

For ends $\alpha \neq \infty$ and θ, draw the line $\alpha\infty$. If this line and the half-line $O\overset{+}{1}$ lie on the same side of the line $\theta\infty$, α is said to be *positive*, which we denote as $\alpha > 0$; otherwise α is said to be *negative*, which we denote as $\alpha < 0$.

2. Addition of ends

Let α and β be two ends. Draw the lines $\alpha\infty$ and $\beta\infty$, the symmetric point O_α of O with respect to $\alpha\infty$ and the symmetric point O_β of O with respect to $\beta\infty$. Draw the perpendicular bisector of $O_\alpha O_\beta$. One may prove that one end of this perpendicular bisector is ∞ (this is equivalent to the theorem that the three perpendicular bisectors on the three sides of $\triangle O O_\alpha O_\beta$ are concurrent at ∞), so that the other is defined to be $\alpha + \beta$.

3. Multiplication of ends

Let $\overset{+}{\alpha}\overset{-}{\alpha}$ and $\overset{+}{\beta}\overset{-}{\beta}$ be two lines perpendicular to $\theta\infty$, where $\overset{+}{\alpha}$ and $\overset{+}{\beta}$ lie on the same side of line $\theta\infty$ as $\overset{+}{1}$ does, and $\overset{-}{\alpha}$ and $\overset{-}{\beta}$ on the same side of line $\theta\infty$ as $\overset{-}{1}$ does. Therefore, according to 1 above we have $\overset{+}{\alpha}, \overset{+}{\beta} > 0$ and $\overset{-}{\alpha}, \overset{-}{\beta} < 0$. Let $\overset{+}{\alpha}\overset{-}{\alpha}$, $\overset{+}{\beta}\overset{-}{\beta}$ intersect $\theta\infty$ at points A, B respectively. Now take a point C on $\theta\infty$ such that the segments $|BC|$ and $|OA|$ are congruent and the direction from B to C is the same as that from O to A. Erect a perpendicular to $\theta\infty$ at C (which

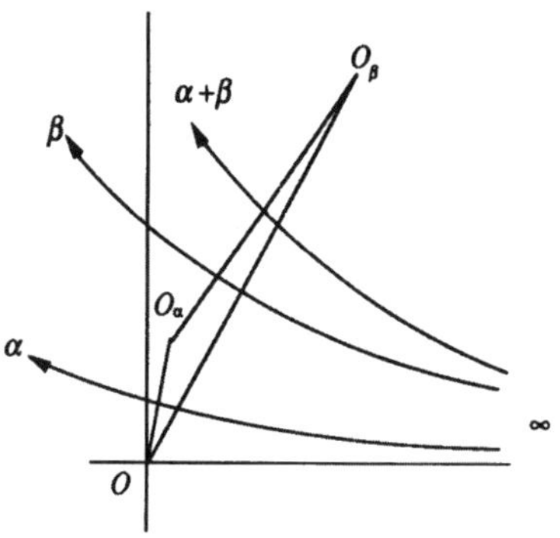

Fig. 6.10

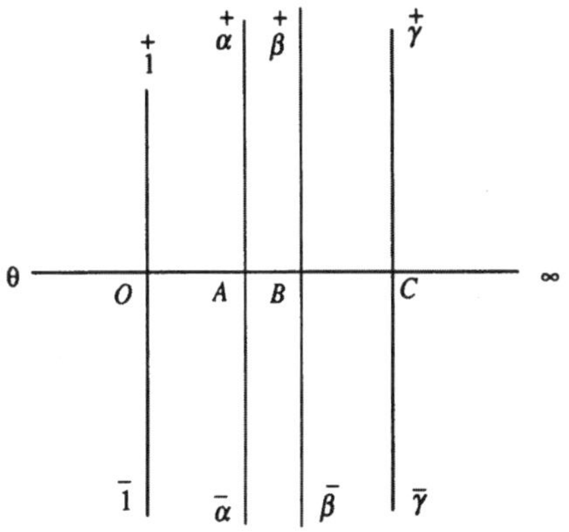

Fig. 6.11

is $\overset{+}{1}\bar{1}$ itself in case $C = O$). Let us denote its two ends by $\overset{+}{\gamma}$ and $\bar{\gamma}$; $C\overset{+}{\gamma}$ and $O\overset{+}{1}$ lie on one and the same side of $\theta\infty$ while $C\bar{\gamma}$ and $O\bar{1}$ lie on the other side of $\theta\infty$, i.e., $\overset{+}{\gamma} > 0$ and $\bar{\gamma} < 0$. Then, we define the multiplication as follows:

$$\overset{+}{\alpha}\overset{+}{\beta} = \overset{+}{\gamma},$$

$$\bar{\alpha}\bar{\beta} = \overset{+}{\gamma},$$

$$\overset{+}{\alpha}\bar{\beta} = \bar{\gamma},$$

$$\bar{\alpha}\overset{+}{\beta} = \bar{\gamma}.$$

In particular, we have

$$\bar{1}\overset{+}{\alpha} = \bar{\alpha}, \quad \bar{1}\bar{\alpha} = \overset{+}{\alpha},$$

$$\overset{+}{1}\overset{+}{\alpha} = \overset{+}{\alpha}, \quad \overset{+}{1}\bar{\alpha} = \bar{\alpha}.$$

Moreover, we define

$$\overset{+}{\alpha}\theta = \theta\overset{+}{\alpha} = \bar{\alpha}\theta = \theta\bar{\alpha} = \theta,$$

$$\overset{+}{\alpha} = -\bar{\alpha}, \quad \bar{\alpha} = -\overset{+}{\alpha}.$$

Hilbert proved the following:

Theorem 1. Introduce a Hilbert coordinate system and define the positiveness, negativeness, addition, and multiplication for ends other than ∞ in the BL non-Euclidean plane as above. Then the ends other than ∞ constitute an ordered number field, i.e., a Desarguesian number system with multiplication commutative that has an order relation and satisfies the axioms N 1–N 17 of number systems in Sect. 1.4. In this number field, the ends θ, $\overset{+}{1}$, $\overset{-}{1}$ play the role of $0, +1, -1$ and are simply denoted as

$$\theta = 0, \quad \overset{+}{1} = +1, \quad \overset{-}{1} = -1.$$

The number field in the theorem is called an *associated number field* with hyperbolic non-Euclidean geometry.

In Schur (1904) there are good geometric explanations of Hilbert's method of introducing the addition and multiplication of ends: $\alpha\beta = \gamma$ if and only if on the infinite line where all ends lie, the following two series of points form a projective correspondence:

$$(\infty 01\beta) \,\bar{\wedge}\, (\infty 0\alpha\gamma).$$

To introduce the coordinates of points, let us first explain some notations as follows.

As in Fig. 6.12, we introduce the addition of oriented segments starting from O on an arbitrary line, say the axis $\theta\infty$, according to the axioms of congruence. For any point A on $\theta\infty$, we denote the oriented segment from O to A by $|\overrightarrow{OA}|$. Then for two arbitrary points A, B on $\theta\infty$, the sum of the oriented segments $|\overrightarrow{OA}|$ and $|\overrightarrow{OB}|$ is still an oriented segment $|\overrightarrow{OC}|$, where C also lies on $\theta\infty$. Now, for an arbitrary oriented segment $t = |\overrightarrow{OA}|$ on $\theta\infty$, draw a half-line from A, perpendicular to $\theta\infty$ and on the same side of $O\overset{+}{1}$. Let us denote its end by λ_t. Then it is easy to see

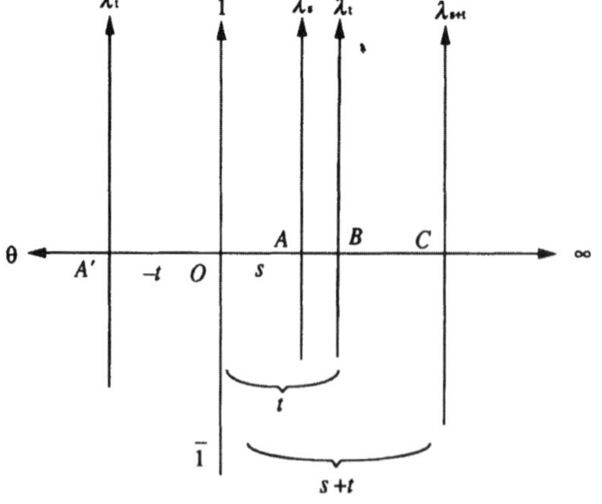

Fig. 6.12

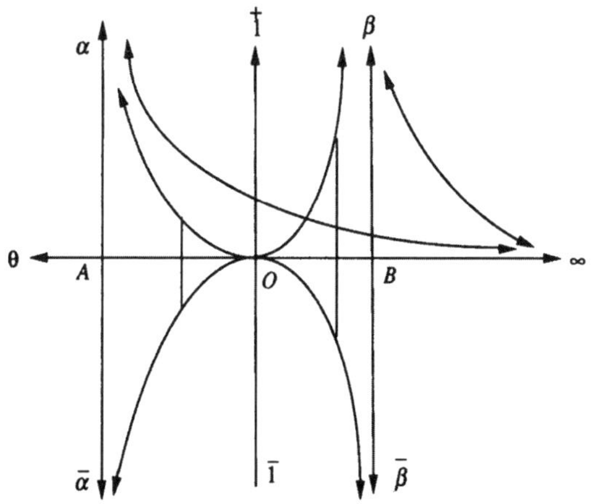

Fig. 6.13

$$\lambda_0 = 1, \quad \lambda_t > 0,$$
$$\lambda_t \cdot \lambda_{-t} = 1,$$
$$\lambda_s \cdot \lambda_t = \lambda_{s+t}.$$

See Fig. 6.13. It is easy to prove that λ induces a one-to-one correspondence between positive ends and oriented segments $t = |\overrightarrow{OA}|$ or their end points A on $\theta\infty$. Let a be an arbitrary positive end. Construct the connecting line $O\alpha$ and the symmetric half-line $O\bar{\alpha}$ of $O\alpha$ with respect to $\theta\infty$; then its end is $\bar{\alpha} = -\alpha$. Construct the connecting line $\alpha\bar{\alpha}$ according to Hilbert (1903), let it meet $\theta\infty$ at a point A and set $|\overrightarrow{OA}| = t$. Then, obviously $\alpha\bar{\alpha}$ is perpendicular to $\theta\infty$, so $\alpha = \lambda_t$.

If another positive end β corresponds to $s = |\overrightarrow{OB}|$ in the above way, then it is easy to know from the axioms of order that $\beta > \alpha$ or $\beta < \alpha$, depending on whether or not the direction from A to B is the same as that of $\theta\infty$. It follows that the correspondence $\alpha \leftrightarrow A$ or $\lambda_t \leftrightarrow t$ is one-to-one.

Now introduce the following ends (according to the operations of ends):

$$\gamma_t = \frac{1}{2}(\lambda_t + \lambda_{-t}),$$
$$\sigma_t = \frac{1}{2}(\lambda_t - \lambda_{-t}),$$
$$\tau_t = (\lambda_t - \lambda_{-t})/(\lambda_t + \lambda_{-t}).$$

Then among these ends the following relations hold:

$$\gamma_t^2 - \sigma_t^2 = 1,$$

$$\gamma_{s+t} = \gamma_s\gamma_t + \sigma_s\sigma_t,$$

$$\sigma_{s+t} = \sigma_s\gamma_t + \sigma_t\gamma_s.$$

If the considered geometry moreover satisfies the axioms of continuity (including Archimedes' axiom and the axiom of completeness), i.e., the case of the usual BL non-Euclidean geometry, then we may consider the oriented segment t as a real number representing its length. Thus $\lambda_t, \gamma_t, \sigma_t, \tau_t$ respectively will be the exponential function exp and the hyperbolic trigonometric functions cosh, sinh, tanh of t. However, we do not assume any axiom of continuity. As λ_t, γ_t, etc. are introduced according to the operations of ends, they do not involve any concept of *transcendental* functions. The entire proof and calculation are not beyond the scope of finiteness and rationality. For this point the reader may refer to Sect. 6.6.

See Fig. 6.14. For an arbitrary point P in the plane, construct a half-line $P\infty$ and let the end of the line determined by $P\infty$ which is opposite to ∞ be α. Moreover, construct the end $\beta = \alpha/2$, the line $\beta\infty$ and the symmetric point P' of P with respect to $\beta\infty$. Then P' lies on the axis $\theta\infty$, so we may suppose that

$$|\overrightarrow{OP'}| = t$$

is an oriented segment. Now set

$$\xi_1 = \sigma_t + \tfrac{1}{2}\alpha^2 \cdot \lambda_{-t},$$

$$\xi_2 = \alpha \cdot \lambda_{-t},$$

$$\xi_3 = \gamma_t + \tfrac{1}{2}\alpha^2 \cdot \lambda_{-t}.$$

Then the triple (ξ_1, ξ_2, ξ_3) of ends is called the Hilbert *homogeneous coordinates* of point P in this Hilbert coordinate system. In particular, the coordinates of the origin O are $(0, 0, 1)$.

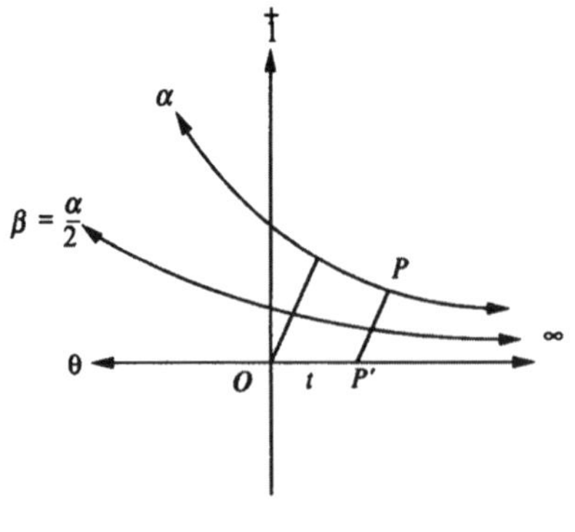

Fig. 6.14

Theorem 2. Take a Hilbert coordinate system in the Bolyai–Lobachevsky non-Euclidean plane. Then the Hilbert coordinates (ξ_1, ξ_2, ξ_3) of an arbitrary point P satisfy the relation

$$\xi_3^2 - \xi_1^2 - \xi_2^2 = 1,$$

$$\xi_3 > 0.$$

Moreover,

$$P \leftrightarrow (\xi_1, \xi_2, \xi_3)$$

induces a one-to-one correspondence between points P in the plane and those triples of ends which are not ∞ and satisfy the above two relations. This correspondence will be written as

$$P = (\xi_1, \xi_2, \xi_3)$$

in what follows.

Theorem 3. Let the Hilbert coordinates of two arbitrary points P, Q in the Bolyai–Lobachevsky non-Euclidean plane be

$$P = (\xi_1, \xi_2, \xi_3), \quad Q = (\eta_1, \eta_2, \eta_3).$$

Then

$$\xi_3\eta_3 - \xi_1\eta_1 - \xi_2\eta_2 > 0.$$

Let

$$P' = (\xi_1', \xi_2', \xi_3'), \quad Q' = (\eta_1', \eta_2', \eta_3')$$

be another two points. Then a necessary and sufficient condition for the segments $|PQ|$ and $|P'Q'|$ to be congruent is

$$|PQ| \equiv |P'Q'| \Longleftrightarrow \xi_3\eta_3 - \xi_1\eta_1 - \xi_2\eta_2 = \xi_3'\eta_3' - \xi_1'\eta_1' - \xi_2'\eta_2'.$$

Theorem 4. The equation of the connecting line $\alpha\beta$ of two non-∞ ends α, β is

$$(\alpha\beta - 1)\xi_1 + (\alpha + \beta)\xi_2 - (\alpha\beta + 1)\xi_3 = 0.$$

The equation of the connecting line $\alpha\infty$ of a non-∞ end α and the end ∞ is

$$\alpha\xi_1 + \xi_2 - \alpha\xi_3 = 0.$$

In the above equations, (ξ_1, ξ_2, ξ_3) are the Hilbert homogeneous coordinates of an arbitrary point on line $\alpha\beta$ or $\alpha\infty$.

Now write the equations of the two lines in the above theorem as

$$u_1\xi_1 + u_2\xi_2 - u_3\xi_3 = 0,$$

in which

$$u_1 = \frac{\alpha\beta - 1}{\beta - \alpha}, \quad u_2 = \frac{\alpha + \beta}{\beta - \alpha}, \quad u_3 = \frac{\alpha\beta + 1}{\beta - \alpha}$$

for line $\alpha\beta$, and

$$u_1 = \alpha, \quad u_2 = 1, \quad u_3 = \alpha$$

for line $\alpha\infty$. In both cases, the above equation is called a *normal form* of line $\alpha\beta$ or $\alpha\infty$ *approaching* α with (u_1, u_2, u_3) as its Hilbert *homogeneous line coordinates*. This is denoted simply as

$$\overrightarrow{\beta\alpha} = (u_1, u_2, u_3) \quad \text{or} \quad \overrightarrow{\infty\alpha} = (u_1, u_2, u_3).$$

Thus, line $\alpha\beta$ approaching another end β may be denoted as

$$\overrightarrow{\alpha\beta} = (-u_1, -u_2, -u_3).$$

For line $\alpha\infty$ approaching the end ∞, we say that

$$-\alpha\xi_1 - \xi_2 - \alpha\xi_3 = 0$$

is its *normal form*. When setting $u_1 = u_3 = \alpha$, $u_2 = 1$ as before, we have

$$\overrightarrow{\alpha\infty} = (-u_1, -u_2, -u_3),$$

where the right-hand side of the equation is the Hilbert *homogeneous line coordinates* of $\overrightarrow{\alpha\infty}$.

It is easy to see that these line coordinates satisfy the relation

$$u_1^2 + u_2^2 - u_3^2 = 1.$$

On the other hand, three arbitrary ends u_1, u_2, u_3 satisfying this relation must be the Hilbert homogeneous coordinates of a line approaching an end. It is also easy to prove that an arbitrary equation

$$v_1\xi_1 + v_2\xi_2 - v_3\xi_3 = 0$$

can be transformed into the normal form, as long as

$$v_1^2 + v_2^2 - v_3^2 > 0.$$

In fact, for any end $\alpha > 0$, let t be such that $\lambda_t = \alpha$. Then $\lambda_{t/2} = \beta$ satisfies the relations that $\beta > 0$ and $\beta^2 = \alpha$, so it can be denoted as $\sqrt{\alpha}$. Then by setting

$$u_i = \pm \frac{v_i}{\sqrt{v_1^2 + v_2^2 - v_3^2}},$$

we immediately get the normal form, where $\pm$ indicates the two different directions approaching the ends.

By now we have established the (Hilbert) coordinate system and represented points and oriented segments by triples of ends satisfying some equality and inequality relations, i.e., by the (Hilbert) homogeneous point coordinates or line coordinates. Therefore, some meaningful geometric relations can be expressed as relations among the corresponding point coordinates and line coordinates. We now list the majority of them.

1. Parallelism
Let the line coordinates of two oriented lines be

$$\vec{l}_u = (u_1, u_2, u_3), \quad \vec{l}_v = (v_1, v_2, v_3)$$

respectively. Then a necessary and sufficient condition for these two oriented lines to be parallel along the prescribed direction is

$$u_1 v_1 + u_2 v_2 - u_3 v_3 = 1.$$

2. Perpendicularity
Let two oriented lines $\vec{l}_u, \vec{l}_v$ be as above. Then the condition for them to be perpendicular is

$$u_1 v_1 + u_2 v_2 - u_3 v_3 = 0.$$

3. Congruence of segments
See Theorem 3 above.

4. Distance between two points
The *distance* d between two points

$$P = (\xi_1, \xi_2, \xi_3), \quad Q = (\eta_1, \eta_2, \eta_3)$$

is defined by the following relation:

$$\gamma_d = \xi_3 \eta_3 - \xi_1 \eta_1 - \xi_2 \eta_2 \ (> 0),$$

where d, as a positive segment on the axis $\theta\infty$, always takes a positive value.

5. Distance from a point to a line
Let the point be

$$P = (\xi_1, \xi_2, \xi_3)$$

and the oriented line be

$$\vec{l} = (u_1, u_2, u_3).$$

Then the *distance* t from point P to line $\vec{l}$ is given by the following relation:

$$\sigma_t = u_1\xi_1 + u_2\xi_2 - u_3\xi_3.$$

When P does not lie on line l, we have $\sigma_t \neq 0$, so that t, as an oriented segment on $\theta\infty$, can take positive or negative value. Whether a positive or a negative value is taken depends on which side of $\vec{l}$ point P lies on, see 7 below.

6. Incidence of a point and a line

Let point P and line $\vec{l}$ be as in 5 above. Then a necessary and sufficient condition for P and $\vec{l}$ to be incident is

$$u_1\xi_1 + u_2\xi_2 - u_3\xi_3 = 0.$$

7. Two sides of a line

Let the oriented line be

$$\vec{l_u} = (u_1, u_2, u_3).$$

Then the two sides of $\vec{l_u}$ are determined by

$$u_1\xi_1 + u_2\xi_2 - u_3\xi_3 > 0$$

and

$$u_1\xi_1 + u_2\xi_2 - u_3\xi_3 < 0$$

respectively. If $P = (\xi_1, \xi_2, \xi_3)$ and the directional position of P with respect to $\vec{l_u}$ is just the same as that of $\overset{+}{1}$ with respect to $O\infty$, then the inequality is taken with > 0; otherwise, the inequality is taken with < 0. These two sides will be called respectively the positive and the negative sides of the oriented line.

8. Degree of an angle

For an angle $\measuredangle O(\alpha, \beta) = \varphi$ at origin O, let us take, without loss of generality, $(\alpha - \beta)/(\alpha\beta + 1)$, as a function T of φ, to measure the degree of φ:

$$T(\varphi) = \frac{\alpha - \beta}{\alpha\beta + 1}.$$

If there is another angle $\measuredangle O(\beta, \gamma) = \psi$:

$$T(\psi) = \frac{\beta - \gamma}{\beta\gamma + 1},$$

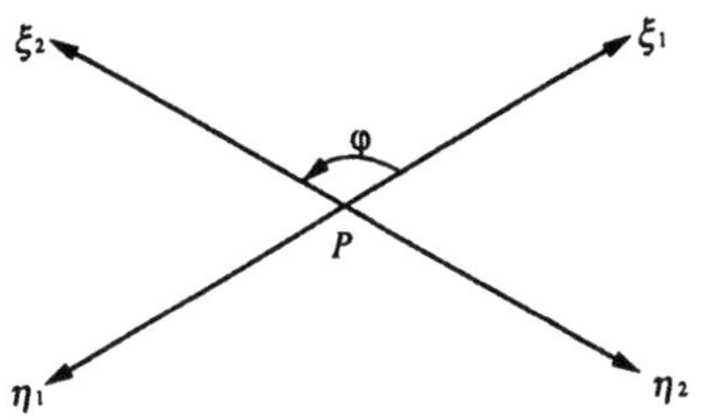

Fig. 6.15

then for the sum $\varphi + \psi = \measuredangle O(\alpha, \gamma)$ of φ and ψ, it is easy to see that

$$T(\varphi + \psi) = \frac{T(\varphi) + T(\psi)}{1 - T(\varphi) \cdot T(\psi)}.$$

See Fig. 6.15. If the vertex of angle φ is at an arbitrary point P with the ends of its two sides being ξ_1, η_1 and ξ_2, η_2 respectively: $\varphi = \measuredangle P(\xi_1, \xi_2)$, then according to Liebmann (1904), the degree of φ may be given by the following function T:

$$[T(\varphi)]^2 = -\frac{\xi_1 - \xi_2}{\xi_1 - \eta_2} \Big/ \frac{\eta_1 - \xi_2}{\eta_1 - \eta_2}.$$

One may prove that the right-hand side of the above formula always takes positive value or 0. As for the value of $T(\varphi)$, it can be positive or negative, determined according to 7 above. Furthermore, if P is the origin O and thus $\eta_1 = -1/\xi_1, \eta_2 = -1/\xi_2$, we immediately get the preceding expression.

If the geometry satisfies the axioms of continuity so that the degrees of angles can be regarded as real numbers, then T is just the usual tangent function:

$$T(\varphi) = \tan \tfrac{1}{2}\varphi.$$

However, we do not assume any axiom of continuity, so T and the previous functions γ, σ, τ are all introduced by using other axioms in the geometry and their properties are purely algebraic.

9. Parallel angles

As in Fig. 6.16, let the ends of a line l be α, β and P be a point not on l with $PA = p$ as its perpendicular distance to l. Then $P\alpha$ is parallel to $A\alpha$ and $\measuredangle P(PA, P\alpha)$ is generally called the parallel angle of P with respect to l,

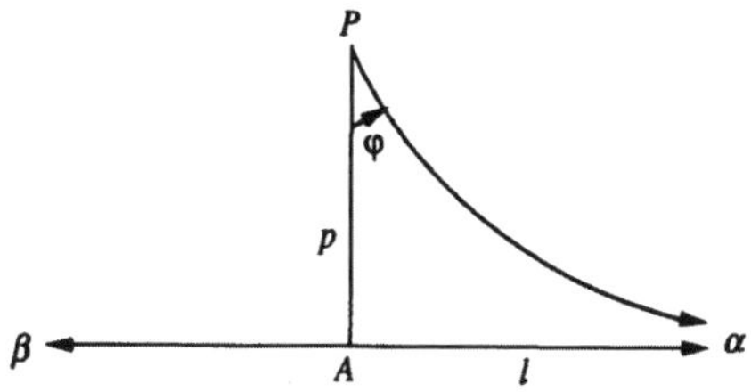

Fig. 6.16

denoted as φ. Thus we have

$$T(\varphi) = \lambda_{-p},$$

where T and φ are purely algebraically defined by using geometric axioms, not necessarily depending on the axioms of continuity as usual.

10. Two arbitrary lines $l_1 = (u_1, v_1, w_1)$ and $l_2 = (u_2, v_2, w_2)$ have the following relations:

$$l_1, l_2 \text{ intersect} \Longleftrightarrow |u_1u_2 + v_1v_2 - w_1w_2| < 1,$$

$$l_1, l_2 \text{ are parallel} \Longleftrightarrow |u_1u_2 + v_1v_2 - w_1w_2| = 1,$$

$$l_1, l_2 \text{ have a common perpendicular (neither intersecting nor parallel)}$$
$$\Longleftrightarrow |u_1u_2 + v_1v_2 - w_1w_2| > 1.$$

Whether the absolute value of the middle expression above takes positive or negative sign depends on the ends in the parallel direction.

Summarizing the above, one sees that in Bolyai–Lobachevskian geometry which takes points and lines as fundamental objects and the axioms H I–H IV (H IV means H IVBL) as a basis, points and lines can be represented by triples of ends via the operations of ends and the Hilbert coordinate system. In this geometry, all fundamental relations and derived relations such as incidence, parallelism, perpendicularity, the separation of order, and the congruence of segments and of angles can be expressed as polynomial equality or inequality relations among the coordinates of points and lines. Therefore, according to Sect. 6.1 we arrive at the following fundamental theorem.

Mechanization theorem. There is a mechanical method for proving theorems in Bolyai–Lobachevsky's hyperbolic plane non-Euclidean geometry.

Some metric concepts such as the lengths of segments and the degrees of angles in the geometry have to be expressed by means of functions like $T, \lambda, \gamma, \sigma, \tau$. But these functions can be expressed in terms of some rational functions of the ends, and the relations among these functions are algebraic, polynomial ones. Hence the identities and the corresponding geometric theorems among them may also be derived and proved mechanically, see Sect. 6.6 for details.

6.4 The mechanization of theorem proving in Riemann's elliptic non-Euclidean geometry

So-called Riemann's elliptic non-Euclidean geometry, simply speaking, is projective geometry with metric concepts. For the sake of simplicity, the following discussions will be restricted to the planar case. We base our presentation on the two articles by Podehl and Reidemeister (1934) and Bachmann and Reidemeister (1937) with some modifications.

First, the fundamental objects of this geometry are also points and lines. Assume that there is no relation of order, and the incidence relation and orthogonal relation are taken as fundamental relations. But different from Podehl and Reidemeister (1934), the congruence relation among pairs of points will be introduced as a derived relation. The fundamental objects and relations are assumed to satisfy the following axioms.

E I Axioms of incidence. The axioms of incidence, the axiom of infinity, and Pappus' (projective) axiom in plane unordered projective geometry are all included, see Sect. 6.2 for details. Now Desargues' (projective) axioms are theorems.

E II Orthogonal axioms (E II1–E II4)

II1. Let a be a line. Then through an arbitrary point there is at least one line b that is distinct from a and perpendicular to a, denoted as $b \perp a$.

II2. Through a point on an arbitrary line a there is one and only one line b perpendicular to a, also denoted as $b \perp a$.

II3. If $b \perp a$, then $a \perp b$.

II4. If one can draw two different lines passing through a point P and perpendicular to a line a, then any line passing through P is perpendicular to a.

In Axiom E II4, point P is called the *pole* of a and a is called the *polar line* of P.

From these axioms, one knows that a point cannot lie on its polar line and the perpendicular of any line through an arbitrary point on the line passes through the pole of this line. If a point B lies on the polar line of A, then point A also lies on the polar line of B. In this case the two points A, B are said to be *conjugate* with each other. The above axioms exclude the existence of isotropic lines, i.e., such lines that are perpendicular to themselves.

Due to Axioms E I, the usual theorems in unordered projective geometry all hold true. Moreover, we can define the *harmonic series of points*, the *harmonic separation* and the *involution* of pairs of points etc. on a line as usual. Now we add the following axiom to the geometry.

II5. The points conjugate with each other on an arbitrary line constitute an involution.

The geometric meaning of this axiom will be explained later.

Definition. A geometry which takes points and lines as its fundamental objects, the incidence relation and orthogonal relation as its fundamental relations, and satisfies the above axioms E I, E II is called an *unordered Riemannian (plane) elliptic orthogonal geometry.*

The above axioms are not independent of each other. For example, Pappus' projective axiom may be derived from other axioms. But in this book, we do not investigate the logical relations among these axioms. What is of significance for us is the following.

Mechanization theorem 1. There is a mechanical method for proving theorems in unordered Riemannian (plane) elliptic orthogonal geometry.

To prove this theorem, let us first make the following preparations.

Introduce a projective coordinate system $(A_1A_2A_3I)$ such that each vertex of the triangle $A_1A_2A_3$ is the pole of the opposite side. This coordinate system is called a *polar reciprocal coordinate system.*

According to Sect. 6.2, for an arbitrary projective coordinate system a necessary and sufficient condition for a point $(x_1 : x_2 : x_3)$ to lie on a line $(u_1 : u_2 : u_3)$ is

$$u_1x_1 + u_2x_2 + u_3x_3 = 0.$$

An arbitrary collineatory transformation that transforms any point $(x_1 : x_2 : x_3)$ to another point $(x'_1 : x'_2 : x'_3)$ may be expressed by means of a set of equations

$$x'_i = \sum_{j=1}^{3} a_{ij} J(x_j), \quad i = 1, 2, 3,$$

where J is an automorphism of the geometry-associated number field $\boldsymbol{K}$ and the determinant $|a_{ij}| \neq 0$. Similarly, the inverse morphism that transforms an arbitrary point $(x_1 : x_2 : x_3)$ to a line $(u_1 : u_2 : u_3)$ may also be expressed by means of a set of equations

$$u_i = \sum_{j=1}^{3} a_{ij} J(x_j), \quad i = 1, 2, 3,$$

where J is also an automorphism of $\boldsymbol{K}$ and the determinant $|a_{ij}| \neq 0$.

The transformation which transforms a point into its polar line and a line into its pole is called a *polar reciprocal transformation* of elliptic orthogonal geometry. Apparently, this transformation is an inverse morphism which not only preserves the incidence relation between points and lines but also transforms the points conjugate with each other into orthogonal lines and the orthogonal lines into points conjugate with each other. Of course, such a transformation may be expressed as a set of equations of the last form above, but in which a_{ij} and the automorphism J have some special properties. If the coordinate system is a polar reciprocal one, then the set of equations can be much simplified, in particular, the automorphism J will become an identity isomorphism. The following theorem originating from Reidemeister (1930) shows the form of polar reciprocal transformations and may be regarded as the main theorem of elliptic orthogonal geometry.

Theorem 1. In a polar reciprocal coordinate system, the polar reciprocal trans-

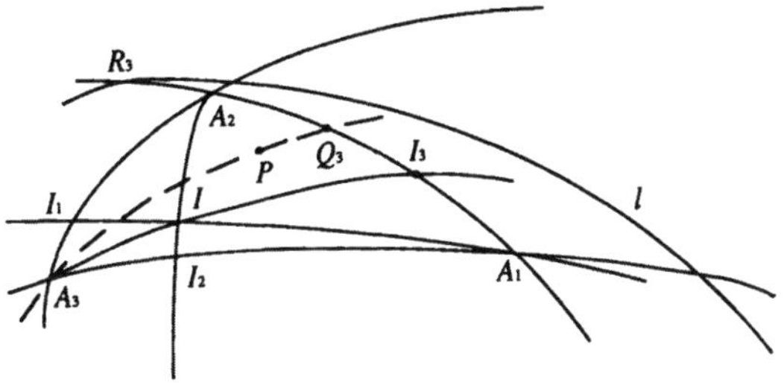

Fig. 6.17

formation may be expressed as

$$u_i = k_i x_i, \quad i = 1, 2, 3,$$

in which $k_1 k_2 k_3 \neq 0$.

Proof. To prove this theorem, let $A_i I$ and $A_{i+1} A_{i+2}$ of the polar reciprocal coordinate system $(A_1 A_2 A_3 I)$ intersect at a point I_i. Define the number field

$$N_i = N(A_{i+1}, I_i, A_{i+2})$$

on $A_{i+1} A_{i+2}$ according to Sect. 6.2 and identify it to the geometry-associated number field $\boldsymbol{K}$, where $i = 1, 2, 3$ and each A_i is defined according to the index i modulo 3. Let the homogeneous coordinates of a point P in this coordinate system be $(a_1 : a_2 : a_3)$ and the polar line of P be l with its homogeneous coordinates $(b_1 : b_2 : b_3)$. Let $A_i P$ and l respectively intersect A_{i+1} and A_{i+2} at points Q_i and R_i. Then Q_i and R_i are conjugate with each other and their homogeneous coordinates are

$$Q_3 = (a_1 : a_2 : 0), \quad R_3 = (-b_2 : b_1 : 0)$$

respectively. Our goal is to prove that there are non-zero numbers $k_i \in \boldsymbol{K}$ determined by the coordinate system $(A_1 A_2 A_3 I)$ such that

$$b_1 : b_2 : b_3 = k_1 a_1 : k_2 a_2 : k_3 a_3,$$

regardless of point P.

For this purpose, we construct a new geometry as follows.

This new geometry consists of two kinds of fundamental objects: "points" and "lines," where "points" are those which are not on $A_1 A_2$ and "lines" are arbitrary ones but distinct from $A_1 A_2$ in the original geometry. Regard a point A not on $A_1 A_2$ and a line a distinct from $A_1 A_2$ in the original geometry as a "point" and a "line" in the new geometry, denoted by "A" and "a" respectively.

Now introduce three fundamental relations, namely, the relation of "incidence," the relation of "parallelism," and the "orthogonal" relation, in the new geometry as follows:

"Point" "A" lies on "line" "a" $\Longleftrightarrow$ A lies on line a;

"Line" "a" is parallel to the "line" "b" $\Longleftrightarrow$ The intersection point of a, b lies on line A_1A_2;

"Line" "a" is perpendicular to "line" "b" $\Longleftrightarrow$ The intersection points of a, b and A_1A_2 are conjugate.

It is not difficult to verify by the original geometric axioms E I, E II that the new geometry satisfies all axioms of unordered orthogonal geometry in Sect. 2.2. For instance, the axiom of orthocenter O 5 in that section may be derived from the involutory axiom E II5 of this section. This also explains the geometric meaning of axiom E II5.

In this new geometry, the "lines" "A_3A_1" and "A_3A_2" are perpendicular. Hence we may take an "orthogonal coordinate system" with "origin" "A_3," the first and the second "axes" "A_3A_1," "A_3A_2" and the "unit points" "I_2" and "I_1" in the new geometry. In this case, from the definition of number systems on the line one easily sees that there are natural isomorphisms

$$N(A_3, I_2) \approx N(A_3, I_2, A_1),$$

$$N(A_3, I_1) \approx N(A_3, I_1, A_2).$$

Furthermore, for any point X distinct from A_2, A_3 on line A_2A_3, when its corresponding numbers in $N(A_2, I_1, A_3)$ and $N(A_3, I_1, A_2)$ are x and x' respectively, one sees that there is a relation $xx' = 1$. From this we know that if the point $P = (a_1 : a_2 : a_3)$ does not lie on A_2A_3 and thus $a_3 \neq 0$, then the "coordinates" of point "P" in the "orthogonal coordinate system" will be "P" $= \left(\dfrac{a_1}{a_3}, \dfrac{a_2}{a_3}\right)$. The equations of the "lines" "A_3Q_3" and "A_3R_3" will be

$$a_2x_1 = a_1x_2,$$

$$b_1x_1 + b_2x_2 = 0.$$

According to Sect. 2.3, the "orthogonal coordinate system" above determines a non-zero "orthogonal rate," denoted by k_3^*. Then, we have

$$k_3^*a_2b_1 - a_1b_2 = 0,$$

for "A_3Q_3" is perpendicular to "A_3R_3."

In the same way, we can construct another new geometry which takes the points other than those on A_1A_3 in the original geometry as its "points," the lines distinct from A_1A_3 as its "lines" and introduce the corresponding fundamental relations such as "incidence," "parallelism," and "orthogonality" as before. Then we may prove as before that there is a non-zero number k_2^* such that

$$k_2^*a_3b_1 - a_1b_3 = 0.$$

Since a_1, a_2, a_3 are not simultaneously zero and neither are b_1, b_2, b_3, from the two equations above we may obtain non-zero numbers k_1, k_2, k_3, for example

$$k_1 = 1, \quad k_2 = k_3^*, \quad k_3 = k_2^*,$$

such that

$$b_1 : b_2 : b_3 = k_1 a_1 : k_2 a_2 : k_3 a_3.$$

This proves Theorem 1. □

Obviously, k_1, k_2, k_3 in Theorem 1 are uniquely determined only up to a non-zero proportional divisor. We say that $k_1 : k_2 : k_3$ is the *ratio of orthogonal rates* determined by the polar reciprocal coordinate system $(A_1A_2A_3I)$.

By Theorem 1, it is easy to give a proof of mechanization theorem 1 as follows.

Take an arbitrary polar reciprocal coordinate system and let the corresponding ratio of orthogonal rates be $k_1 : k_2 : k_3$ by Theorem 1, where $k_1k_2k_3 \neq 0$. In this coordinate system, a necessary and sufficient condition for a point $P = (x_1 : x_2 : x_3)$ and a line $l = (u_1 : u_2 : u_3)$ to be incident is

$$x_1u_1 + x_2u_2 + x_3u_3 = 0.$$

Let

$$l_u = (u_1 : u_2 : u_3), \quad l_v = (v_1 : v_2 : v_3)$$

be two lines. If the pole of l_u is $P = (x_1 : x_2 : x_3)$, then by Theorem 1 we have

$$x_1 : x_2 : x_3 = \frac{1}{k_1}u_1 : \frac{1}{k_2}u_2 : \frac{1}{k_3}u_3.$$

Since a necessary and sufficient condition for l_u, l_v to be parallel to each other is that l_v passes through pole P of l_u, that is

$$\frac{1}{k_1}u_1v_1 + \frac{1}{k_2}u_2v_2 + \frac{1}{k_3}u_3v_3 = 0$$

or

$$k_2k_3u_1v_1 + k_3k_1u_2v_2 + k_1k_2u_3v_3 = 0,$$

the fundamental relations in the geometry can all be expressed as polynomial equality relations among the coordinates. According to the basic principle of Sect. 6.1, theorem proving in this geometry is mechanizable. □

As in Sects. 2.2–2.4, we may introduce some derived concepts such as midpoint, symmetry, and reflection in unordered elliptic orthogonal geometry and introduce a new axiom.

Axiom of symmetric axes. For two arbitrary lines a_1, a_2, there is a symmetric axis l such that a_1 and a_2 are symmetric with respect to l.

We say that an unordered elliptic orthogonal geometry satisfying this axiom of symmetric axes is an *unordered (plane) elliptic metric geometry*, in which the *congruence* relation may be introduced as a derived relation and the metric concept as a derived concept. We may add the relation of separation as a fundamental relation and introduce some axioms of separation as we did in Sect. 6.2 so as to define various ordered elliptic orthogonal geometry, metric geometry, and the usually so-called Riemann's elliptic non-Euclidean geometry. Just like the mechanization theorem above, we may prove the following:

Mechanization theorem 2. Theorem proving is mechanizable in unordered elliptic metric geometry as well as various ordered elliptic orthogonal geometry, metric geometry, and non-Euclidean geometry.

The proof of this theorem is quite easy and thus omitted. As a supplement we give a theorem below, also without proof.

Theorem 2. In a polar reciprocal coordinate system, let

$$\begin{aligned} A &= (x_1 : x_2 : x_3), \quad & B &= (y_1 : y_2 : y_3), \\ A' &= (x_1' : x_2' : x_3'), \quad & B' &= (y_1' : y_2' : y_3') \end{aligned}$$

be four points. Then a necessary and sufficient condition for two pairs of the points to be congruent

$$(AB) \equiv (A'B')$$

is

$$\begin{aligned} &\frac{(k_1x_1y_1 + k_2x_2y_2 + k_3x_3y_3)^2}{(k_1x_1^2 + k_2x_2^2 + k_3x_3^2)(k_1y_1^2 + k_2y_2^2 + k_3y_3^2)} \\ &= \frac{(k_1x_1'y_1' + k_2x_2'y_2' + k_3x_3'y_3')^2}{(k_1x_1'^2 + k_2x_2'^2 + k_3x_3'^2)(k_1y_1'^2 + k_2y_2'^2 + k_3y_3'^2)}. \end{aligned}$$

6.5 The mechanization of theorem proving in two circle geometries

All (plane) geometries considered so far in this book take only points and lines as their fundamental objects. Of course, this is not a restriction necessarily required. Several geometries that take other figures, in particular circles, as their fundamental objects have also appeared. Among circle geometries, two were more extensively investigated: one is the so-called Möbiusian geometry or inversive geometry which studies such properties as the tangency and intersection of circles in the complex plane and the other is the so-called Laguerrean geometry or equilong geometry which studies such properties as the oriented tangency of oriented lines and oriented circles in the complex plane (cf. Klein 1926: sects. 49 and 67, Morley and Morley 1933, and others). These geometries were treated axiomatically in van der Waerden and Smid (1935), and later Ewald (1956a) and Uhl (1964) made some modifications. We point out that theorem proving in these geometries is also mechanizable and give a very brief introduction below. For further details, the author will write a technical article.

Möbiusian circle geometry

The fundamental objects for this geometry consist of two kinds: points and circles, and the fundamental relation consists of the incidence relation: points lying on circles. The *tangency* or *orthogonality* of two circles may be introduced either as a fundamental relation or as a derived relation. In the axiom system, one axiom associated with the name of Miquel plays a very important role which is similar to the role of Pappus' axiom in projective geometry. Now we state it as follows.

Miquel's axiom. Let

$$\begin{matrix} A & B & C & D \\ A' & B' & C' & D' \end{matrix}$$

be eight points. Take arbitrarily two from the first four points and take two others from the last four points in such a way that they are not on the same columns as the two points previously taken. Such quadruplets of points are six in number. If the four points in five of the six quadruplets all lie on the same circle while these circles and points are pairwise distinct, then the four points in the sixth also lie on the same circle.

Miquel's axiom has various weaker forms.

We call all circles that are tangent together at the same point a *pencil of circles*. Van der Waerden has used the following method to introduce a number system so as to get coordinatization in the plane.

Take a fixed point W in the plane of circle geometry. Now construct a projective plane $\mathbf{P}$ as follows.

A "point" in $\mathbf{P}$ is either a point distinct from W in the original plane or an arbitrary pencil of circles with W as its common tangent point. The former is called a "principal point" and the latter a "secondary point" of $\mathbf{P}$.

A "line" in $\mathbf{P}$ is either a circle passing through the point W in the original plane, called a "principal line," or a "line" passing through all "secondary points," called a "secondary line," denoted by g.

The definition of the incidence relation between "points" and "lines" in $\mathbf{P}$ is evident.

It is easy to prove that under the incidence relation, the "points" and "lines" in $\mathbf{P}$ satisfy all axioms of the usual unordered projective geometry. In particular, Pappus' axiom can be derived from Miquel's axiom. So in the corresponding projective geometry one may introduce an associated number field $\boldsymbol{K}$. After fixing a projective coordinate system, in which one side of the coordinate triangle is a secondary line g, we may prove that circles in the original plane are all "conics" in the new projective plane and the quadric part of their equations (corresponding to the other two sides of the coordinate triangle) have a fixed form. From this it is easy to elicit the conclusion that theorem proving in this circle geometry is mechanizable.

Laguerrean circle geometry

This geometry takes the geometry constituted by such figures as oriented lines and oriented circles in ordinary plane geometry as a concrete model. After axiomatization, Laguerrean geometry has *spears* (Speere, oriented lines) and *cycles* (Zykel, oriented circles) as its fundamental objects. The fundamental relations consist of the *incidence* between spears and cycles, the *tangency* of cycles and the *parallelism* of spears. The parallel relation of spears may also be defined as a derived relation, namely that they are not incident to the same cycle. Likewise the tangency of cycles may be defined as a derived relation, namely that they are exactly incident to one and the same spear. Sometimes the incidence of spears and cycles is also said to be the *tangency* of spears and cycles. The number of spears which are incident to two cycles can only be 0, 1, or 2. There is no concept of points, which can at most be introduced as a derived concept in this geometry. See Fig. 6.18.

In the axiom system, there is an axiom which occupies a similarly important position as Miquel's axiom does in Möbiusian geometry, denoted still as Miquel's *axiom*. Let us state it as follows.

Miquel's axiom. Let a spear W be tangent to (i.e., incident to) three cycles $\alpha_1, \alpha_2, \alpha_3$. Suppose the cycles α_2, α_3 have spear A_1 as one of their common tangent spears; similarly, the cycles α_3, α_1 and α_1, α_2 respectively have common tangent spears A_2 and A_3. Furthermore, the cycles $\alpha_1, \alpha_2, \alpha_3$ are tangent to three other spears P_1, P_2, P_3 respectively, where the latter are not parallel to each other. Let all spears above be distinct from each other and the cycles r_1, r_2, r_3 be tangent to the spears $A_1P_2P_3$, $A_2P_3P_1$, $A_3P_1P_2$ respectively. Then r_1, r_2, r_3 have a common tangent spear B.

We say that the collection of cycles which are tangent to each other and tangent to a fixed spear S is a *tangent family* with S as its common tangent

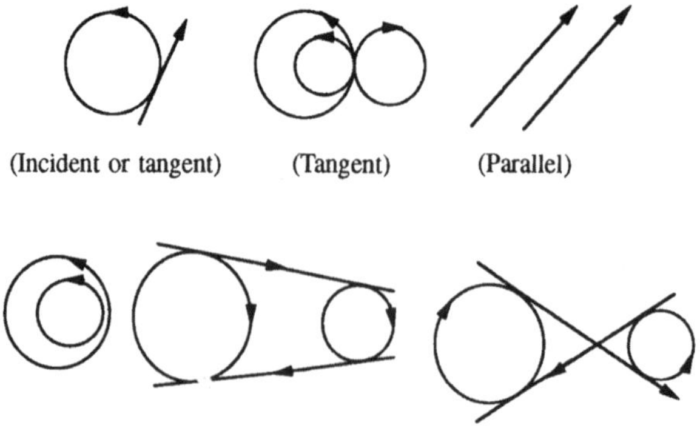

Fig. 6.18

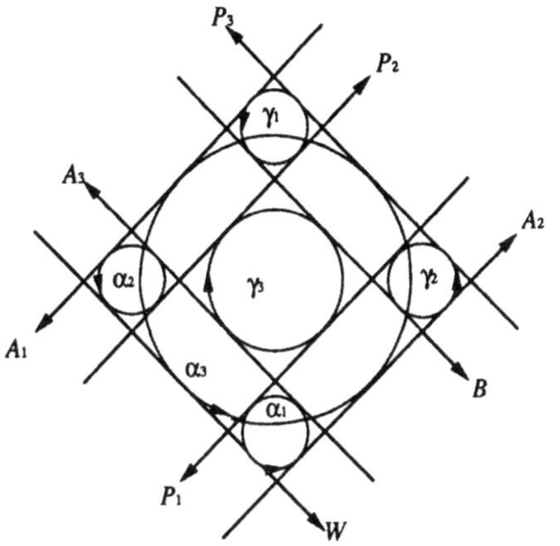

Fig. 6.19

spear. Similarly as for Möbiusian geometry, we may construct a projective plane in a dual way as follows.

Take a fixed spear W in the original plane of the circle geometry, called a *base spear*. Define a projective plane **P**, on which the "points" consist of three kinds: an arbitrary spear not parallel to W, called a "principal point"; an arbitrary tangent family with W as its common tangent spear, called a "secondary point"; a single point, called a "base point" and denoted by G. The "lines" in **P** also consist of three kinds: an arbitrary cycle z tangent to W, called a "principal line," on which the "points" consist of the corresponding "principal points" of the spears tangent to z but not parallel to W and the corresponding "secondary point" of the tangent family defined by W and z; an arbitrary *spear family* consisting of a family of parallel spears, called a "secondary line," on which the "points" consist of the corresponding "principal points" of the spears in the spear family and a "base point" G; a single "line," called a "base line," on which the "points" consist of the corresponding "secondary points" of the tangent family with W as their common tangent spear and the "base point" G.

In van der Waerden and Smid (1935), Smid proved that the geometry determined by these "points" and "lines" satisfies a number of axioms of unordered projective geometry. In particular, Pappus' axiom can be proved by using Miquel's axiom. Therefore, in this geometry one may introduce a coordinate system, in which a cycle not tangent to W will be expressed as a conic. From this it is easy to elicit the corresponding mechanization theorem that theorem proving in Laguerrean geometry is mechanizable.

6.6 The mechanization of formula proving with transcendental functions

Mechanical proving of identities involving trigonometric functions

In the ordinary geometry of ancient China, the measurement of length, area, and volume occupied an important position, but angles and their measure were seldom mentioned. In modern Riemannian geometry, ds^2 or the length of arcs is taken as a fundamental concept, from which the others such as the measure of angles are derived. In contrast with these two, in ordinary geometry of the usual Euclidean form angles and segments are two important metric concepts that almost occupy an equal position. Especially for a triangle the consideration of relations between sides and angles has cardinal significance. These relations are all expressed by means of some trigonometric functions such as sin, cos, tan, which are transcendental with respect to the angles and of which the investigation is often not easy. Nevertheless, if we take a Descartes coordinate system, let the initial side of an angle A coincide with the positive ray of the abscissa axis and take an arbitrary point $P = (x, y)$ on the terminate side, then these trigonometric functions all become rational functions of x, y and $\overline{OP} = r$, i.e.,

$$\sin A = y/r, \quad \cos A = x/r, \quad \tan A = y/x,$$

where

$$r^2 = x^2 + y^2, \quad r > 0.$$

Hence, in principle these transcendental functions can be treated via the relations among rational functions or even polynomials. This makes it possible to express the concept of angles and transcendental relations among segments and angles in other forms or even to avoid using them. If we really need to consider the so-called trigonometric identities led by these relations, we may also give a mechanized proof according to Sect. 6.1 by using the methods of Chaps. 3–5. We shall explain this below.

There are various polynomial relations among trigonometric functions. Let us list some of them as follows.

1. Relations among the six trigonometric functions of an angle:

$$\sin^2 A + \cos^2 A = 1, \tag{1}$$

$$\tan A \cos A = \sin A, \tag{2}$$

$$\cot A \sin A = \cos A, \tag{3}$$

$$\sec A \cos A = 1, \tag{4}$$

$$\csc A \sin A = 1, \tag{5}$$

$$\sec^2 A = 1 + \tan^2 A, \tag{6}$$

$$\csc^2 A = 1 + \cot^2 A. \tag{7}$$

2. Relations for the negative of an angle:

$$\left.\begin{array}{ll} \sin(-A) = -\sin A, & \cos(-A) = \cos A, \\ \tan(-A) = -\tan A, & \cot(-A) = -\cot A, \\ \sec(-A) = \sec A, & \csc(-A) = -\csc A. \end{array}\right\} \tag{8}$$

3. Relations for the sum and difference of angles:

$$\sin(A \pm B) = \sin A \cos B \pm \cos A \sin B, \tag{9}$$

$$\cos(A \pm B) = \cos A \cos B \mp \sin A \sin B, \tag{10}$$

$$\tan(A \pm B)(1 \mp \tan A \tan B) = \tan A \pm \tan B. \tag{11}$$

4. Relations for some special angles such as 0, $\pi/2$, π etc. (where n is an arbitrary integer):

$$A = n\pi \Longleftrightarrow \sin A = 0, \tag{12}$$

$$A = 2n\pi \Longleftrightarrow \cos A = +1, \ \sin A = 0, \tag{13}$$

$$A = (2n+1)\pi \Longleftrightarrow \cos A = -1, \ \sin A = 0, \tag{14}$$

$$A = \frac{\pi}{2} + n\pi \Longleftrightarrow \cos A = 0, \tag{15}$$

$$A = \frac{\pi}{2} + 2n\pi \Longleftrightarrow \sin A = 1, \ \cos A = 0, \tag{16}$$

$$A = -\frac{\pi}{2} + 2n\pi \Longleftrightarrow \sin A = -1, \ \cos A = 0. \tag{17}$$

This kind of relations for special angles may be augmented in different cases.

We can also consider the formulas for half an angle:

$$\sin^2 \frac{A}{2} = \frac{1 - \cos A}{2}, \tag{18}$$

$$\cos^2 \frac{A}{2} = \frac{1 + \cos A}{2}, \tag{19}$$

$$\tan^2 \frac{A}{2} = \sin \frac{A}{1 + \cos A} = \frac{1 - \cos A}{\sin A}. \tag{20}$$

Of course, none of the above relations is independent of the others and they can be elicited from the original definition of sin, cos, tan, etc.

If a trigonometric identity to be proved involves the angles A and B, then the trigonometric functions of angles may be expressed in terms of x, u, etc., so that the original identity is transformed to a polynomial relation according to (1)–(17) above. Whether or not this polynomial relation is satisfied can be proved by the mechanical method mentioned in Sect. 6.1. The following is a concrete example.

Example 1. Prove that for the three angles A_1, A_2, A_3 of a triangle $A_1A_2A_3$, there is an identity

$$\sin 2A_1 + \sin 2A_2 + \sin 2A_3 = 4 \sin A_1 \sin A_2 \sin A_3.$$

Solution. For the three angles of a triangle we have

$$A_1 + A_2 + A_3 = \pi.$$

By using the formulas of angular sum (9), (10) and (14), we obtain from this relation that

$$\cos A_1 \cos A_2 \cos A_3 - \cos A_1 \sin A_2 \sin A_3 - \sin A_1 \cos A_2 \sin A_3 \\ - \sin A_1 \sin A_2 \cos A_3 + 1 = 0$$

and

$$\sin A_1 \cos A_2 \cos A_3 + \cos A_1 \sin A_2 \cos A_3 + \cos A_1 \cos A_2 \cos A_3 \\ - \sin A_1 \sin A_2 \sin A_3 = 0.$$

Similarly, the identity to be proved may be transformed to

$$\sin A_1 \cos A_1 + \sin A_2 \cos A_2 + \sin A_3 \cos A_3 - 2 \sin A_1 \sin A_2 \sin A_3 = 0.$$

Now set

$$\begin{aligned} \sin A_1 &= u_1, & \sin A_2 &= u_2, \\ \cos A_1 &= x_1, & \cos A_2 &= x_2, \\ \sin A_3 &= x_3, & \cos A_3 &= x_4. \end{aligned}$$

Then the original problem is reduced to the following one.

The set of hypothesis equations may be taken as

$$\begin{aligned} &F_1 \equiv x_1^2 + u_1^2 - 1 = 0, \\ &F_2 \equiv x_2^2 + u_2^2 - 1 = 0, \\ &F_3 \equiv (x_1x_2 - u_1u_2)x_3 + (u_1x_2 + u_2x_1)x_4 = 0, \\ &F_4 \equiv -(u_1x_2 + u_2x_1)x_3 + (x_1x_2 - u_1u_2)x_4 + 1 = 0, \end{aligned}$$

and the conclusion equation is

$$g \equiv x_3x_4 - 2u_1u_2x_3 + u_2x_2 + u_1x_1 = 0.$$

The problem amounts to proving that $g = 0$ is a formal consequence of $F_1 = 0, \ldots, F_4 = 0$.

Since the coefficient determinant of x_3, x_4 in F_3, F_4 is 1, we may solve the two equations for x_3, x_4 and obtain

$$\begin{aligned} &F_3' \equiv x_3 - (u_1x_2 + u_2x_1) = 0, \\ &F_4' \equiv x_4 + (x_1x_2 - u_1u_2) = 0. \end{aligned}$$

Hence, the problem is further reduced to proving that $g = 0$ is the formal consequence of $F_1 = 0,\ F_2 = 0,\ F_3' = 0,\ F_4' = 0$.

Using the general method, we reduce g successively by F_4', F_3', F_2, F_1 with respect to x_4, x_3, x_2, x_1 and get

$$\begin{aligned} g \equiv & -(u_1x_2+u_2x_1)(x_1x_2-u_1u_2)-2u_1u_2(u_1x_2+u_2x_1) \\ & + u_2x_2+u_1x_1 \quad \text{mod} \quad (F_4', F_3') \\ \equiv & -u_1x_1x_2^2-u_2x_1^2x_2-u_1u_2(u_1x_2+u_2x_1)+u_2x_2+u_1x_1 \quad \text{mod} \quad (F_4', F_3') \\ \equiv & -u_1x_1(1-u_2^2)-u_2x_2(1-u_1^2)-u_1u_2(u_1x_2+u_2x_1) \\ & + u_2x_2+u_1x_1 \quad \text{mod} \quad (F_4', F_3', F_2, F_1) \\ \equiv & \ 0 \quad \text{mod} \quad (F_4', F_3', F_2, F_1). \end{aligned}$$

That is, $g = 0$ is a formal consequence of $F_4' = 0,\ F_3' = 0,\ F_2 = 0,\ F_1 = 0$. In other words, for the three angles of $\triangle A_1A_2A_3$ the original identity holds.

The above method proceeds in a mechanical way and is applicable to various trigonometric identities. For instance, using the same mechanical method we may prove:

If three angles A_1, A_2, A_3 satisfy

$$A_1 + A_2 + A_3 = (2n+1)\pi,$$

then the identity

$$\cos 2A_1 + \cos 2A_2 + \cos 2A_3 + 4\cos A_1 \cos A_2 \cos A_3 = 0$$

does not hold.

Mechanical proving of theorems in non-Euclidean geometries and identities involving hyperbolic functions

In the usual literature, the treatment of the two non-Euclidean geometries considered above often requires using transcendental functions such as trigonometric, exponential, and hyperbolic functions; and so does the proof of the corresponding theorems and formulas require using methods of expanding power series and limit approximation. Take Bolyai–Lobachevsky's plane hyperbolic non-Euclidean geometry as an example. For the sake of simplicity, let the associated constant of this geometry be $k = 1$ (cf. Greenberg 1973). As in Fig. 6.20, take two oriented perpendiculars with foot at O in the plane as axes. From a point P draw two lines PU and PV perpendicular to the two axes respectively. The quadrilateral $OUPV$ is all of right angles at O, U, V, commonly called a Lambert quadrilateral. Let the oriented distances be denoted as

$$\overline{OU} = u, \quad \overline{OV} = v,$$
$$\overline{UP} = w, \quad \overline{VP} = z.$$

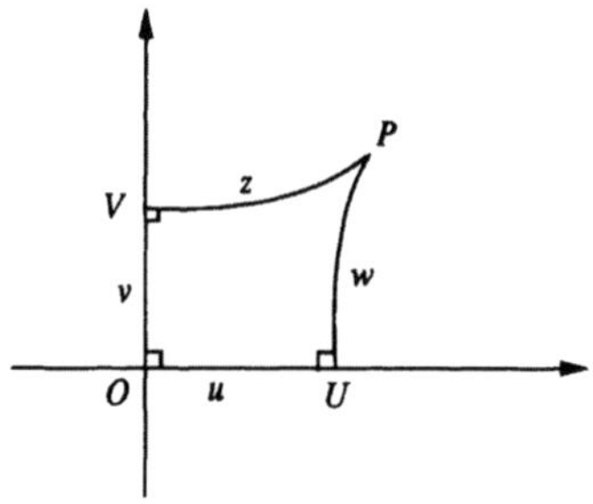

Fig. 6.20

Then, among u, v, w, z there are two transcendental relations as follows:

$$\tanh w = \tanh v \cosh u,$$
$$\tanh z = \tanh u \cosh v.$$

Set

$$x = \tanh u, \quad y = \tanh v,$$
$$T = \cosh u \cosh w, \quad X = xT, \quad Y = yT.$$

There are different manners to introduce coordinates for point P. For example,

$$(u, v) = \text{axis coordinates of } P,$$
$$(u, w) = \text{Lobachevsky coordinates of } P,$$
$$(x, y) = \text{Beltrami coordinates of } P,$$
$$(T, X, Y) = \text{Weierstrass coordinates of } P.$$

As these coordinates all use transcendental functions, some relations and properties with geometric meaning have to be expressed by means of transcendental functions, and so do the theorems and formulas. We give a few examples below.

1. Distance between two points

In an axis coordinate system the distance $\overline{P_1 P_2}$ between two points $P_1(u_1, v_1)$ and $P_2(u_2, v_2)$ may be determined by

$$\cosh \overline{P_1 P_2} = \frac{1 - \tanh u_1 \tanh u_2 - \tanh v_1 \tanh v_2}{\sqrt{1 - \tanh^2 u_1 - \tanh^2 v_1} \cdot \sqrt{1 - \tanh^2 u_2 - \tanh^2 v_2}},$$

while in a Beltrami coordinate system it can be simplified to

$$\cosh \overline{P_1 P_2} = \frac{1 - x_1 x_2 - y_1 y_2}{\sqrt{1 - x_1^2 - y_1^2} \cdot \sqrt{1 - x_2^2 - y_2^2}},$$

where, however, the transcendental and irrational functions still appear.

2. Hyperbolic Kou-Ku theorem

Let C be the right angle and a, b, c be the lengths of the three sides of a triangle ABC, where c is the opposite side of the right angle C. Then the relation corresponding to the Kou-Ku theorem in ordinary geometry is

$$\cosh c = \cosh a \cosh b.$$

3. Hyperbolic cosine law

For an arbitrary triangle ABC with three angles A, B, C and lengths a, b, c of their corresponding sides, there is a cosine formula

$$\cosh c = \cosh a \cosh b - \sinh a \sinh b \cos c.$$

When C is a right angle, this formula is simplified to the Kou-Ku relation.

Other cases such as the intersectional angle of two lines, the distance from a point to a line, circles, maximal circles, equidistant lines and even the equations of lines as well as the conditions for parallelism, perpendicularity, incidence etc., all cannot be deviated from transcendental functions, mainly hyperbolic functions. We do not give further examples.

The occurrence of these transcendental functions makes that the traditional proof of theorems in non-Euclidean geometries often has to be completed by complicated calculations of trigonometric formulas. One method is to list various formulas of hyperbolic functions such as

$$\cosh^2 x - \sinh^2 x = 1,$$

$$\sinh(x \pm y) = \sinh x \cosh y \pm \cosh x \sinh y,$$

so that the proof of identities and geometric theorems can proceed in a mechanical way as previously done for the proof of trigonometric formulas. However, since the occurring hyperbolic functions all need to be defined by using analytical methods, it makes that at least in theory the geometry has to be restricted to the case of having the real or complex number field as its basic number field. This excludes the possibility of considering geometries like non-Euclidean geometry over more general fields, while geometries over general fields occupy an important position in modern mathematics.

But, the method of Hilbert (1903) by introducing the operations of ends for hyperbolic non-Euclidean geometry completely avoids using the concept of continuity to establish the axiom system of this geometry so that transcendental functions become unnecessary, as Hilbert wrote at the end of that paper: "The familiar formulas of Bolyai–Lobachevskian geometry can then also be derived with no difficulty and the development of this geometry has been thus completed with the aid of Axioms I–IV alone."

The assertion that the trigonometric formulas of non-Euclidean geometry may be derived easily without the aid of the axioms of continuity by Hilbert was confirmed a few years afterwards (cf. the books of Liebmann, Gerretsen,

Szász, Szmielew, and others). This has already been explained somewhat in Sect. 6.3. We now restate the relevant part as follows.

Let hyperbolic plane non-Euclidean geometry satisfy the system of axioms H I–H IV (where H IV means H IVBL) as shown in Sect. 6.3. By applying the operations of ends, one gets a geometry-associated number field $\boldsymbol{K}$ of characteristic 0. Introduce the orthogonal axes $\theta\infty$ and $\overset{-+}{1\,1}$ so as to determine a Hilbert coordinate system. Then the coordinates of a point P are represented by a triple (ξ_1, ξ_2, ξ_3) of numbers in $\boldsymbol{K}$, where

$$\xi_3^2 - \xi_1^2 - \xi_2^2 = 1, \quad \xi_3 > 0,$$

and an arbitrary oriented line is also represented by a triple (u_1, u_2, u_3) of numbers in $\boldsymbol{K}$, where

$$u_1^2 + u_2^2 - u_3^2 = 1.$$

If the two ends of an oriented line are α and β with direction from β to α, then

$$u_1 = \frac{\alpha\beta - 1}{\alpha - \beta}, \quad u_2 = \frac{\alpha + \beta}{\alpha - \beta}, \quad u_3 = \frac{\alpha\beta + 1}{\alpha - \beta} \quad \text{if} \quad \alpha, \beta \neq \infty,$$

$$u_1 = -\alpha, \quad u_2 = -1, \quad u_3 = -\alpha \quad \text{if} \quad \beta = \infty,$$

and

$$u_1 = \beta, \quad u_2 = 1, \quad u_3 = \beta \quad \text{if} \quad \alpha = \infty.$$

Dually, if (u_1, u_2, u_3) are fixed, then the two ends α, β are

$$\alpha = \frac{u_2 + 1}{u_3 - u_1}, \quad \beta = \frac{u_2 - 1}{u_3 - u_1} \quad \text{if} \quad u_1 \neq u_3,$$

$$\alpha = \infty, \quad \beta = u_1 \quad \text{if} \quad u_1 = u_3,\ u_2 = 1,$$

and

$$\beta = \infty, \quad \alpha = -u_1 \quad \text{if} \quad u_1 = u_3,\ u_2 = -1.$$

Furthermore, the condition for a point (ξ_1, ξ_2, ξ_3) to lie on a line (u_1, u_2, u_3) is

$$u_1\xi_1 + u_2\xi_2 - u_3\xi_3 = 0.$$

In Sect. 6.3, we have also introduced some functions λ_t, γ_t, σ_t and τ_t which will be redenoted as $\lambda(t)$, $\gamma(t)$, $\sigma(t)$ and $\tau(t)$ in what follows. These functions are defined over the geometry-associated number field $\boldsymbol{K}$ with ends as their values. Here

$$\gamma(t) = \tfrac{1}{2}(\lambda(t) + \lambda(-t)),$$
$$\sigma(t) = \tfrac{1}{2}(\lambda(t) - \lambda(-t)),$$
$$\tau(t) = (\lambda(t) - \lambda(-t))/(\lambda(t) + \lambda(-t)).$$

The introduction of these functions depends on the operations of ends, but not on any concept of continuity. They are purely algebraic and satisfy some relations that are similar to those that the real transcendental functions $\exp t$, $\cosh t$, $\sinh t$, $\tanh t$ satisfy, which we will not list here. In the same section, we have also introduced the function

$$T(\varphi) = \frac{\alpha - \beta}{\alpha\beta + 1}$$

of the angle φ between two oriented lines which pass through the origin and approximate to two ends α and β. This function has some properties that are similar to those of the usual real transcendental function $\tan \frac{1}{2}\varphi$. From this function, we can purely algebraically introduce two other functions

$$S(\varphi) = \frac{2T(\varphi)}{1 + [T(\varphi)]^2}, \quad C(\varphi) = \frac{1 - [T(\varphi)]^2}{1 + [T(\varphi)]^2},$$

which correspond to the usual real transcendental functions $\sin\varphi$ and $\cos\varphi$ respectively.

Various geometric properties and relations with geometric meaning such as the parallelism and perpendicularity of two lines, the distance between two points, the distance from a point to a line and the angle between two lines may all be concretely expressed by means of these functions, as we have stated already in Sect. 6.3. From this it follows that the mechanical method mentioned in Sect. 6.1 can immediately be applied to proving theorems and formulas in this non-Euclidean geometry, without need of any assumption on continuity as pointed out by Hilbert. As examples, let us prove the above-mentioned hyperbolic Kou-Ku theorem and the cosine formula as follows.

Example 2 (Kou-Ku theorem). To simplify the calculation, we suppose that the vertex of the right angle C of $\triangle ABC$ is at the origin O of a Hilbert coordinate system, i.e.,

$$C = (0, 0, 1).$$

Let A, B not lie on the axis $\theta\infty$ with

$$A = (\xi_1, \xi_2, \xi_3),$$
$$B = (\eta_1, \eta_2, \eta_3).$$

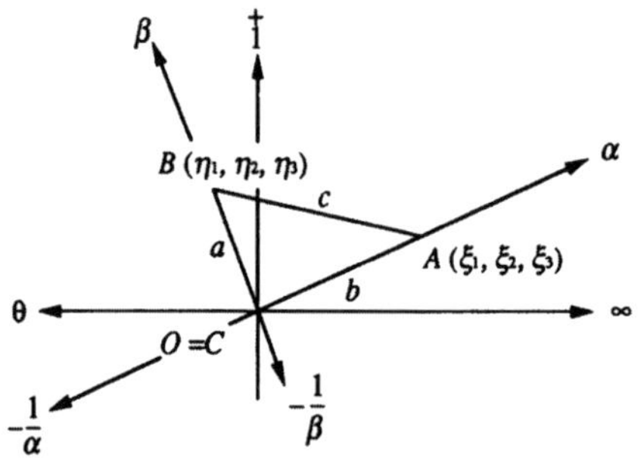

Fig. 6.21

Then

$$\xi_3^2 - \xi_1^2 - \xi_2^2 = 1, \quad \xi_3 > 0,$$
$$\eta_3^2 - \eta_1^2 - \eta_2^2 = 1, \quad \eta_3 > 0.$$

According to the distance formula (see Sect. 6.3), we have

$$\gamma(a) = \eta_3,$$
$$\gamma(b) = \xi_3,$$
$$\gamma(c) = \xi_3\eta_3 - \xi_1\eta_1 - \xi_2\eta_2.$$

Now, let one end of the line OA be α. Then the other is $-1/\alpha$, so the coordinates (u_1, u_2, u_3) of $\overrightarrow{O\alpha}$ are

$$u_1 = -\frac{2\alpha}{\alpha^2+1}, \quad u_2 = \frac{\alpha^2-1}{\alpha^2+1}, \quad u_3 = 0.$$

Similarly, if one end of the line OB is β, then the other is $-1/\beta$, so the coordinates (v_1, v_2, v_3) of $\overrightarrow{O\beta}$ are

$$v_1 = -\frac{2\beta}{\beta^2+1}, \quad v_2 = \frac{\beta^2-1}{\beta^2+1}, \quad v_3 = 0.$$

The conditions for the points A, B to lie on lines OA and OB respectively are

$$-2\alpha\xi_1 + (\alpha^2-1)\xi_2 = 0,$$
$$-2\beta\eta_1 + (\beta^2-1)\eta_2 = 0.$$

Furthermore, the condition for C to be a right angle is

$$u_1v_1 + u_2v_2 - u_3v_3 = 0,$$

i.e.,

$$4\alpha\beta + (\alpha^2-1)(\beta^2-1) = 0.$$

The formula to be proved becomes

$$\gamma(c) = \gamma(a)\gamma(b).$$

Now introduce the variables

$$x_1, x_2, x_3, \ldots, x_{10}, x_{11},$$

corresponding successively to

$$\xi_1, \xi_2, \xi_3, \eta_1, \eta_2, \eta_3, \gamma(a), \gamma(b), \gamma(c), \alpha, \beta,$$

in which

$$\alpha, \beta \neq \theta, \infty,$$

due to the hypothesis that A, B are not on the axis $\theta\infty$. Therefore,

$$x_{10} \neq 0, \quad x_{11} \neq 0.$$

Moreover,

$$x_3 > 0 \quad (\text{i.e. } \xi_3 > 0),$$

$$x_6 > 0 \quad (\text{i.e. } \eta_3 > 0).$$

The remaining hypothesis relations are

$$F_1 \equiv x_3^2 - x_1^2 - x_2^2 - 1 = 0,$$

$$F_2 \equiv x_6^2 - x_4^2 - x_5^2 - 1 = 0,$$

$$F_3 \equiv x_7 - x_6 = 0,$$

$$F_4 \equiv x_8 - x_3 = 0,$$

$$F_5 \equiv x_9 - x_3x_6 + x_1x_4 + x_2x_5 = 0,$$

$$F_6 \equiv x_2x_{10}^2 - 2x_1x_{10} - x_2 = 0,$$

$$F_7 \equiv x_5x_{11}^2 - 2x_4x_{11} - x_5 = 0,$$

$$F_8 \equiv (x_{10}^2 - 1)(x_{11}^2 - 1) + 4x_{10}x_{11} = 0.$$

The conclusion relation to be proved is

$$G \equiv x_9 - x_7x_8 = 0.$$

Proceeding now according to the method of Chap. 4, we first well-order $F_1, \ldots,$ F_8. From $F_6 = 0$, $F_7 = 0$, $F_8 = 0$, we obtain

$$x_{10}x_{11}(x_1x_4 + x_2x_5) = 0.$$

Under the non-degeneracy conditions

$$x_{10} \neq 0, \quad x_{11} \neq 0,$$

(which are already true by hypotheses) we have

$$F_9 \equiv x_1 x_4 + x_2 x_5 = 0.$$

Adjoining F_9 to the polynomial set $\{F_1, \ldots, F_8\}$, we get

$$\Lambda = \{F_1, \ldots, F_8, F_9\}.$$

A quasi-basic set of Λ is

$$\Phi: \quad F_1, F_9, F_2, F_3, F_4, F_5, F_6, F_7.$$

Reducing G with respect to Φ, the remainder is simply verified to be 0, or

$$G = F_5 - F_3 F_4 - x_3 F_3 - x_6 F_4 - F_9.$$

Hence, $G = 0$ is a formal consequence of $F_1 = 0, \ldots, F_8 = 0$, and the Kou-Ku theorem is proved.

Example 3 (Cosine formula). We still suppose the vertex C of $\triangle ABC$ is at O but do not suppose that C is a right angle. Similarly, let A, B not lie on the axis 0∞ and denote the angle between $\overrightarrow{O\alpha}$ and $\overrightarrow{O\beta}$ by φ. Then we have

$$T(\varphi) = \frac{\alpha - \beta}{1 + \alpha\beta},$$

$$C(\varphi) = \frac{(1+\alpha\beta)^2 - (\alpha - \beta)^2}{(1+\alpha\beta)^2 + (\alpha - \beta)^2}.$$

Similar to Example 2, let $\xi_1, \ldots, \alpha, \beta$ be denoted by $x_1, \ldots, x_{10}, x_{11}$. In addition, we introduce the variables x_{12}, x_{13}, x_{14} such that

$$x_{12} = C(\varphi), \quad x_{13} = \sigma(a), \quad x_{14} = \sigma(b).$$

Then in the hypothesis relations, $F_1 = 0, \ldots, F_7 = 0$ are the same as before, while $F_8 = 0$ becomes

$$F_8' \equiv [(1 + x_{10}x_{11})^2 + (x_{10} - x_{11})^2]x_{12} - [(1 + x_{10}x_{11})^2 - (x_{10} - x_{11})^2] = 0,$$

and in addition we have

$$F_9' \equiv x_{13}^2 - x_7^2 + 1 = 0,$$
$$F_{10}' \equiv x_{14}^2 - x_8^2 + 1 = 0.$$

The set of polynomials $\Sigma = \{F_1, \ldots, F_7, F_8', F_9', F_{10}'\}$ corresponds to the set of hypothesis relations (excluding the part of inequalities). The cosine formula

$$\gamma(c) = \gamma(a)\gamma(b) - \sigma(a)\sigma(b)C(\varphi)$$

to be proved can be rewritten as

$$G' \equiv (x_9 - x_7x_8) - x_{12}x_{13}x_{14} = 0.$$

Due to the indefiniteness of signs, we relax in proving that

$$G'' \equiv [(x_9 - x_7x_8) + x_{12}x_{13}x_{14}][(x_9 - x_7x_8) - x_{12}x_{13}x_{14}]$$
$$\equiv (x_9 - x_7x_8)^2 - x_{12}^2x_{13}^2x_{14}^2 = 0.$$

Using the general method of Chap. 4, we well-order Σ so as to enlarge it to another set Λ of polynomials. Then by reducing G'' with respect to the quasi-basic set of Λ, we know that under the non-degeneracy conditions

$$x_2 \neq 0, \quad x_5 \neq 0, \quad x_{10} \neq 0, \quad x_{11} \neq 0,$$

the remainder of G'' with respect to the quasi-basic set of Λ is 0, i.e., $G'' = 0$ is a formal consequence of $F_1 = 0, \ldots, F_7 = 0, F_8' = 0, F_9' = 0, F_{10}' = 0$. From the original hypotheses, we know that these non-degeneracy conditions are all naturally satisfied, so the cosine formula $G'' = 0$, without considering the sign, is proved. We may prove the original cosine formula $G' = 0$ by considering $\xi_3 > 0, \eta_3 > 0$ and using the method of Chap. 5.

Mechanical proving of formulas involving other transcendental functions

The method explained above can be applied to prove not only theorems and formulas in ordinary geometry and hyperbolic non-Euclidean geometry, but also theorems and formulas involving trigonometric and hyperbolic functions in sphere geometry or elliptic non-Euclidean geometry. Moreover, for other types of transcendental functions, when we do not need to know the exact definition and real meaning of continuity but only to find the formal identity relations among them, we can also apply methods similar to those above to achieve mechanical proving. Such methods may also be applied to prove relevant geometric theorems in which transcendental functions such as elliptic functions in the theory of cubic curves and the θ function in the theory of some algebraic curves and surfaces appear. This makes the use of the great part of transcendental functions in the research of various geometries unnecessary, or it can be reduced to problems treatable by purely algebraic methods. We hope to be able to systematically treat this topic in the near future.

References

Artin, E. (1927): Über die Zerlegung definiter Funktionen in Quadrate. Abh. Math. Sem. Univ. Hamburg 5: 100–115.

Artin, E. (1950): Geometric algebra. Interscience, Chichester.

Artin, E., Schreier, O. (1927): Algebraische Konstruktion reeller Körper. Abh. Math. Sem. Univ. Hamburg 5: 85–99.

Bachmann, F. (1959a): Aufbau der Geometrie aus dem Spiegelungsbegriff. Springer, Berlin Göttingen Heidelberg.

Bachmann, F. (1959b): Axiomatischer Aufbau der ebenen absoluten Geometrie. In: Henkin, L., Suppes, P., Tarski, A. (eds.): The axiomatic method. North-Holland, Amsterdam, pp. 114–126.

Bachmann, F., Reidemeister, K. (1937): Die metrische Form in der absoluten und der elliptischen Geometrie. Math. Ann. 113: 748–765.

Baer, R. (1944): The fundamental theorems of elementary geometry: an axiomatic analysis. Trans. Amer. Math. Soc. 56: 99–129.

Baker, H. F. (1929): Principles of geometry, vol. 1, foundations. Cambridge University Press, Cambridge.

Chasles, M. (1889): Aperçu historique sur l'origine et le développement des méthodes en géométrie, 3rd edn. Gauthier-Villars, Paris.

Coxeter, H. S. M. (1947): Non-Euclidean geometry. University of Toronto Press, Toronto.

Davis, P. J. (1927): La géométrie de René Descartes. Gauthier-Villars, Paris.

Degen, W., Profke, L. (1976): Grundlagen der affinen und euklidischen Geometrie. Springer, Berlin Heidelberg New York.

Descartes, R. (1974): The Schwarz function and its applications. Mathematical Association of America, Washington.

Ewald, G. (1956a): Axiomatischer Aufbau der Kreisgeometrie. Math. Ann. 131: 354–371.

Ewald, G. (1956b): Über den Begriff der Orthogonalität in der Kreisgeometrie. Math. Ann. 131: 463–491.

Gerretsen, J. C. H. (1942a): Die Begründung der Trigonometrie in der hyperbolischen Ebene. Indag. Math. 4: 133–139, 169–173, 199–206.

Gerretsen, J. C. H. (1942b): Zur hyperbolischen Geometrie. Indag. Math. 4: 207–213.

Greenberg, M. J. (1973): Euclidean and non-Euclidean geometries: development and history. Freeman, San Francisco.

Gröbner, W. (1949): Moderne algebraische Geometrie. Springer, Wien Innsbruck.

Hartshorne, R. (1967): Foundations of projective geometry. Harvard University Press, Cambridge, MA.

Henkin, L., Suppes, P., Tarski, A. (eds.) (1959): The axiomatic method. North-Holland, Amsterdam.

Hermann, G. (1926): Die Frage der endlich vielen Schritte in der Theorie der Polynomideale. Math. Ann. 95: 736–788.

Hessenberg, G. (1905a): Beweis des Desarguesschen Satzes aus dem Pascalschen. Math. Ann. 61: 161–172.

Hessenberg, G. (1905b): Begründung der elliptischen Geometrie. Math. Ann. 61: 173–184.

Hilbert, D. (1899): Grundlagen der Geometrie, 1st edn. Teubner, Stuttgart (8th edn. 1956).

Hilbert, D. (1900): Über den Zahlbegriff. Jahresber. Deut. Math. Verein. 8.

Hilbert, D. (1901): Mathematische Probleme. Arch. Math. Phys. 1: 44–63, 213–237 [also in Hilbert, D. (1935): Gesammelte Abhandlungen, vol. 3. Springer, Berlin, pp. 290–329].

Hilbert, D. (1903): Neue Begründung der Bolyai-Lobatschefskyschen Geometrie. Math. Ann. 57: 137–150 [also in Hilbert, D. (1956): Grundlagen der Geometrie, 8th edn. Teubner, Stuttgart, pp. 159–177].

Hilbert, D. (1933): Gesammelte Abhandlungen, vol. 2. Springer, Berlin.

Hodge, W. V. D., Pedoe, D. (1947): Methods of algebraic geometry, vol. 1. Cambridge University Press, Cambridge.

Hodge, W. V. D., Pedoe, D. (1952): Methods of algebraic geometry, vol. 2. Cambridge University Press, Cambridge.

Jacobson, N. (1974): Basic algebra, vol. 1. Freemann, San Francisco.

Kerékjártó, B. (1969): Les fondements de la géométrie, vol. 1. Gauthiers-Villars, Paris.

Klein, F. (1893): Vergleichende Betrachtungen über neuere geometrische Forschungen. Erlangen 1872. Math. Ann. 43: 63–100.

Klein, F. (1926): Vorlesungen über höhere Geometrie. 3rd edn. Springer, Berlin.

Klein, F. (1928): Vorlesungen über nicht-euklidische Geometrie. Springer, Berlin.

Klein, F. (1939): Elementary mathematics from an advanced standpoint, vol. 2, geometry. Macmillan, New York.

Kline, M. (1972): Mathematical thought from ancient to modern times. Oxford University Press, New York.

Knuth, D. E. (1969): The art of computer programming, vol. 2. Addison-Wesley, Don Mills.

Kreisel, G. (1958): The mathematical significance of consistency proofs. J. Symb. Logic 23: 155–182.

Liebmann, H. (1904): Über die Begründung der hyperbolischen Geometrie. Math. Ann. 59: 110–128.

Liebmann, H. (1905): Elementargeometrischer Beweis der Parallelenkonstruktion und neue Begründung der trigonometrischen Formeln der hyperbolischen Geometrie. Math. Ann. 61: 185–199.

Morley, F., Morley, F. V. (1933): Inversive geometry. Bell, London.

Pasch, M., Dehn, M. (1926): Vorlesungen über neuere Geometrie, 2nd edn. Springer, Berlin.

Podehl, E., Reidemeister, K. (1934): Eine Begründung der ebenen elliptischen Geometrie. Abh. Math. Sem. Univ. Hamburg 10: 231–255.

Reidemeister, K. (1930): Grundlagen der Geometrie. Springer, Berlin.

Ritt, J. F. (1932): Differential equations from the algebraic standpoint. American Mathematical Society, New York.

Ritt, J. F. (1950): Differential algebra. American Mathematical Society, New York.

Robinson, A. (1955): On ordered fields and definite functions. Math. Ann. 130: 257–271.

Robinson, A. (1956): Further remarks on ordered fields and definite functions. Math. Ann. 130: 405–409.

Robinson, A. (1963): Introduction to model theory and to the mathematics of algebra. North-Holland, Amsterdam.

Salmon, G. (1879a): A treatise on conic sections, 6th edn. London.

Salmon, G. (1879b): Higher plane curves, 3rd edn. Dublin.

Schur, F. (1903): Zur Propositionslehre. Math. Ann. 57: 205–208.

Schur, F. (1904): Zur Bolyai-Lobatschefskyschen Geometrie. Math. Ann. 59: 314–320.

Seidenberg, A. (1954): A new decision method for elementary algebra. Ann. Math. 60: 365–371.

Szasz, P. (1958): Unmittelbare Einführung Weierstrasscher homogener Koordinaten in der hyperbolischen Ebene auf Grund der Hilbertschen Endenrechnung. Acta Math. Acad. Sci. Hung. 9: 1–28.

Szasz, P. (1959a): Direct introduction of Weierstrass homogeneous coordinates in the hyperbolic plane, on the basis of the end-calculus of Hilbert. In: Henkin, L., Suppes, P., Tarski, A. (eds.): The axiomatic method. North-Holland, Amsterdam, pp. 97–113.

Szasz, P. (1959b): Begründung der analytischen Geometrie der hyperbolischen Ebene mit den klassischen Hilfsmitteln, unabhängig von der Trigonometrie dieser Ebene. Acta Math. Acad. Sci. Hung. 8: 139–157.

Szmielew, W. (1959): Some metamathematical problems concerning elementary hyperbolic geometry. In: Henkin, L., Suppes, P., Tarski, A. (eds.): The axiomatic method. North-Holland, Amsterdam, pp. 30–52.

Szmielew, W. (1961): A new analytic approach to hyperbolic geometry. Fund. Math. 50: 129–158.

Tarski, A. (1948): A decision method for elementary algebra and geometry, 1st edn. Rand Corporation, Santa Monica (2nd edn., University of California Press, Berkeley, 1951).

Tarski, A. (1959): What is elementary geometry. In: Henkin, L., Suppes, P., Tarski, A. (eds.): The axiomatic method. North-Holland, Amsterdam, pp. 16–29.

Trager, B. H. (1976): Algebraic factoring and rational function integration. In: Proceedings of the 1976 ACM Symposium on Symbolic and Algebraic Computation, pp. 219–226.

Uhl, A. (1964): Ein axiomatischer Aufbau der Laguerregeometrie. Springer, Berlin Göttingen Heidelberg New York.

Van der Waerden, B. L. (1930a): Moderne Algebra, vol. 1. Springer, Berlin.

Van der Waerden, B. L. (1930b): Eine Bemerkung über die Unzerlegbarkeit von Polynomen. Math. Ann. 102: 738–739.

Van der Waerden, B. L. (1931): Moderne Algebra, vol. 2. Springer, Berlin.

Van der Waerden, B. L. (1945): Einführung in die algebraische Geometrie. Springer, Berlin.

Van der Waerden, B. L. (1955): Algebra, 4th edn. Springer, Berlin Göttingen Heidelberg New York.

Van der Waerden, B. L., Smid, L. J. (1935): Eine Axiomatik der Kreisgeometrie und der Laguerregeometrie. Math. Ann. 110: 753–776.

Veblen, O., Young, J. W. (1910): Projective geometry, vol. 1. Ginn, Boston.

Veblen, O., Young, J. W. (1918): Projective geometry, vol. 2. Ginn, Boston.

Wang, H. (1959): A survey of mathematical logic. Science Press, Beijing.

Wang, H. (1960): Toward mechanical mathematics. IBM J. 4: 2–22.

Wang, H. (1965): Formalization and automatic theorem proving. In: Proceedings of the IFIF Congress, vol. 1, pp. 51–58.

Wang, H. (1981): Popular lectures on mathematical logic. Science Press, Beijing.

Wu, W.-t. (1978): On the decision problem and the mechanization of theorem-proving in elementary geometry. Sci. Sin. 21: 159–172.

Wu, W.-t. (1979): Mechanical theorem-proving in elementary differential geometry. Sci. Sin. Math. Suppl. (I): 94–102.

Wu, W.-t. (1982a): Mechanical theorem-proving in elementary geometry and differential geometry. In: Proceedings of the 1980 Beijing DD-Symposium, vol. 2. Science Press, Beijing, pp. 1073–1092.

Wu, W.-t. (1982b): Toward mechanization of geometry – some comments on Hilbert's "Grundlagen der Geometrie". Acta Math. Sci. 2: 125–138.

Subject index